CONSERVATION OF TIMBER BUILDINGS

CONSERVATION OF TIMBER BUILDINGS

F. W. B. Charles

HUTCHINSON

London Melbourne Sydney Auckland Johannesburg

Hutchinson & Co. (Publishers) Ltd
An imprint of the Hutchinson Publishing Group
17–21 Conway Street, London W1P 6JD

Hutchinson Publishing Group (Australia) Pty Ltd
PO Box 496, 16–22 Church Street, Hawthorne, Melbourne,
Victoria 3122

Hutchinson Group (NZ) Ltd
32–34 View road, PO Box 40–086, Glenfield, Auckland 10

Hutchinson Group (SA) (Pty) Ltd
PO Box 337, Bergvlei 2012, South Africa

First published 1984

Set in Bembo by BAS Printers Ltd
Over Wallop, Hampshire

Printed and bound in Great Britain by
Butler & Tanner Ltd, Frome and London

British Library Cataloguing in Publication Data
Charles, F.W.B.
The conservation of timber buildings.
1. Half-timbered buildings – Great Britain.
I. Title
721'0448 NA7175

Library of Congress Cataloguing in Publication Data
Charles, F.W.B. (Frederick William Bolton)
Conservation of Timber Buildings
Bibliography: p.
Includes index.
1. Wooden-frame buildings — Conservation and restoration. I. Title.
TH'3411.C43 1984 694'.028'8 84-12984
ISBN 0 09 145090 X

Frontispiece acknowledgement
Middle Littleton Barn. Courtesy Martin Charles

CONTENTS

Only note that there are no tokens of poverty about them: no tumble-down picturesque; which to tell you the truth, the artist usually availed himself of to veil his incapacity for drawing architecture. Such things do not please us, even when they indicate no misery. Like the mediævals we like everything trim and clean, and orderly and bright; as people always do when they have any sense of architectural power.

William Morris, *News from Nowhere*

FOREWORD

When an architect has developed a superskill through dedicating his professional and even personal life to conservation as Freddie Charles has done with his wife's constant support, we must be grateful. When this is done in such a lively, informative and slightly provocative way we can be delighted as we should be with this book.

'The approach is practical rather than philosophical and the final argument was the architectural one', in the author's own words. In some cases this inevitably leads to arguments with antiquarians, art historians and archaeologists who often find it difficult to understand that a building must stand up and resist live and dead loads and be protected against the numerous hazards and causes of decay that affect wood. Indeed, there are two main types of building: those made of some type of masonry and those made of wood, and the ethics of conservation of wood structures have not yet been as fully developed as those of masonry buildings. This is because in Europe wood buildings generally serve the practical pragmatic Northern races, whereas masonry structures serve the Latin races who are more inclined to theorize, by and large. In fact, some marriage of both approaches is necessary and an agreed theory of conservation can help us to find interdisciplinary harmony. The recommended technique is to examine the practical alternatives and then to see which is 'least bad' according to theory.

The Japanese, having some of the oldest and certainly the world's largest wooden buildings, find in practice that these must be dismantled every 150 years to renew rotten pieces of wood and remake joints. Theorists in England often have too limited approach. Archaeologists and art historians find it hard to accept that buildings have to be dismantled, for they would value the wattle-and-daub or other infill more highly as long as it was in place and undisturbed. Failing to appreciate the action of the timber structural system, they would prefer to see it glued together if possible, rather than jointed and pegged. The structural integrity and firmness of the timber framing must, however, be the first priority and this point is abundantly well illustrated by Freddie Charles, who, in fact, is an outstanding expert in building archaeology, although his approach is limited to England.

As the infill of a framed building is almost inevitably finished with plaster and probably limewashed, critics of reconstruction should take heart that these materials weather quite quickly compared with brick and stone, so there are fewer problems connected with patina. There is, of course, nothing to equal the feel and patina of well worn wood. It is amazing what abuse a framed building will put up with as long as its joints hold, as is well shown by the case of the Ancient High House.

Freddie Charles' description of investigating buildings, with the preliminary survey followed by a structural survey is fascinating, yet he admits that the work is full of surprises and gives us a frank description of some of the difficulties he met, and overcame in practice. My surprise was how often he was able to use lump sum contracts. Part of his skill lies in his knowledge of the sequence in which timber buildings were put together. His chapters on framing and development of timber building are essential reading for all students of vernacular architecture, for without this knowledge it is impossible to interpret a framed building.

Historic buildings may give their conservators insuperable problems if, as sometimes happens, timber of sufficient scantlings cannot be provided, so the contribution of those timber merchants who can supply British hardwoods should not be underestimated. Indeed I believe the Government should interest itself in this question and reserve good woodlands for conservation.

Freddie Charles and his wife Mary, by saving many outstanding buildings such as Cheylesmore Manor House and those in Spon Street Coventry, have given a large number of persons great pleasure. For this work and their account of it in this book they deserve our thanks.

Bernard M. Feilden, CBE
DUniv., FSA, FRSA, FRIBA, AADipl. (Hons)
Architectural Conservation Consultant

PREFACE

This book arose directly out of a series of lectures given at the Institute of Advanced Architectural Studies at York. The lectures were mainly case studies of the repair of timber-framed building. They included some of those of Coventry's Spon Street Scheme, by which certain buildings near the town centre, which otherwise would have been demolished, were dismantled and re-erected in Spon Street. They also included the Bear Steps group in the heart of Shrewsbury; Chester House, which became a branch library in the High Street of Knowle; and the great roof of Middle Littleton Tithe Barn near Evesham. All of these have been published, though not in the context of conservation, and are mentioned only in passing in this book.

Instead, Shell Manor near Droitwich, Cheylesmore Manor in Coventry, Much Wenlock Guildhall and the Wellington Inn in Manchester have been selected as more appropriate for the present purpose. These buildings are of much the same size and, after compensating for price increases, their contracts were of similar value. There, however, all similarity ends. Not only are the buildings different in design, date, construction and every detail but, from the first day of the survey to the final handing over, the courses of the contracts diverged more and more. Nor is it possible to describe every problem of any one building. Rather, particular aspects are brought out in each, so that in taking all four together something approaching a comprehensive picture of a restoration scheme may emerge.

The case studies are preceded, as in the lectures, by an account of the subjects of which everybody concerned with the conservation of timber buildings should have some knowledge. Not least important is the history of the timber-frame tradition, a topic that touches on academic theory as well as archaeological research, and being relatively new to scholastic circles is prone to controversy – no doubt a healthy sign of growth. But if the views of one who has studied the buildings as a practitioner, rather than a scholar, do not agree with theories of diffusion, distribution, hybridization and the rest that have been applied to structural carpentry, I can only plead that I can see no essential difference in the process of designing buildings and structures as practised today from that of the carpenter of five hundred or even a thousand years ago. Building is technical at whatever level, and while technical limits can be overreached (as at Beauvais Cathedral) the builder normally only undertakes what is within his capability. It is the dialectic of social demand and technical response that brings changes in design, rather than cultural migrations or idealistic attributions to the carpenter. The major changes that did take place and their dates are of course generally accepted by all researchers; it is only why they happened that is at issue.

Conservation is no less controversial. My 'philosophy', if it warrants that name at all, is once again based on practice, above all on observation or rediscovery of what the old carpenters did to create the building in the first place, and, through close survey, on deciding what must be done to preserve or restore its essential structure. Seeing how today's carpenters set about it, particularly in handling great timbers, sometimes in what might seem to be hopelessly restricted conditions, has been almost as informative.

For today's carpenters lack nothing in comparison with their forebears, except opportunity. With the demands of a restoration contract, entirely different from those of the normal nail-bashing job, it soon becomes clear that craftsmanship has not declined, let alone disappeared. Indeed, in the case of Gunold Greiner, not even independence has been lost. And my first acknowledgement is to him for the invaluable few years' experience of working with him in the reconstruction of the Bromsgrove House at Avoncroft.

Also in direct descent from the master carpenter of old is the contracting firm of Parker and Morewood in which Ronald Morewood himself, latterly with his two sons, is the job carpenter, as well as the one who orders his timber, organizes the contract and keeps his accounts. So far as I know, no other contractor, because of today's gulf between administration and worker, can compete in any respect with that firm. Nevertheless, several other carpenters deserve acknowledgement: Gerald Dauncey of Spicers, Worcester; Michael Goode of Sandys, Stafford; Arthur Berry of Pettifers, Pershore; John Baskerville of Bridgemans, Lichfield; and Kenneth Higgins of Sapcotes, Birmingham. From all of these I have learnt to appreciate and admire today's carpenters as a breed of men who never deserved to lose their craft independence and possession of the 'mysteries' of carpentry. It is still noticeable, however, that whenever a really tricky operation is to be done, and the time for doing it has been agreed with the architect so that he may watch it (he should be helping), it is always already done by the time he comes on the job!

Next to the carpenters is what, on looking back, seems to have been a constant stream through the office of assistant architects – certainly they were inadequately paid! It is possible only to acknowledge those who have contributed directly to this book. They are John Giles, who did the published site sketches, among many others, of Shell Manor; John Greaves Smith, one or two of whose superb perspectives also appear; and Russell Jones, who drew the framing plans of Chapter 3. To the rest who surveyed, drew, supervised jobs and learnt with us, we can only say thank you.

To find the right place – or right words – for acknowledging the labours of my wife and coworker is not easy. Better perhaps just to say that she has done *everything* except the writing.

Also as concerns the book and office together, not only has our secretary Sue Lester typed continuously and remorselessly, but her memory, mental alacrity and uncanny ability to find the right page of the text, let alone whole drafts that have gone amiss in the welter of an architect's office, leave us speechless with admiration and gratitude.

The drawings, as already noted, are mostly my wife's, the photographs mostly mine, but those which stand out as clearly professional are those of my son Martin. His pictures of our work would also, on their own, make a book.

Of 'outsiders', I would thank Charles Venables, whose knowledge of oak after a lifetime of buying and selling native timber is second to none, and who has kindly checked Chapter 2 for me. I thank also Alexander Scott, a lifelong member of the same firm, with whom I have often visited the forests to select oaks for our special requirements and have watched experiments in cleaving, and who has gone to great trouble in teaching me the science of tree growth, timber characteristics, conversion and so on. He also made important comments on Chapter 2, which have been (I hope correctly) incorporated.

Of the many eminent architectural historians, archaeologists and others who have indirectly contributed to this book, both through their works and personally, I would mention only two by name – Walter Horn and Abbott Cummings, one from each seaboard of America and about as remotely opposite to each other in their special interests and scholarship. To Walter we are inexpressibly grateful for what he has taught us, particularly about medieval builders and their great tithe barns, and to Abbott for his generously imparted knowledge of the seventeenth century, by no means confined to the timber-framed house of New England, and to both for their warm and enduring friendship. To their English peers we owe almost as much, except perhaps in the last respect which may be simply because they don't have to find somewhere to stay on this side of the Atlantic. But it would be invidious to try to name them. I only, and most sincerely, say thank you for each individual effort on our behalf.

And the same must be said to Jack Bowyer for my introduction to Hutchinson in the first place and to Doug Fox of that firm for his patience and confidence.

Lastly, Bernard Feilden not only paid me the compliment of agreeing to write the Foreword but also has been always generous with his time in answering my questions, mainly about his pioneering restoration of York Minster, and in his sympathetic interest in the somewhat humbler kind of restoration we attempt in our office.

INTRODUCTION

There are as many ways of building with timber as there are species of trees, differences of climate and contrasts of terrain. The broad categories of construction and architectural style are of course clearly established geographically – horizontal-log construction in the northern or mountainous coniferous forests, timber framing in the temperate deciduous zones, the post-and-beam or trabeated systems of big conifers in the warmer and subtropical regions, and all those structures of uncountable sizes and shapes in tropical Asia and Africa.

Wherever trade and wealth brought the rise of civilization, wooden architecture expressed its unique culture. Vernacular building, at the bottom of the social hierarchy, has still closer links with local and regional conditions; each type of building is ingeniously designed for the particular way of life of each community and to meet the special conditions of each region, even including earthquakes. There cannot be a country in the world today that is not looking at – or having looked at on its behalf – the humblest dwellings of its people, and learning in the process a great deal of their skills, customs and art, in many cases for the first time.

The variations are so numerous that it is vain to look for any such thing as a universal prototype. It may only be postulated that digging posts into the ground and tying a roof on top of them is likely to have been the most wide spread system from the day when man first built a tribal settlement, and that the other method, that of leaning poles against each other, as for a wigwam or charcoal burner's hut, must have begun on the day – if not the day before – he came down from the trees.

Aeons after these events, the same methods may still be seen or recognized in standing buildings, archaeological remains or documentary descriptions. The greatest timber buildings in existence, the Japanese temples and shrines, represent the former method of construction but at the very advanced stage when means had been found to stand the posts *on* the ground instead of in it, so making possible 'permanent' wooden architecture. Their elegant framework of tall, square or round posts and die-straight beams acquired its character through the use of the giant cedar tree – the sugi. The fabulous buildings of King Solomon, according to the Bible's description, no doubt had similar framework as this was also of cedar, but their great trees came from the ancient forest of Lebanon. Only the roofs must have been different, for surely not even Solomon could have conceived of such roofs as those of the Far East.

In the tamer and more temperate zone of north-western Europe, the Norwegian stave churches also directly represent earthfast post construction, but again the posts stand clear of the ground. The timber is pine and their construction is of palisade form, perhaps surprising to find in one of the regions of horizontal-log building. The only comparable structure in England, Greensted Church in Essex, has contiguous uprights of oak, without the main posts or 'staves' of the Norwegian churches.

In the deciduous zones, archaeology has shown an immense variety of post buildings from Roman times onwards, and even before. They were oblong, bow-shaped, oval, circular, sometimes aisled with a central row of posts or with two rows dividing a central nave from side aisles. The pattern of post-holes has given rise to an equally varied range of theoretical reconstructions. Their significance in this context, however, is that even the most important buildings, the halls of royal palaces, were being built in England with posts set in the ground until well into the twelfth century. For the lesser buildings the system continued right through the Middle Ages, and indeed is still in use.

There must also have been a mass of buildings of the second method, leaning posts against each other, but these leave no trace for the archaeologist. Not only do circular structures belong to this category, but also 'tongs' or 'A-frames' forming oblong buildings. The larger of these, following the principle of framing their trusses, or cross-frames, on the ground and then rearing them, keeping them upright by braced members spanning two such frames, comprise the unique category of crucks. Since archaeology must draw a blank in excavations, however intense the search for them, other clues – most obviously how trees may grow into the shape capable of yielding crucks – must be followed instead. Such trees do not grow in any of the coniferous forests. Nor are the more sheltered deciduous regions likely to produce them, at least in sufficient quantity to establish a building tradition, for in the forest or woodlands even oaks may grow as straight as firs. Thus only the hillier parts of the western seaboard of northern France and Britain must have been true cruck regions, as they were no less of round houses. That is not to say bent timbers may not be found or used almost anywhere, even in Japan. But they are not crucks.

To revert to the post structures, it may be said that the stage at which permanent buildings were demanded, probably not at the same time throughout the oak regions of Europe, and perhaps last of all in England, produced a qualitative change in wood construction as dramatic in its own context as that in the higher realm of masonry architecture, when Gothic emerged out of Romanesque. This was the idea of the framed wall. In no other method is the frame so closely and consistently integrated with the building's external casing, nor so economical in providing a structure which is capable of both supporting the roof, with which it is also structurally unified, and keeping out weather and providing within its structure all the secondary elements of panels, door openings and windows, complete with mullions.

This timber-frame system creates the entirely prefabricated building. The pattern of timbers and panels varies from region to region as well as from age to age, but common to them all are the mortice-and-tenon joints and *halved* timbers. Hence no doubt the system's more traditional name, half-timber.

Buildings with framed walls range from churches, church steeples and porches to tithe barns, market halls, dovecotes and of course domestic halls and their solar and service wings, and they all share architectural style, initially Gothic and latterly Renaissance and Jacobean. The great medieval roofs, open to the rafters, were the supreme achievement. These too were of the same technique and style (though no two were ever identical) whether supported by a framed or solid substructure. The local carpenters would acquire the newest method and latest fashion whenever the new church or its new roof, designed no doubt by the master carpenter for the cathedral, was under construction – and that was a pretty continuous occupation. When at last the open roof had gone out, at least for domestic buildings, the carpenter's virtuosity was transferred to the timbering of upper floors, designed of course to be admired as the ceilings of the rooms below.

In such context the timber-frame tradition can hardly be termed 'vernacular' but, since it encompassed practically all wooden construction from the cathedral roof downwards, the definition, as in stone architecture, must rather relate to buildings than to materials and method. Timber framing may be said to have extended, or descended, to 'vernacular' social level when the lesser halls and some peasant houses were being built in it, roughly, it seems from documents, from about 1450. By the beginning of the seventeenth century, every property owner in town or country could probably boast a timber-framed house, if he was not already living in a brick or stone one. And after that, from about 1700, he used those materials, or a plaster rendering, to cover up the timber, and so bring at least the street front of his house up-to-date.

Lastly, some very late cottages are often a mixture of timber framing and post construction, having posts and rails half-lapped across each other instead of morticed-and-tenoned. Such little buildings are particularly prominent in Denmark where wooden farmhouses and barns were still being erected at the end of the last century. They are best dateable by knowledge of how long such flimsy erections are capable of standing.

Such is the scope of timber building. Yet its history was scarcely even a subject of its own before the Second World War. Now it runs to a considerable corpus of articles and books, mostly by archaeologists and historians. Scientists, especially physicists (for radiocarbon dating) and dendrochronologists, have also played a vital role. More recently dendrologists studying the trees themselves, rather than only their annual rings, have contributed to the ever-increasing store of knowledge and theories of construction and development.

The conservator, on the other hand, or conservation office, including the specialist in architectural history and expert on the principles and methods in preservation and restoration, is yet to be established, at any rate in Britain. The Society for the Preservation of New England Antiquities sets a model with its multidisciplinary staff and practical experience in the restoration exclusively of timber structures. Such an organization, probably with official status, must sooner or later come into being in England, working closely with the two international bodies ICCROM and ICOMOS*, though even they do not yet have specialists in timber.

In the Appendix is reproduced a paper by Bernard Feilden that sets out important principles for the conservation of timber structures; no doubt these principles will eventually be established internationally.

As a further comment on the principles of conservation, it may be noted that William Morris laid down his revolutionary viewpoint in furious opposition to the nineteenth-century 'restorers'. This was his Manifesto of 1877, establishing the Society for the *Protection* of Ancient Buildings and most closely concerned with the imminent threat to Tewkesbury Abbey. While everything he stood for is no less valid today, the reinterpretation of his precepts for modern conditions and especially for framed buildings, as opposed to cathedrals and minsters, is long overdue. The more obvious conditions which have changed are, first, that the buildings we are concerned with have had

*ICCROM, the International Centre for the Study of the Preservation and the Restoration of Cultural Property, Ospizio di San Michele, Via di San Michele 13, 00153, Rome, was set up by UNESCO in 1959, and ICOMOS, the International Council on Monuments and Sites, Hotel de Saint Aignan, 75 Rue du Temple, 75003 Paris, grew out of the Second Congress of Architects and Technicians of Historic Monuments at Venice in May 1964. Their aims and functions are set out in Feilden (1979), p. 58 *et seq.*

to survive another century of every bad practice that existed then – as well as a few more. Thousands of timber-framed buildings have been lost through normal wastage, replacement, war destruction and, most devastating of all, post-war redevelopment. Only in the last few years have the lesser buildings once again been sought after, but by members of a very different level of society with still more different values from those of their builders and previous inhabitants. The result, 'gentrification', is generally as unbeguiling as the term, and may be even more destructive than the slipshod repairs or ill-conceived over-restoration that characterized Victorian practice, not least at this level.

Second, a revolution in architectural thinking has taken place. This is the direct result of the 'modern movement' of architecture, of which Morris himself has been ascribed a pioneer. It has radically altered the approach to design, giving first place to respect for structure and 'honesty' in the use of materials, instead of plagiarism of past styles and falsification of appearances. Thus the serious attitude to ancient buildings (as opposed to sentimental or commercially exploitive ideas by which old buildings can still be, and are, appallingly abused) is that they have meaning, not only historically and socially, but also as the embodiment of much of today's architectural ethic. Moreover, modern materials can assist in their essential expression – which is structure – without resort to copying the old or disguising their own nature. In short, there is now an aesthetic unity between ancient and modern, so long as each is allowed to be its undisguised self.

What the restorers, on the one hand, and Morris, on the other, stood for has been most succinctly summed up by E. P. Thompson:

> The fashionable architects attempted to impose a superficial Gothic style upon their work, copying interesting Gothic features, often disregarding both structure and modern requirements. Philip Webb and Morris and their group, on the other hand, were concerned with the handling of materials by the medieval builder and craftsman, with substance and structure rather than with 'style'.

And that precisely has been our concern in restoring ancient timber-framed buildings – with what degree of success or failure will be seen in the following chapters.

PART ONE

BACKGROUND

STRUCTURAL TYPES

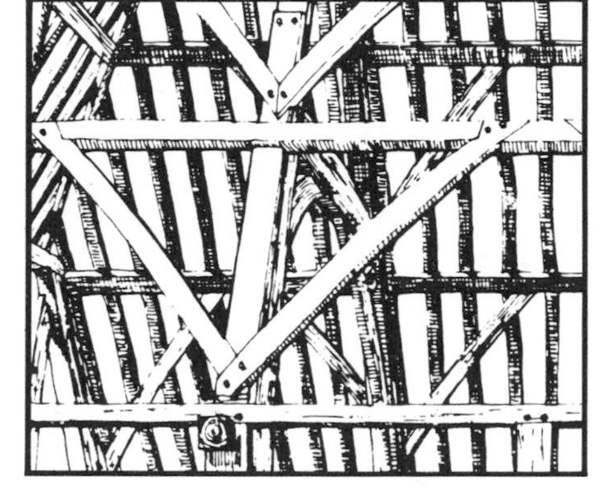

There are four types of structure generally associated with the timber-frame tradition: 'box-frame', 'post-and-truss', 'cruck' and 'base-cruck'.[1]★ Such a range of terms – and it could be extended almost without limit – is less daunting than it might appear at first sight. We are in fact dealing with only two roof types – the rafter or 'single' roof and the purlin or 'double' roof – each imposing its own clear principles of construction on the substructure, and both of them developed to a high technical level long before the timber-frame tradition came on the scene (Figure 1(a) and (b)).

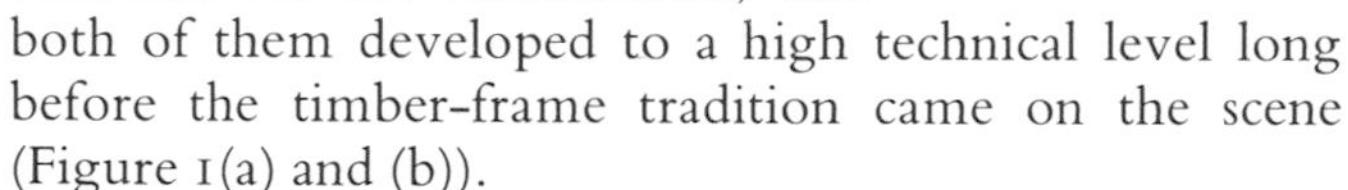

Rafter roofs

The origin of the rafter roof must go back to man's discovering that by tying three straight sticks together he could make a rigid frame. The beauty of the system, when the time came for erecting A-frames large enough for dwellings, was that additional props, ties, struts and braces could be put in wherever and whenever a weakness appeared.

★Notes and references are collected at the back of the book.

But the greatest advantage was that, regardless of size, no timber had to be larger than an easily felled and handled pole of birch, poplar, willow or conifer.

The next stage, placing the frames on to walls to obtain greater headroom with less span, resulted in the kind of structure whose remains are daily discovered by archaeologists – post-hole buildings or post structures.[2] The roof skeleton of the simplest of these must still have been set up on the plot first in order to determine where the posts should be placed. For the posts would have to conform to the vagaries of the wall-plates or a 'ring-beam', perhaps made up of oak branches, by which the whole roof would be supported in the finished building (Figure 2).

In the erection process, the ring-beam sections would be inserted beneath the A-frames, and tied to each other and to the underside of the rafters. The post positions could then be marked on the plot wherever they would give the most efficient support to the ring-beam. Next the roof skeleton would be moved to one side or, if it were too large to be lifted bodily, could be dismantled, all the components having been marked beforehand for final reassembling in exactly the same manner. The holes

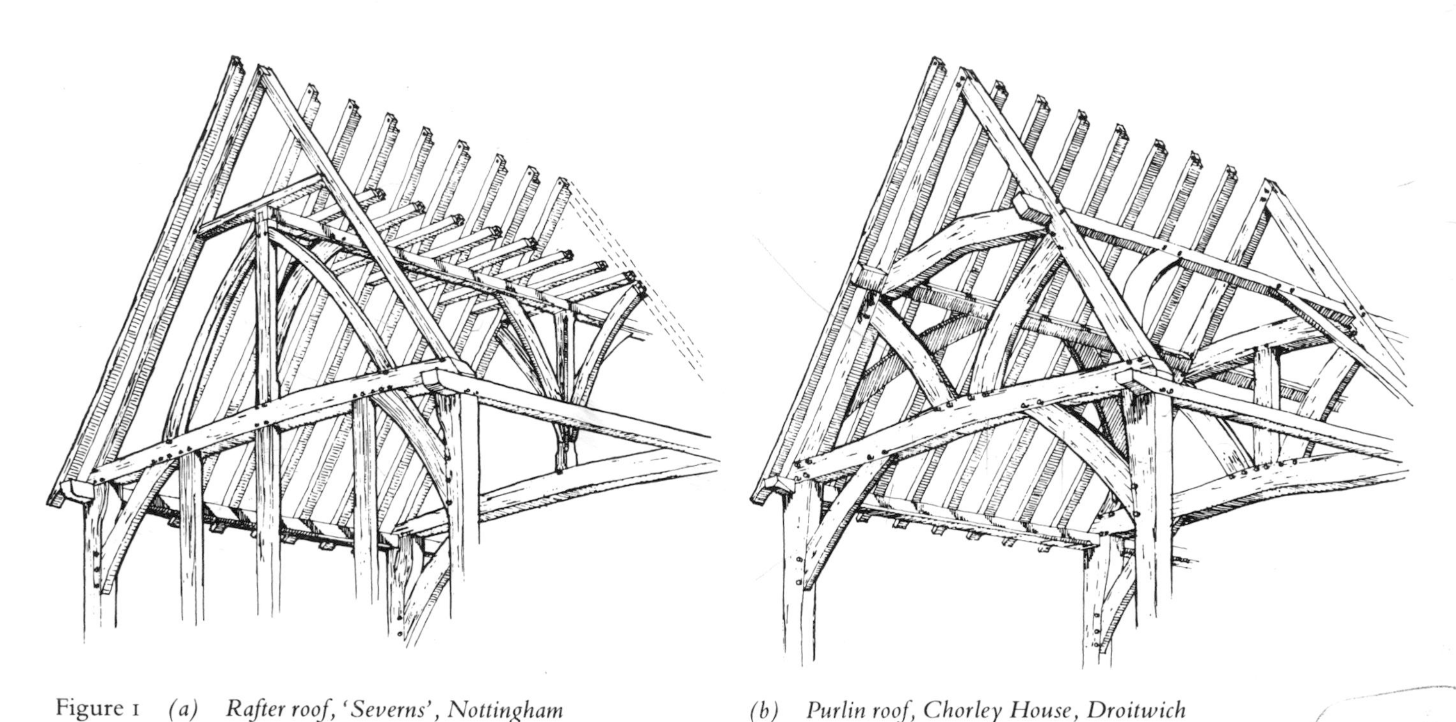

Figure 1 *(a) Rafter roof, 'Severns', Nottingham* *(b) Purlin roof, Chorley House, Droitwich*

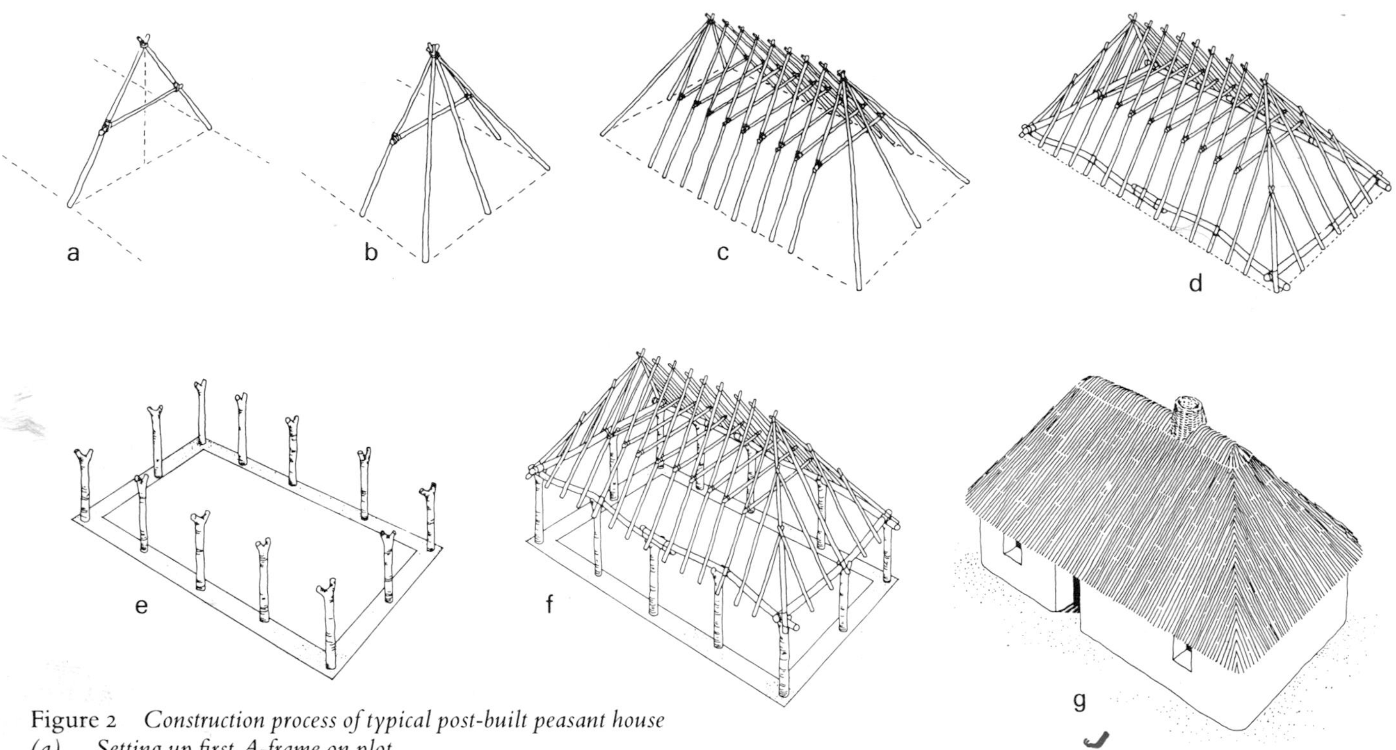

Figure 2 *Construction process of typical post-built peasant house*
(*a*) *Setting up first A-frame on plot*
(*b*) *Forming hip*
(*c*) *Completed roof skeleton*
(*d*) *Ring-beam inserted beneath rafters*
(*e*) *Posts set up as determined by alignment of ring-beam*
(*f*) *Roof lifted on to the posts*
(*g*) *Completed house*

would be dug, and the posts, each of appropriate length (equal to the depth of hole plus the height to the eaves plus the varying height above that to the ring-beam) would be set up and secured by back-filling. With the reassembling of the roof structure, now upon the posts, the external cladding of cob, wattle-and-daub, turf or stone would result in a building not very dissimilar in appearance, however different the technique, from the scores of traditional cottages still to be seen in some remoter parts of Britain.

This method was equally suitable for circular, oval, oblong or square buildings or for any shape in between, and it must have lasted for millenia. Indeed the principle of designing the roof first and then arranging the substructure to fit the roof applies as much to the vast exhibition hall of today as to the Iron Age hut; the difference is that while the former must first be designed on the drawing board to small scale, the latter was designed on the ground to full scale.

Perhaps the final stage of the rafter roof combined with earthfast posts was reached with the construction of the largest in the series of halls of the royal Saxon Palace at Cheddar in about 1120 AD (Figure 3).[3] The span of the nave was 28 feet and of the aisles 16 feet, giving a total width of 60 feet, greater than that of any standing tithe barn. But perhaps the Cheddar Hall was, after all, a little over-ambitious. For at some time after its erection the aisles had to be narrowed (Figure 4). There are no *purlin* roofs of this scale and date, at any rate in England. There are, however, rafter roofs only slightly later and of similar or greater clear span. One of these is at Lincoln Cathedral, another at Beverley Minster and yet another, of nearly 40 feet, at the Abbey of Fontevraud in the Loire valley.

In the earliest buildings of the timber-frame tradition, rafter roofs are all lap-jointed, following the lead of post construction. Examples are the well-known Essex barns.[4] These are aisled, and one of their structural characteristics is long straight braces – generally referred to as 'passing'

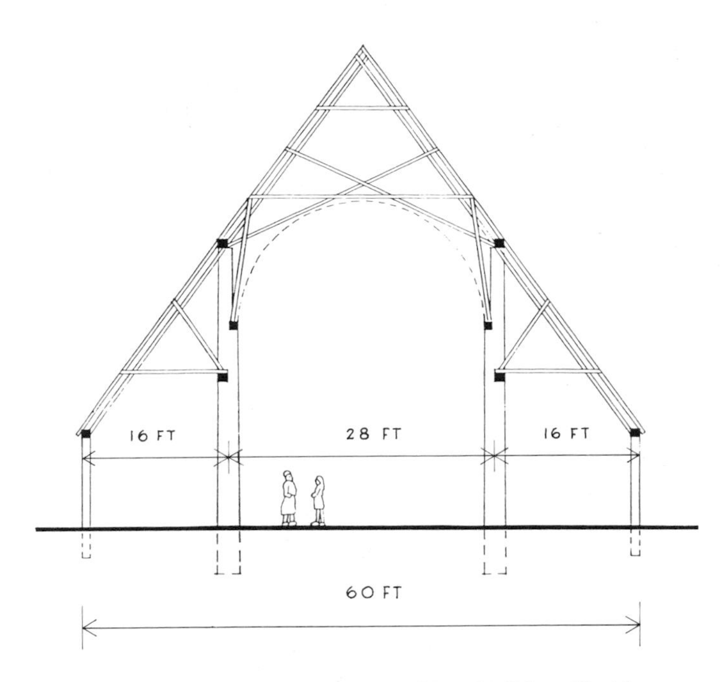

Figure 3 *Conjectural reconstruction of East Hall I at Cheddar – cross-section*

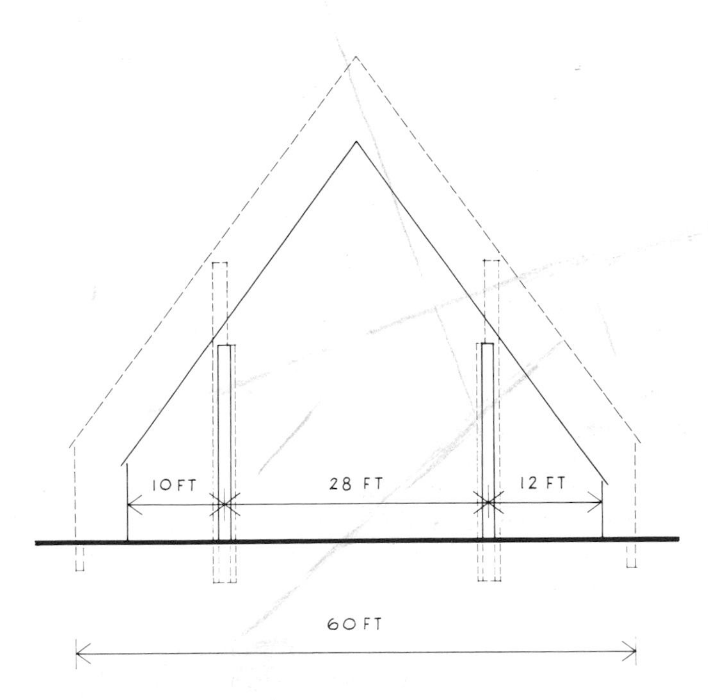

Figure 4 *Section of the East Hall II at Cheddar compared with East Hall I (shown dotted). The difference of 10 feet in width reduces the roof height and length of posts by 9 feet, making a substantial difference in the size of posts needed for the second hall as compared with the earlier, I*

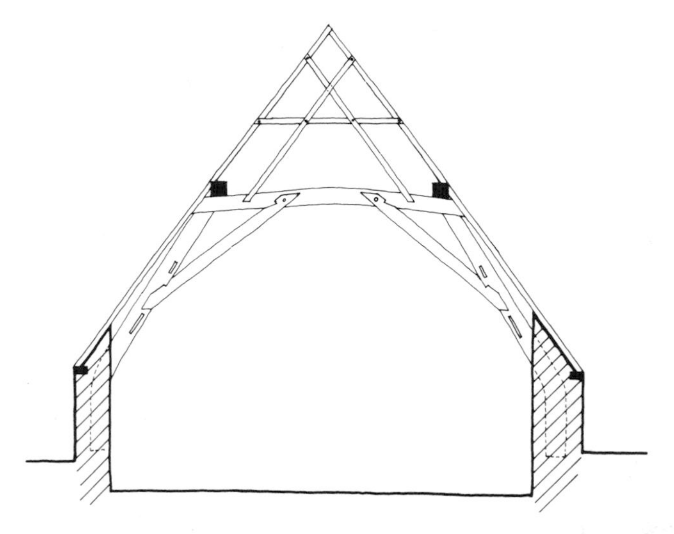

Figure 5 *Typical cross-frame of Siddington Barn, Gloucestershire, showing lap-joints*

braces – which sometimes extend from the outer posts of the aisles right across the nave and far up the opposite slope of the roof, half-lapped and generally notched over every member they cross. Siddington Barn, in the very different Cotswold region, and perhaps even older, has the same jointing method but with 'base-crucks' instead of aisle-posts (Figure 5). The free-standing belfry at the Hereford and Worcester church of Pembridge is, or was, also a lap-jointed structure, though alterations in later centuries have left only empty sockets on the surviving posts (Figure 6). Many Danish and Norwegian churches also have lap-joints. Variations in their design are practically unlimited but there is generally a sound reason for the more subtle variations. The secret-notch lap, for instance, safeguards splitting along the grain of the tongue as the continuous outer skin preserves its continuity (Figure 7). Again, the tongues of some joints are tapered so that cross-grain stress is less likely to snap off the tongue at the shoulder. Experts in both this country and Denmark claim that buildings can be dated by such variations.[5]

Domestic examples of rafter roofs with half-lap joints are even more numerous and widespread than the barns and church buildings. One of them, Fyfield Hall, also in Essex, was recently found to have earthfast posts for its main cross-frame, wholly lap-jointed, while the wall-frames were mortice-and-tenoned.[6] Though there must be special ground conditions to enable posts set in the earth to last six or seven centuries, there are probably other such transitional buildings awaiting identification – possibly

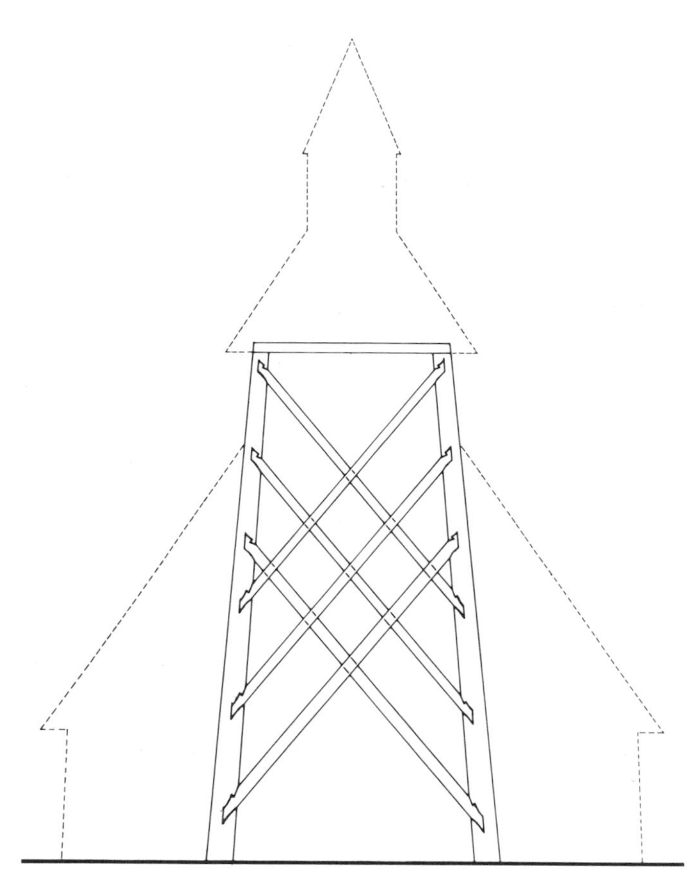

Figure 6 *Elevation of original frame of Pembridge Belfry, near Hereford (existing tower shown in dotted line)*

the great hall of the Bishop's Palace at Hereford (Figure 8).[7]

Thus lap-jointing denotes an early date. But it might as readily denote a very late date! Inferior houses and lesser buildings were still built in the old way until the eighteenth or even nineteenth century – for post construction and its methods persisted right through, and after, the period of the timber-frame tradition though for progressively lesser and shorter-lived structures. And the system continues today; there are even catalogues of so-called 'pole' barns and cowsheds.

It was the introduction of the framed wall around 1100 to 1200, at first probably for buildings required to be 'permanent', that revolutionized structural methods and gradually eliminated lap-joints. This was the box-frame system, properly confined to that class of buildings with rafter roofs of which the load is taken through the rafters on to the lateral as well as end walls. A characteristic, directly deriving from post construction, and which persisted throughout the period of the timber-frame tradition in the lowland counties, is the close-studding of the wall-frames and absence of rails, exactly as open-palisading in post construction.

Development was extremely rapid, at any rate in comparison with the ages required for the evolution of the rafter roof. The next stage had to be the introduction of cross-beams, as otherwise the posts or studs, no longer restrained by the ground, would have spread at wall-plate

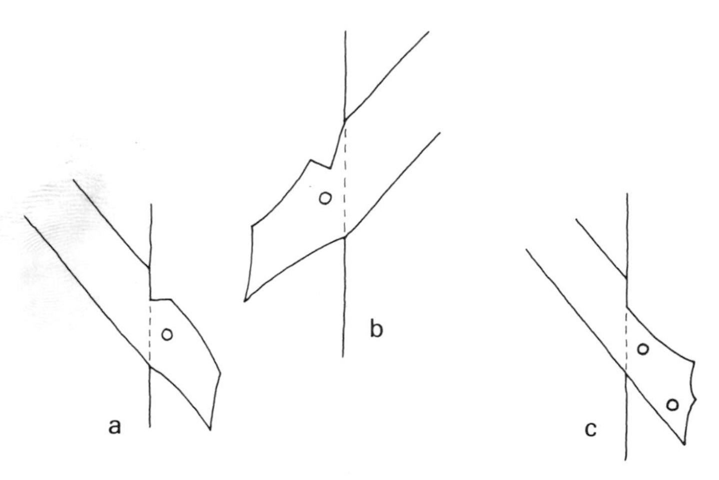

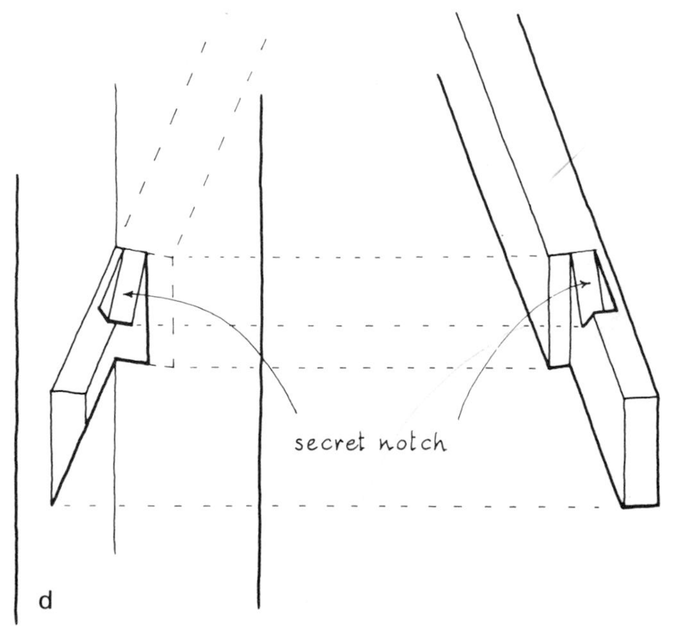

Figure 7 *Examples of lap joints*
(a), (b) Notched half-lap joints noted by Elna Möller at Pembridge Belfry and compared with
(c) A Danish example of the thirteenth century
(d) Secret-notch half-lap joint at Cressing Temple Barn, Essex. Courtesy Cecil Hewett

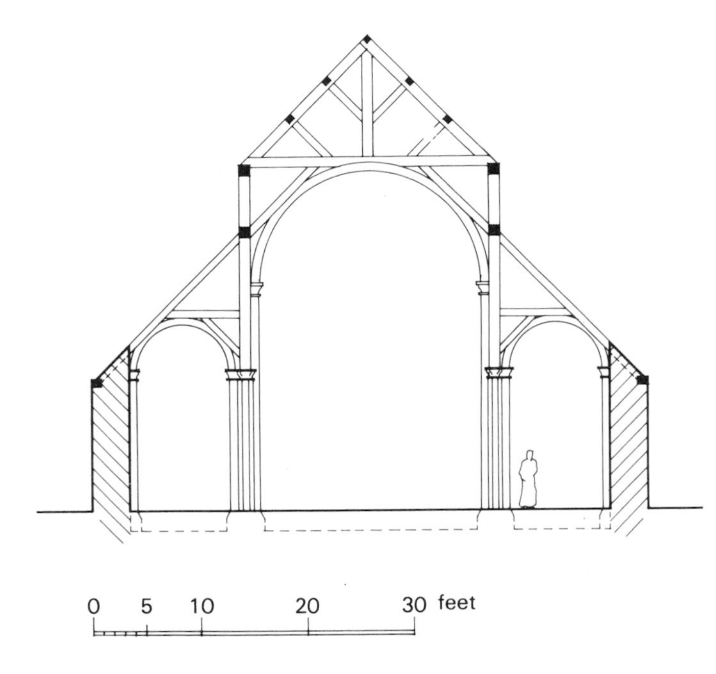

Figure 8 *Section through Great Hall of Bishop's Palace at Hereford*

level under the thrust of the roof. The cross-beams for this purpose were thus wall-members, not roof-members. But it was also convenient to tenon the rafter feet into the cross-beams wherever they coincided. Hence the first appearance of trusses in rafter roofs. It was also an advantage to incorporate intermediate or principal posts within the wall-frame which could be braced to the cross-beams and so resist transverse racking.

The final step was to make use of the cross-beam as a means of additionally supporting the roof, so permitting not only a wider span but also greater resistance to spread. This was done by setting a crown-post on the centre of the beam to support a collar-plate – perhaps better termed 'crown-plate' (see Figure 1(a)). Its function was to support each rafter couple at midspan of its collar. The effect is to alter the stresses in the rafters so that the tendency to spread at their feet is reduced. Under certain conditions of loading the direction of stress could be reversed, but in proportion the rafters would tend to spring apart at their heads (Figure 9). The best joint to resist this is clearly the bare-face dovetailed tenon (Figure 10), but strangely this has so far turned up only in France.[8] In this country

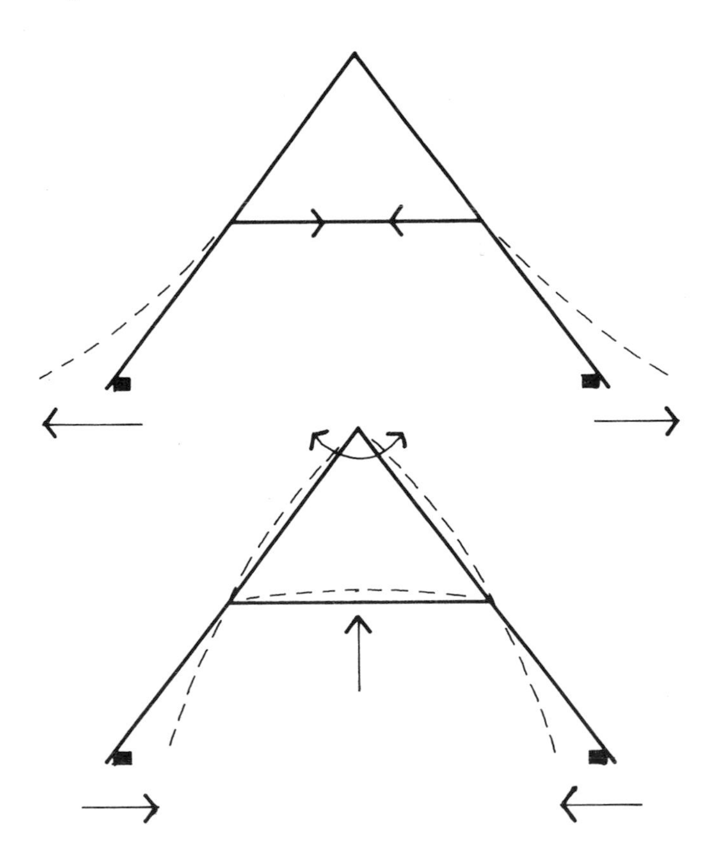

Figure 9 *Stresses in components of A-frame rafter roof*

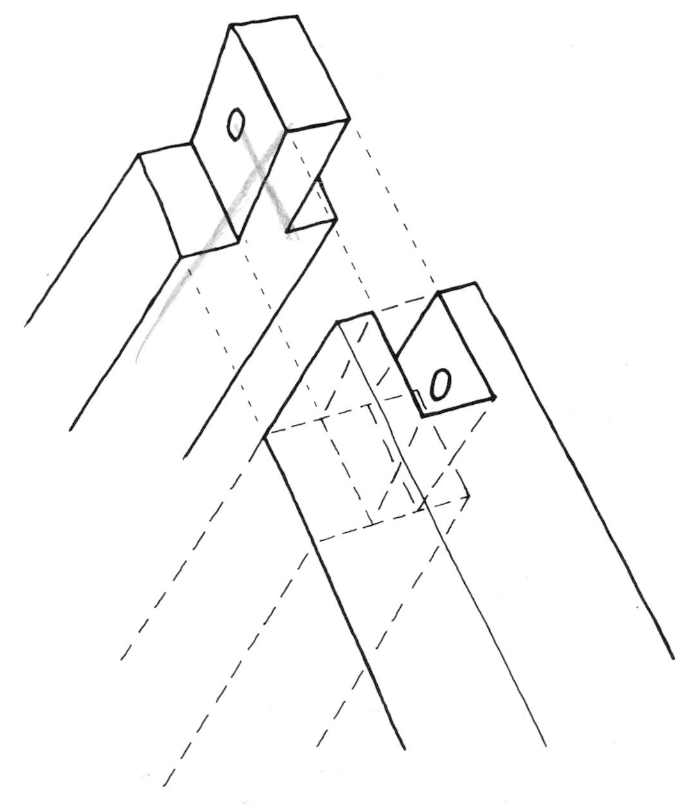

Figure 10 *Ridge joint with dovetailed tenon*

Figure 11 *Typical crown-post of a Wealden House*

Figure 12 *Wealden House at Benenden, Kent*

the ordinary bare-face tenon or even half-lap continued to be used, the peg being relied upon to resist shear.

The crown-post truss was still not a true triangulated truss, the tie-beam merely providing a strong enough footing for the crown-post on to which a large proportion of the roof load was exerted via the crown-plate. The central purpose of the post, to act as a column, at least as seen by the designer, is reflected in its generally square, polygonal or cylindrical form (Figure 11). It is also in most medieval halls richly moulded and carved and elaborated by a complex of braces from tie-beam to post and from post to collar and plate. That both tie-beam and crown-post are in fact tension members would surely have been incomprehensible to the medieval designer.

Thus with the introduction of framed walls not only did structural technique change but architectural style as well. Now carefully selected swept timbers were halved to make matching pairs; the wall-frames and roof were structurally unified; the moulding of timbers, sometimes extremely elaborate, was designed to give a continuous profile to the central open 'truss' of hall or upper chamber; and the arch effect was accentuated by a steeply cambered tie-beam modulated to the braces and jowled posts. The style became Gothic.

No house type more elegantly achieves all this, as well as architectural unity of different elements, than does the Kentish Wealden (Figure 12).[9] In this type, fortunately still numerous, the two-bay hall, open to the roof, is flanked by a two-storey jettied bay at each end. The complete plan is enveloped by an enormous hipped roof, sometimes with gablets at the terminal trusses of the hall to let out the smoke. The jettied ends form a recess across the hall with a system of double wall-plates, of which the outer is longitudinally braced from the adjacent corner posts of the jettied bays. This recess is the feature by which the Wealden has come to be defined, to the exclusion of their no less important rafter roofs and hipped ends. Thus Wealdens, though of a debased sort and often better termed 'half-floored' halls (see Figure 91) may be found a hundred miles and more from Kent.

Purlin roofs

All other types of structure have purlin roofs. Their structural principle is that the load of the roof is conveyed by horizontal timbers (the ridge-pole, purlins and wall-plates) to trusses and cross-frames, the two being structurally integral – in the modern term, a portal-frame. The structure is thus essentially bay-divided, and the end walls, or frames, must be gabled to support the purlins and ridge-piece (should there be one). In some cases they are half-hipped, the wall being built only high enough to take the purlins; the ridge is cantilevered over the adjacent truss. Longitudinal racking is resisted by wind-bracing by which the right angle in the plane of the roof between principal rafter and purlin is triangulated.

The purlin roof is Mediterranean, the Doric temple being the most often cited rationalization or skeuomorph in stone or marble of a timber structure with purlin roof.[10] Vitruvius describes the type with multiple purlins and close-spaced trusses. No doubt the Romans adapted it for more northerly climates by increasing the pitch. It was this that resulted in the transformation of the Mediterranean roof to the kind we are concerned with. The trusses had to be taller and so fully framed; thus the bays were lengthened to economize in their number. The number of purlins was also reduced to one or two to each slope, so that rafters, absent from the multiple-purlin roof, had to be laid over them – hence the 'double' roof.

These changes also necessitated longitudinal bracing. In France, in contrast with the familiar wind-bracing of English roofs, the system was to span the length of the building with what is virtually a lattice-girder. This consists of a top and lower ridge-piece braced to each other and to the king-posts, standing on the collar or less often the tie-beam. The same system was used in the larger French rafter roofs, above the collars, and still survives as a normal method in housebuilding (Figure 13).

The purlin roof, once the problems of handling and lifting large timbers had been mastered, had several advantages over the rafter roof. First, the number of joints to be cut was substantially reduced, since there was only one frame element to each bay as compared with seven or eight more or less complicated rafter couples. Second, during erection, instead of having the whole space above the wall-head cluttered with rafters awaiting assemblage or rearing, the structure could be completed through the length of the building and then the rafters laid on singly. With two gangs at work, the one following the other, the time required for construction would be practically halved.

Thus the introduction of the purlin roof was a major step in structural technique and articulation of operations. Not surprisingly it superseded the rafter roof even in its south-eastern homeland once the open hall had gone out of fashion. A convenient date for this change is *c.*1550, though it must have been proceeding through a century or more.

There were almost as many designs of medieval trusses as there were roofs – even within one roof all the trusses

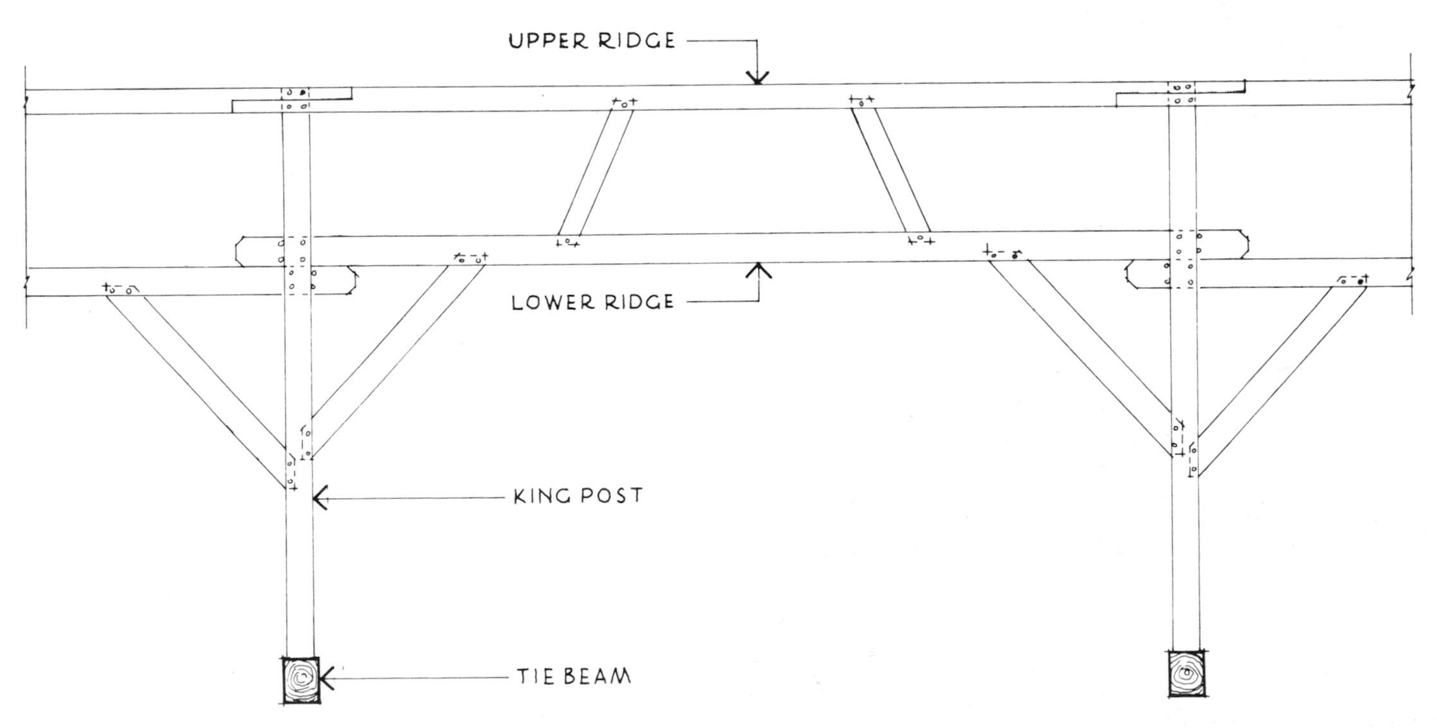

Figure 13 *Roof of barn reconstructed at Abbey of St Wandrille, Normandy, with king-posts and braced 'lattice' ridge-piece*

might be different (see the description of Chorley House, page 218). All of them were designed according on the one hand to the way the roof structure would be assembled, and on the other to the stressing and loading each component could be *seen* to have to resist. Medieval trusses are thus highly unscientific – and no less elegant. In general their timbers are substantial, particularly in the depth of the halved-face, the tie-beam again being looked on as a beam carrying a load rather than as a tie without need for compressive strength. The principals are generally also (and rightly) of heavy section, having to take the weight of the purlins. By contrast, the purlins in many of the earlier roofs are noticeably slender, and even may not be primarily intended as load-bearing members. In at least one example their purpose seems rather to have been that of distance-pieces in the erection of the roof, each successive truss being reared to the awaiting purlins and wind-braces of the completed bay. At Amberley Court, near Hereford, the rafters lie a good 6 inches above the purlin (see Figure 274).

Another early form is the compound principal, made up of two members of common rafter dimensions held apart by tenoned blocks (Figure 14). The purlins of the same depth as the space between the principals are threaded through the rafters or, to be more exact, are set on the inner rafter against the block, and the outer rafter is then laid on to it. At the apex the inner rafters are cross-lapped and, since the roof pitch is invariably more than 45 degrees, the space circumscribed by all four of them is diamond shaped, and the ridge-piece must be the same in section. Compound rafters are again reminiscent of post construction and rafter roofs, but they are not necessarily

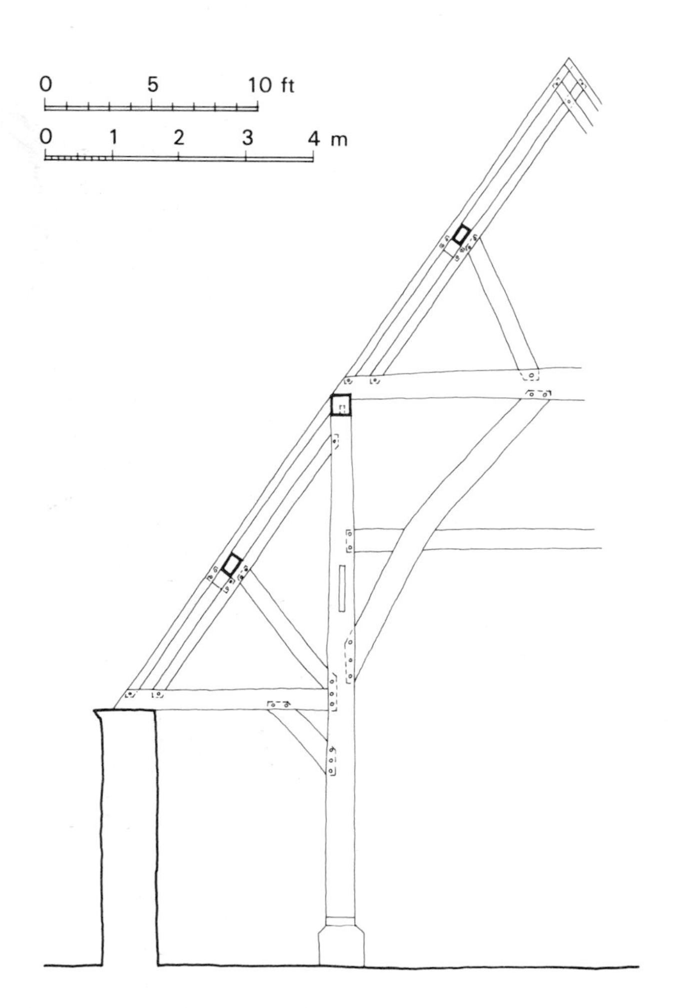

Figure 14 *Roof truss at Bredon Tithe Barn, near Tewkesbury showing compound-rafter system and strainer-beam*

Figure 15 *Roof of Chorley House, with arcading effect of wind-braces*

Figure 16 *Guesten Hall roof, originally at Worcester Cathedral. Courtesy John Greaves Smith*

earlier than the solid principal; the contrast points rather to different traditions, the former of course to post construction.

Many medieval roofs have intermediate arch-braced trusses (without tie-beam) within each bay, generally standing directly on the wall-plates. As well as reducing the purlin span, they provide a footing for an additional pair of wind-braces, so that the whole effect of such a roof is of Gothic arcading down its whole length (Figure 15). With double or triple purlins, as in the Guesten Hall roof from Worcester Cathedral, the wind-braces often become the dominating element of the roof's design – sometimes a sheer extravaganza (Figure 16).[11]

A technical point to note is that in all medieval roofs of quality (and there are not very many without it) the principals stand above the plane of the wind-braces and purlins, acting as common rafters (Figure 17). In the later, less architectural ones, the common rafters lie over the principals, and their spacing may be independent of the bay lengths. It was also a mark of good design in medieval roofs to conceal the joints. Thus purlins may appear to be a single timber as long as the building – in barns, 100 feet and more! In fact each purlin is only the length of a bay, or of two bays at the most, scarfed to the next within the thickness of the principal, either by splaying the tenons, if they are tenoned-purlins, or rebating their ends to form a square-butt scarf (Figure 18).

Figure 17 *Chorley House roof, showing principal rafters projecting above plane of purlins and wind-braces*

A very common medieval roof throughout the midlands is the so-called clasped-purlin type. This permits the truss to be built up on the tie-beam. First the struts are fixed to support the collar, the collar being shaped at each end to form a seating for the purlin. Then the principals, reduced in section exactly at the same level, are placed on to the tenons cocked upwards at the ends of the collar, making the seating into a recess or trench (Figure 18). All the trusses can thus be completed before threading the purlins through them and dropping them in. Occasionally there is a tenon at the cut-back of the principal to engage a mortice in the underside of the purlin. The system represents a further step towards articulation of construction. Instead of having to complete each bay before the next truss could be erected, assemblage even of the primary roof structure now becomes a follow-on of operations, each of which could be completed from end to end of the structure.

Tenoned- and clasped-purlins persisted until, again, the open roof was superseded. Then the common system of trenching the purlin into the back of the principal became general, as no doubt it always had been for the more utilitarian buildings. By this time, too, the length of the bays had been increased so that the purlins had to be of larger section. Wind-braces were used more functionally and sparingly, but in the superior buildings they were still elegantly shaped and chamfered.

In all types of roofs in which the purlins ride over the principals the joints are of course visible, though not always easy to identify. The straight splayed-scarf, relying on pegs to secure it, is the simplest. The tabled-scarf prevents longitudinal slip, and the addition of squint-butts also locks them longitudinally. But the bridled-scarf, probably the commonest in the better roofs, is the most effective, preventing sideways slip and, by pegging, longitudinal movement. Its only weakness is its liability to shear vertically if any subsidence of the adjacent cross-frame, by no means unusual, causes one purlin to exert too much pressure on the floor of the other's mortice. Bredon Tithe Barn has the most refined bridle-scarfs of all (Figure 19). The bridles are very long – about 20 inches – and tapered, as if to transfer the load of the one member (in this case, arcade-plates as well as purlins) to the other *gradually*; the bridle is of course also strengthened at the point of shear. Every such joint has to be made with greatest accuracy, which may have been another reason for its design. Needless to say, the structural concept is again wholly unscientific.

Figure 18 *Purlin-to-principal joints*

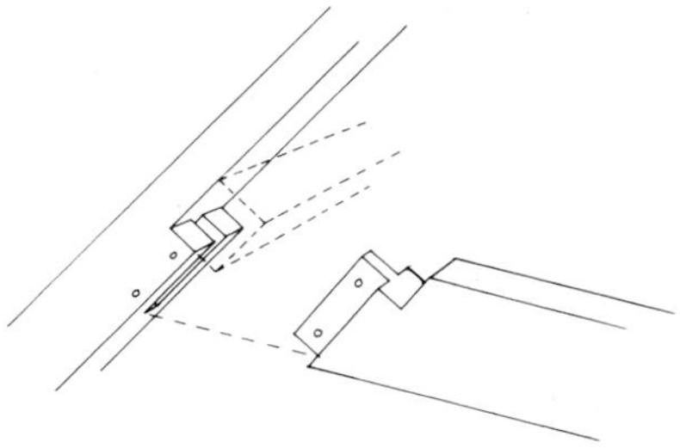

(a) *Clasp-purlin with reduced principal*

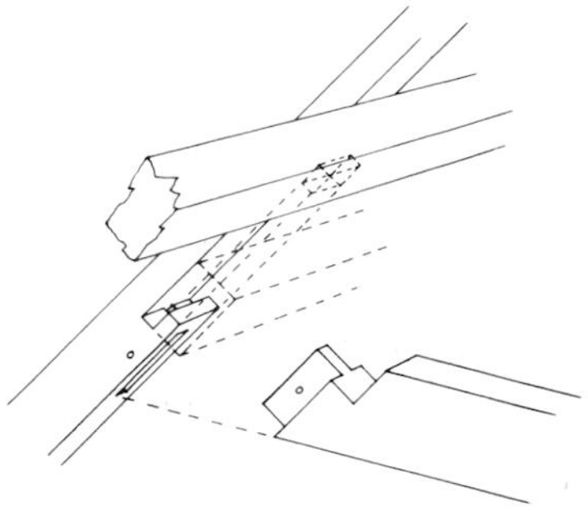

(b) *Clasp-purlin with principal tenoned into purlin*

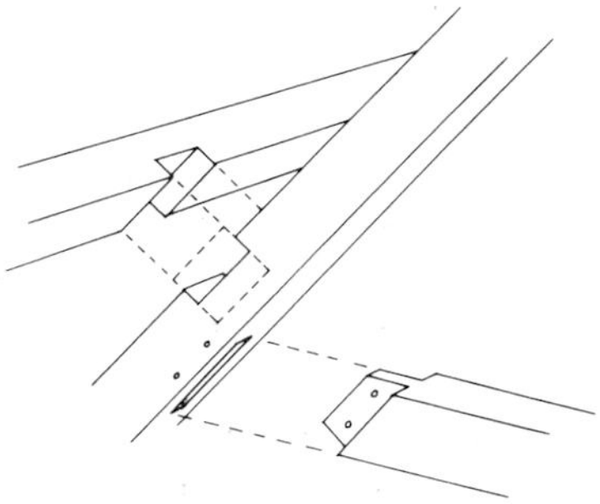

(c) *Trenched-purlin*

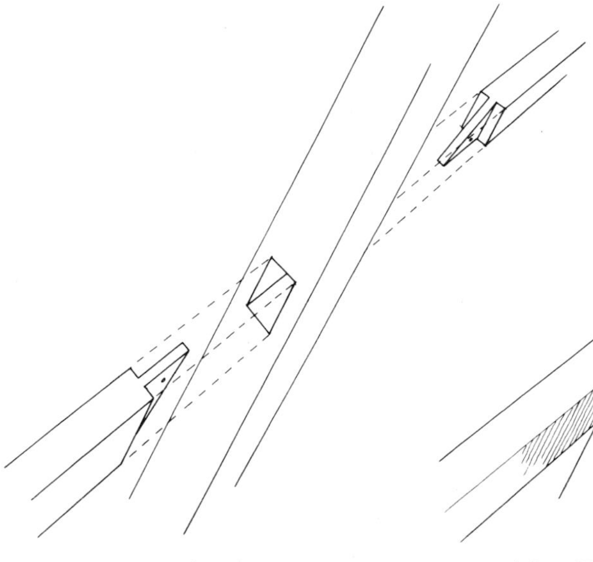

(d) *Tenoned-purlin*

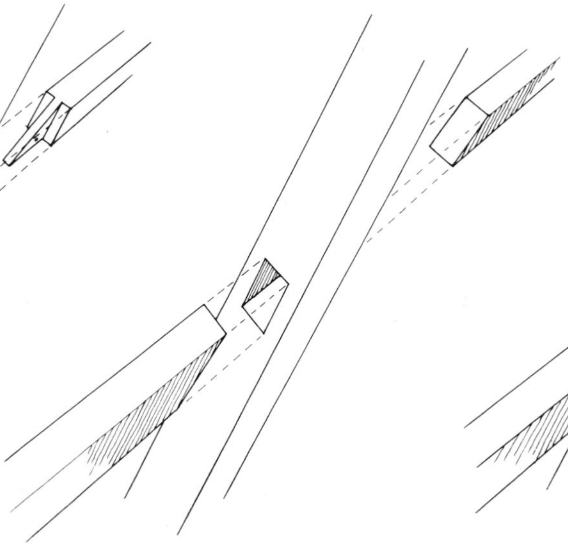

(e) *Through splayed-scarf purlin*

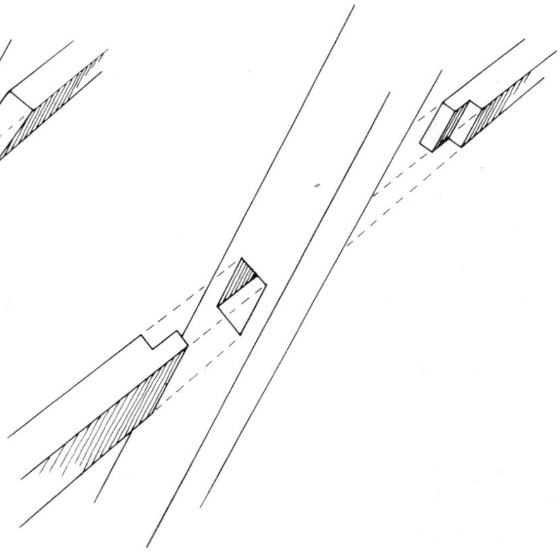

(f) *Through halved-scarf purlin*

Figure 19 *Bridled-scarfs*

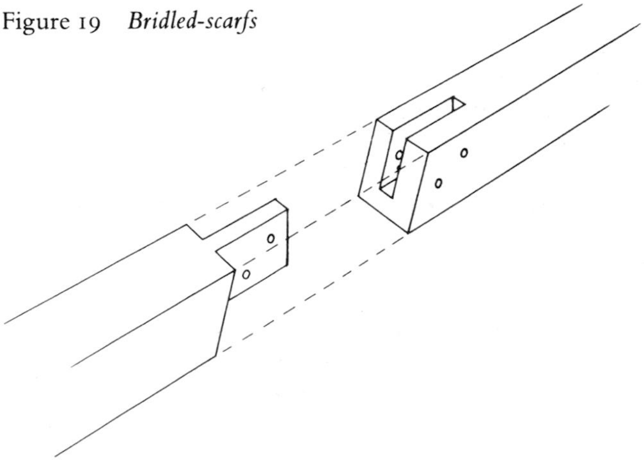

(a) *Typical joint*

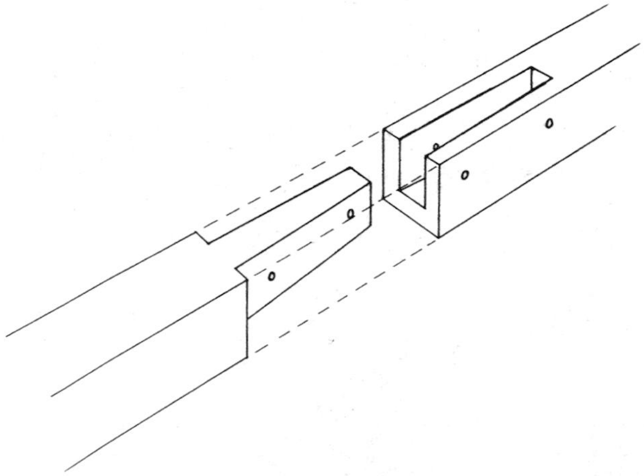

(b) *Tapered joint found only at Bredon Barn*

Post-and-truss

The combination of the purlin roof and wall-frame produced the largest category of all timber-framed buildings – post-and-truss construction. The system is the most unified in its structural concept and erection process. The framing of such a building is a continuing operation from upper corner post to diagonally opposite lower corner post and, unlike box-frame structures, each bay is triangulated, including the roof. The result could be said to be over-strong. The wall-frames are perfectly capable of supporting the roof load, yet the roof is carried by the cross-frames; the walls, theoretically at least, are thus merely weather screens. Over-strength, however, was hardly a concern of the medieval carpenter, which may be one good reason why we still have so many of his structures.

The construction and erection of a hypothetical post-and-truss building is fully described in Chapter 3.

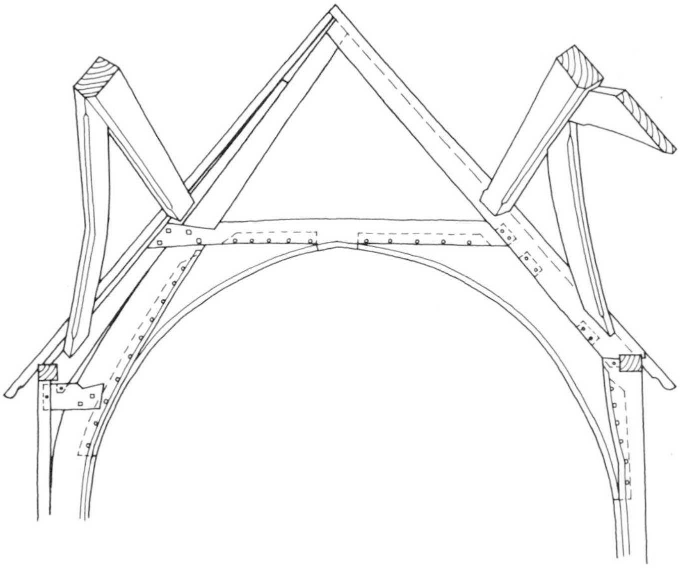

Figure 21 *Composite drawing to show similarity between post-and-truss and cruck construction in typical cross-frame*

Crucks

The structural principle of crucks is no different from that of post-and-truss. The cruck cross-frame represents a two-

Figure 20 *Interior of Leigh Court Tithe Barn, near Worcester. Courtesy Walter Horn*

Figure 22 *Great Coxwell Barn, Berkshire, showing main aisle-post cross-frames and cruck-like intermediate 'posts'*

Figure 23 *Bedern Hall, York, with rafter roof and cruck-like intermediate trusses*

pin portal, designed essentially for framing on the ground and rearing. It consists of two blades which, roughly conforming to the profile of the vertical wall and traditional pitched roof, are continuous from ground or plinth to ridge – or very close to the ridge. The blades are connected by one or more cross-members, the whole framing having to be sufficiently strong to withstand the stresses it must be subjected to while being reared (Figure 20).

Formally, as well as in principle, the cruck and post-and-truss may bear a close resemblance. In the hypothetical illustration of Figure 21 the main difference is in the typical shaping of the members, those of the cruck being sharply elbowed in comparison with the more curved sweeps of the post-and-truss.

There are examples of cruck and post-and-truss cross-frames within the same building, the cruck usually within the structure, the post-and-truss frames at the gable ends. But despite obvious kinship there is little likelihood that one developed from the other. Rather there was a gradual supersession of crucks by post-and-truss construction towards the end of the Middle Ages; the use of the former was confined after about 1500 to lesser structures and cottages, though in the more upland areas cruck construction of a rough kind persisted even into the nineteenth century.

If we accept that rearing is part of the definition of the cruck, there are many 'false' crucks, their curved members being better described as curved principals.[12] In the Low Countries and Rhinelands are those high roofs containing two or three attics each supported on inclined posts, elbowed a few feet above the cross-beam into which they are tenoned. Similar, if less ambitious, attics occur in this country in eighteenth- and nineteenth-century buildings, including Georgian houses, in which the principals are tenoned into floor beams 3 or 4 feet below the wall-plate, thus allowing headroom below the collar and through-access longitudinally which would otherwise have been blocked by tie-beams.

The two best examples of medieval false crucks in this country are at Great Coxwell Tithe Barn[13] and the Bedern Hall at York (Figures 22 and 23). At the former they are intermediate trusses between the main aisle-post cross-frames; at the latter, a scissor-braced rafter roof is 'reinforced' by a pair of huge swept braces midway down the length of the hall – incidentally, an upper hall. Both of these buildings are *c.*1300.

When the true cruck was first used cannot be discovered by archaeology (however ardent, even feverish, the search) for the simple reason that no reared structure leaves identifiable evidence in the ground. Claims based on inclined holes as indicating crucks can therefore be discounted, especially since the lower part of a cruck-blade, if it ever had penetrated the ground, should have left a vertical hole! Our only aids therefore are extrapolation, documentary evidence or standing buildings.[14]

The hog-back tombstones of north-western England have been cited as reproducing the cruck form, suggesting that the cruck was of Viking origin.[15] This, however, is contradicted, first by the absence of convincing evidence that cruck buildings were ever hog-backed, and second by the fact that one of the tombstones clearly shows a king-post truss. The Viking examples do no more than indicate by their tile pattern that they represent roofs.

More promising is a piece of literary evidence suggesting that by the sixth century at least one structure with the form and strength of the true cruck already existed. The following is from a book, still to be published, on the origins of crucks.[16] The author writes:

> In his *Manners and Customs of the Ancient Irish*, Eugene O'Curry discusses the story of the construction of a wooden oratory for St Moling, the sixth-century founder of the monastery now known as St Mullins in Carlow, as reported in the ancient Gaedhelic life of the Saint, the Book of St Moling:
>
> 'The oratory was built by a great artificer, called Gobban Saor, who, at his wife's instigation, demanded as his fee for the construction of the church "its full of rye". St Moling when appraised of this replied, "Invert it and turn its mouth up, and it shall be filled for thee . . . ." So Gobban applied machinery and force to the oratory, so that he turned it upside down, and not a plank gave the smallest way beyond another.'

Only a true cruck could have stood such treatment.

Hardly less intriguing are certain standing oratories and churches of Western Ireland. Near the coast in the very region in which cruck-shaped oaks would, and once clearly did, grow (page 48) are several little stone buildings of cruck form. The best known is Gallerus Oratory (Figure 24). As in all structures of corbelled stonework, the profile tends to be shaped in the form of an elbowed cruck rather than a segmental arch. Also in rectangular, as opposed to circular buildings ('beehives') of corbelled construction, there is a weakness at the ridge midway between the end walls. At Kevin's Kitchen at the monastery of Glendalough this point is supported by an internal elbowed stone arch. Elsewhere the roofs have collapsed and it is believed that wooden supports, of cruck form, must have held them up until the wood rotted.[17]

But the more convincing evidence of true crucks is in the architectural details of the now ruined churches. The church of Kilmalkedar (Figure 25),[18] only a few miles from Gallerus Oratory, has external crucks reproduced in stone on its west front, the finial suggesting the crossing of the cruck-blades at the apex. Other 'carpentry' features are the projecting eaves course along the side walls with a carved corbel stone at one end, representing, exactly as in cruck construction, the wall-plate and beam-end of the truss. Lastly, inside the church, blind 'arcading' in the side walls has pilasters with capitals, but instead of arches there are lintels. These, instead of being set *on* the capitals, as in any self-respecting stone structure, are incongruously 'jointed' into the sides of them, as in carpentry. The earliest of these 'cruck' churches is on St Macdara's Isle, dated by the authorities to *c.*700.[19] Timber crucks must, however, have been in existence long before that in order to establish their form and details as an architectural style worthy to be followed in masonry.

Figure 24 *Gallerus Oratory, Dingle Peninsula, County Kerry, Eire, showing corbelled construction*

That is as far back as the evidence goes, but at least it is far enough to demolish the theory, once put forward, that cruck construction is the carpenter's version of the Gothic arch.[20] More likely it antedates timber framing even by millennia. In support of that, the use of half-lap joints in all but the most 'architectural' of the crucks; the fact that it is a form – *the* form – designed essentially for rearing, the only method by which heavy structures could be raised when, or where, there was no scaffolding or lifting tackle; and finally its archaic appearance, a point perhaps not valid to the dispassionate archaeologist, but the most telling in architecture; suggest extremely remote origins.

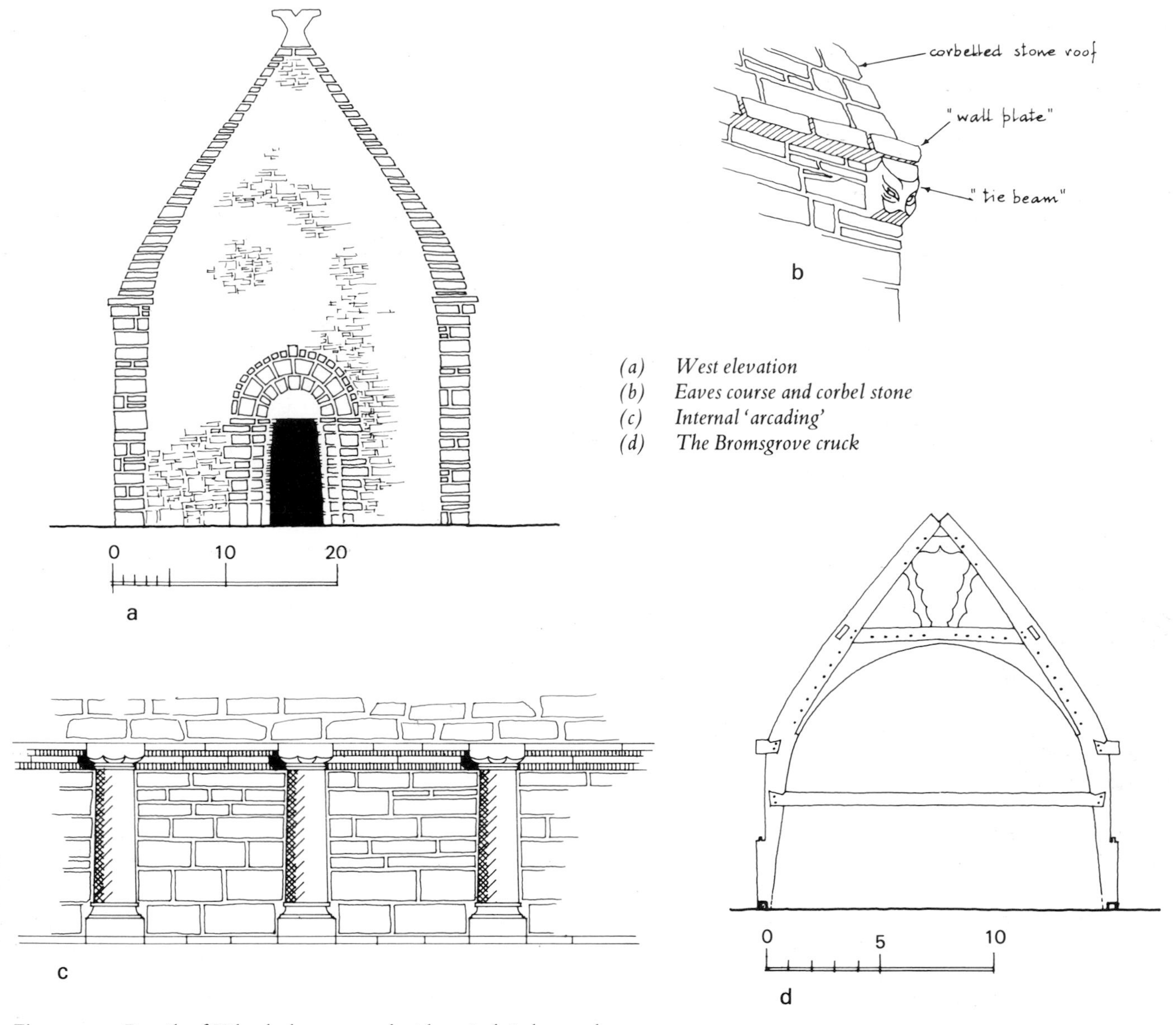

(*a*) *West elevation*
(*b*) *Eaves course and corbel stone*
(*c*) *Internal 'arcading'*
(*d*) *The Bromsgrove cruck*

Figure 25 *Details of Kilmakedar compared with typical timber cruck*

Construction and type

In the simplest of crucks the cross-frames consist of only three members – a pair of blades and a cross-beam – though in practice tapered principal rafters may also have to be pegged to the back of the blades to even out the roof-slope from ridge to wall-plate. The ridge-pole is supported in the forked apex of the blades, half-lapped over each other, the purlins in trenches cut out of the blades, and the wall-plates on the projecting ends of the cross-beams. Longitudinally, as in all bay-divided structures with purlins, every bay has to be braced. The timbers can be set down directly on the building plot instead of on a framing floor and they need not have been prefabricated or numbered. They must only fit accurately to preplanned fixed points – the cruck-feet, the extremities of the cross-beam and the apex.

If the timbers are large – and the deeper the cross-beam the better it will resist torsion when the truss is being reared – as many as four pegs may be used to secure the half-laps. Each is driven into the 'upper' face and skewed in a different direction. The pegs are not on the same horizontal line but rather located at random to avoid splitting the cross-beam along the grain. They often do not penetrate the lower face of the blades.

In the process of rearing, a means had to be devised to prevent the heavy truss from slipping forward as the point of lift approached its centre of gravity, just as a ladder must be held at the bottom rung as the other end is lifted. Many cruck-blades will be found to have a dead-mortice, or rectangular hole, 4 or 5 inches deep, a few inches above the foot on its upper face. In this was a projecting peg. A rope was looped round the foot of the blade and the purpose of the peg was to make sure that the rope could not ride up as it took the strain. An actual peg has survived in Middle Littleton Tithe Barn and may be seen in the restored building, though before that it had been bedded in the stonework for seven hundred years (Figure 26)!

There are many variations, especially at the head of the blades in the way the blades meet and in the design of the ridge. The simplest method, as noted, is the lapped crossover joint with a diagonal ridge-piece set in the projecting ends of the blades; but the blades may as readily be connected by a yoke, so that they abut each other without any joint, and a V-cut is made for the ridge-piece. Or, yet again, the heads may not meet at all but are joined by a saddle, sometimes with a post set on top of it to obtain more height, and the ridge-piece may again be diagonally set or have a flat soffit resting on the saddle or post (Figure 27(a) and (b)). Both the yoke and saddle methods allow

Figure 26 *Middle Littleton Tithe Barn, near Evesham when cruck peg was exposed during restoration*

Figure 27 *Leigh Court Tithe Barn – variation of cruck-head joint and ridge piece. Courtesy Walter Horn*

(a) Saddle with flat ridge

(b) Short post and 'V'-set ridge

Figure 28 *Reconstruction of typical cruck hall*

greater height and wider span for blades of the same length – the latter clearly being the most flexible in this respect. Cruck-blades were also sometimes heightened by scarfing-on additional pieces, but only in the rather poorer barns. More than one system may also be found in the same building.

Some researchers[21] are tempted to date crucks by these variations, and even postulate a technical evolution of cruck construction; however, having regard to the ages which carpenters had in which to learn the tricks, this is little more than an archaeological pastime. Even the present regional pattern presented by such details must be accepted with some reserve, for wastage, especially of the lesser buildings, must have altered it considerably.

It was again the introduction of the framed wall and techniques of the timber-framed tradition, especially the emphasis on halving and hence mirror images, that transformed the simple cruck into a work of architecture. The cruck frames were now reared on sill-beams and the use of the mortice-and-tenon, instead of lap-joints, resulted in arch-braced trusses, affording much greater headroom than could be obtained with cross-beams, which had to be at wall-plate level. Architecturally the halved blades, huge multipegged braces and steeply cambered collar, often shouldered and shaped to complete the pointed arch, produced crucks that were indeed Gothic and in every way comparable with the halls of the post-and-truss or box-frame construction (Figure 28).

But there were problems in combining such disparate systems as the half-timber wall with the cruck. First, the joint between the wall-post or stud and the back of the blade was never finally resolved. The blade in some instances has a step cut into it for the post to sit on, while in others the post is tenoned and pegged to the back of the blade, or in yet others both post and blade are tenoned into the sill-beams. The joints at wall-head level where the cruck, wall-plate, wall-post, cruck-spur and, most often, principal rafter all converge upon each other are even more complicated (Figure 29(a) and (b)). The framing and panelling of the end walls with the cruck-blades interfering with the normal rectangularity was also awkward, at least in appearance.

An anomaly of a slightly different kind is that the large swept or elbowed brace in the side walls of most West Midland crucks, apparently for the purpose of propping the truss when it had been reared, fulfils no such purpose. For the wall-frame was assembled independently, probably when all the crucks were up and otherwise secured. Such braces thus merely announce the nature of the underlying structure. In short, the cruck and

Figure 29 *Leigh Court Tithe Barn*
(a) Cruck joints

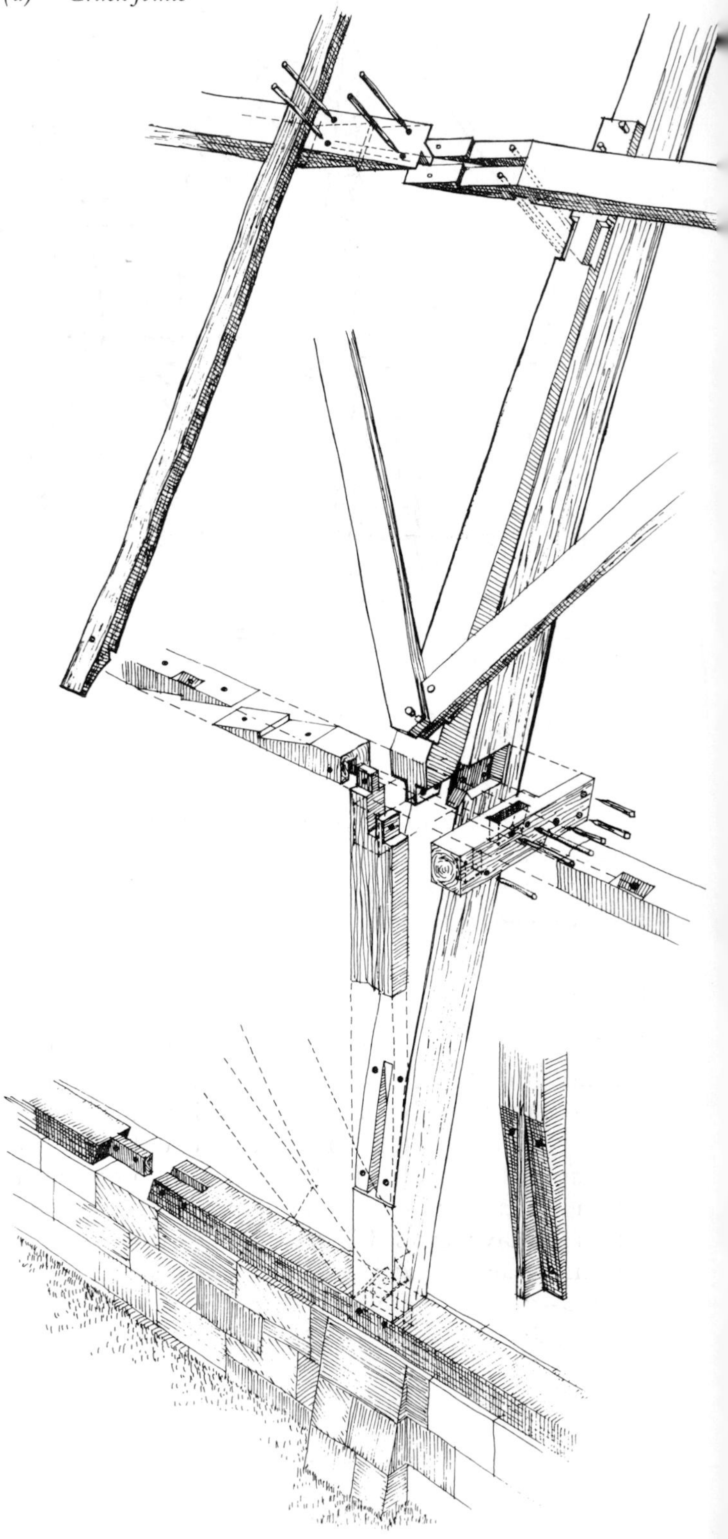

timber-frame tradition never made an entirely comfortable marriage.

The rearing of the cruck structure is best described by drawings (Figure 30).

Base-cruck

Finally we come full circle to the base-cruck, which has elements of all the structural forms we have discussed. Its cruck connection is obvious, the only difference being that the blades are terminated half-way up the roof where they are connected by a collar-beam. Thus with shorter blades than the simple cruck of the same height the span may be considerably increased (Figure 31). The medieval base-cruck hall is also practically identical with those of aisled post-and-truss construction, the only difference here being that the main open truss, or trusses, have swept or elbowed instead of straight posts, so clearing the floor of obstructions. In such halls – invariably the more grand of such buildings – the spere-truss and closed end-trusses have aisle-posts directly supporting the arcade-plates. Lastly, above the collar-beam (or nave tie-beam) there may be either a purlin or a rafter roof (Figure 32).

Thus all of these methods of construction had been invented or developed by the time these halls, as well as numerous tithe barns with the same combination of elements, were built. And they are all early in date. West Bromwich Manor House is *c.*1200.[22] Eastington Hall may be even earlier. Siddington Barn looks as if it may be even of the eleventh century (see Chapter 11). These structures – and there are others – are distinguished both by half-lap joints, instead of mortice-and-tenon, in their upper roof, and by straight scissor- or passing-braces of uniform slender strength. Siddington Barn is half-lapped throughout, even in its immensely heavy substructure.

The distribution of base-crucks, though much fewer in number than simple crucks (a hundred or so, compared with over three thousand of the latter so far recorded) is very much wider. But for those beyond the true cruck region the term 'false' may again have to be applied, as they were not reared – at any rate so far as is at present known. Tickerage in Sussex (Figure 31) is an example of an assembled timber-framed base-cruck,[23] while the Old Deanery at Salisbury is one with stone walls, the latter with scissor-braces of unusually large section but wholly mortice-and-tenoned.[24]

Figure 30 *Series of figures illustrating the design and rearing of a typical medieval cruck hall, based on the demolished Lower Norchard Cottage at Peopleton (Chapter 11) and the 'Bromsgrove Cruck' of which only the central truss of the hall was salvaged and is now at the Avoncroft Museum*

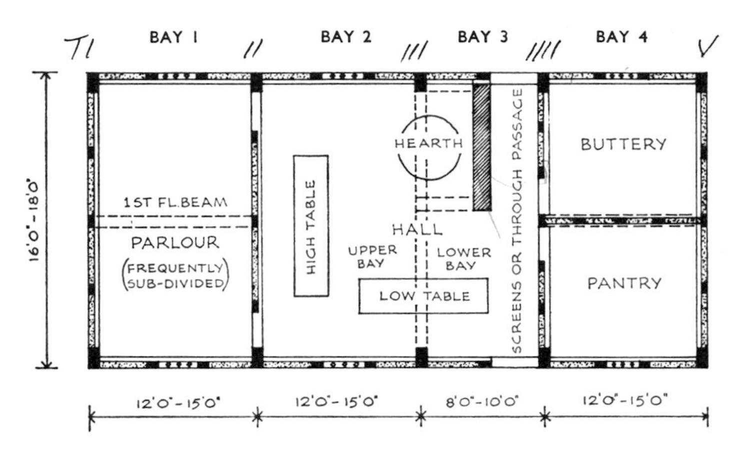

(a) The plan shows the typical two-bay hall, with the lower bay (3) as the slightly shorter and containing the through-passage. Also typical is the position of the later inserted fireplace, backing on to the passage

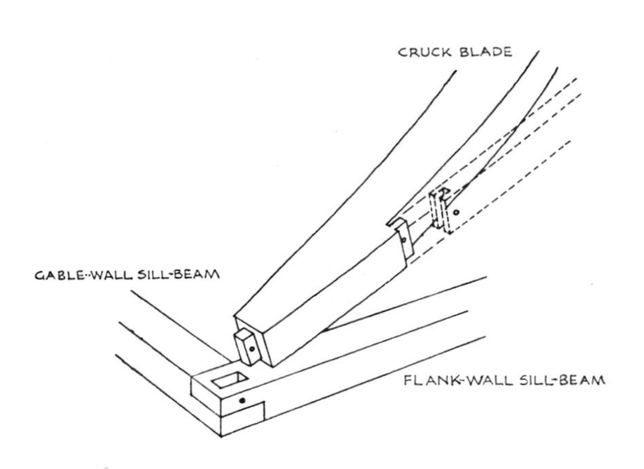

(c) Shows the method of securing the foot of the post to the blade and of rearing the truss on to the sill-beam

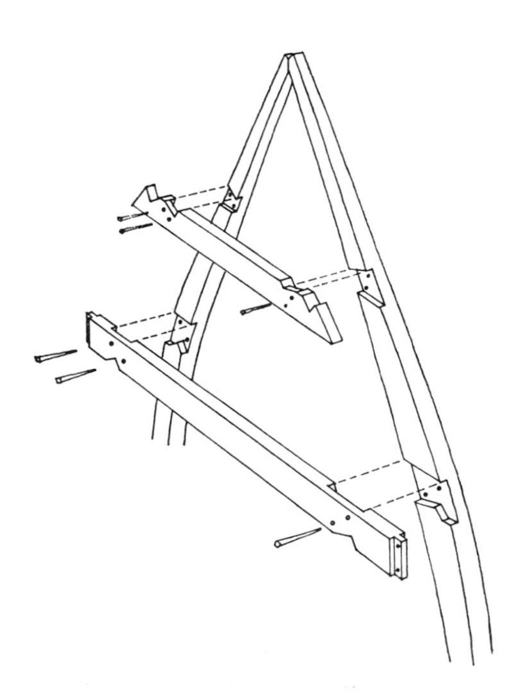

(b) Shows T I, the first to be reared, with the jointing of collar and cross-beam, the latter projecting beyond the blades to be tenoned into the wall-post which carries the plate

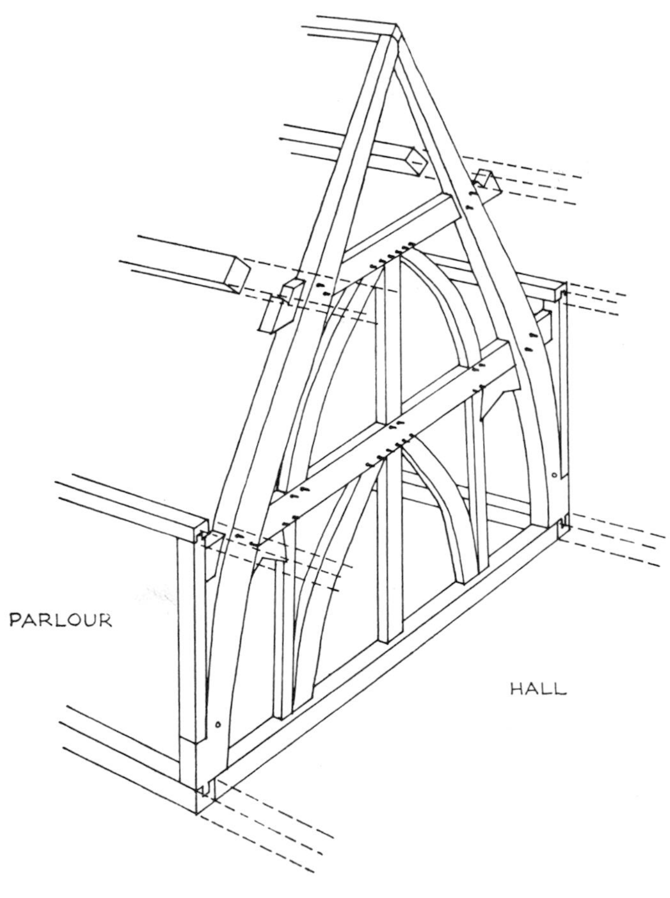

(d) Shows the 'upper' wall of the hall, an arrangement of cross-beams, centre-post and braces, typical also of other structural types in an area that includes even the home counties

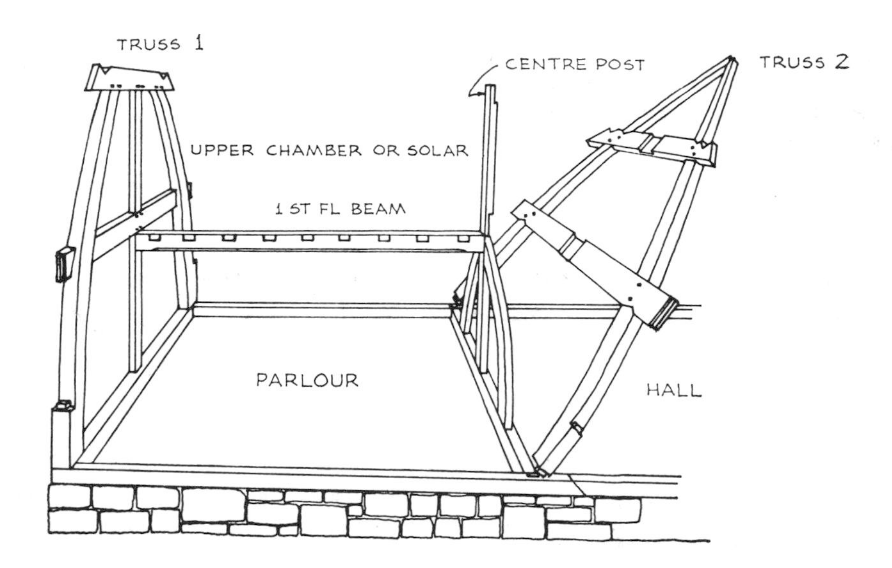

(e) Shows the structural function of the centre-post in the rearing of this truss, T II

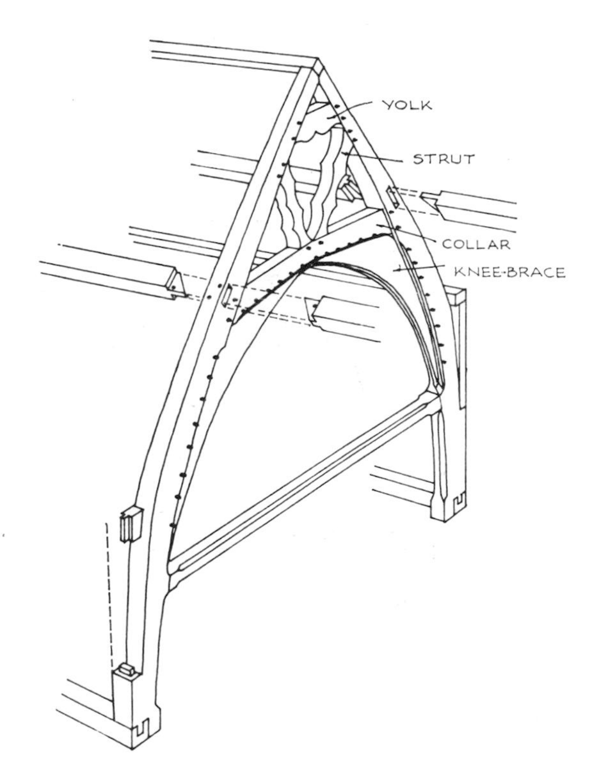

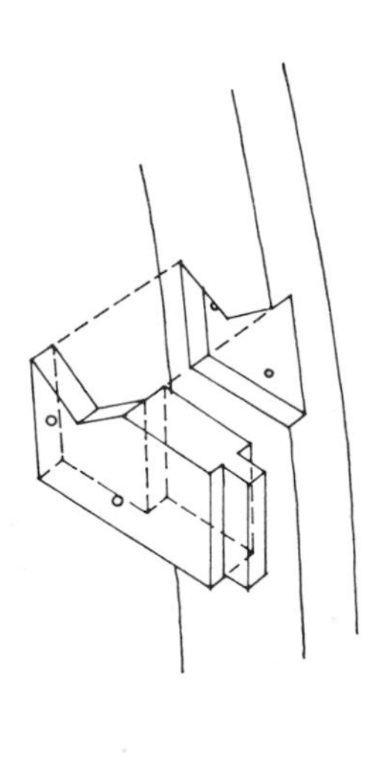

(f) Shows the central arch-truss demarcating the hall bays, always the most 'architectural' of the trusses. Points to note are: the mantle-beam serves the dual purpose of restraining 'spread' and as a means of hanging pots over the open hearth immediately beneath it; the blades are acutely enough angled to allow for tenoned-purlins, though in many cases the purlins are trenched into tapered principals lying on the back of the blades

(g) Shows the cruck-spur, notch-lapped or dovetail-lapped, always to the upper face of the blades

By no means every cruck structure follows the rules set by this composite example. There are variations in the sequence of rearing and anomalies in the construction of the trusses. In general, the poorer the quality of the building the more frequent the anomalies and the later its date

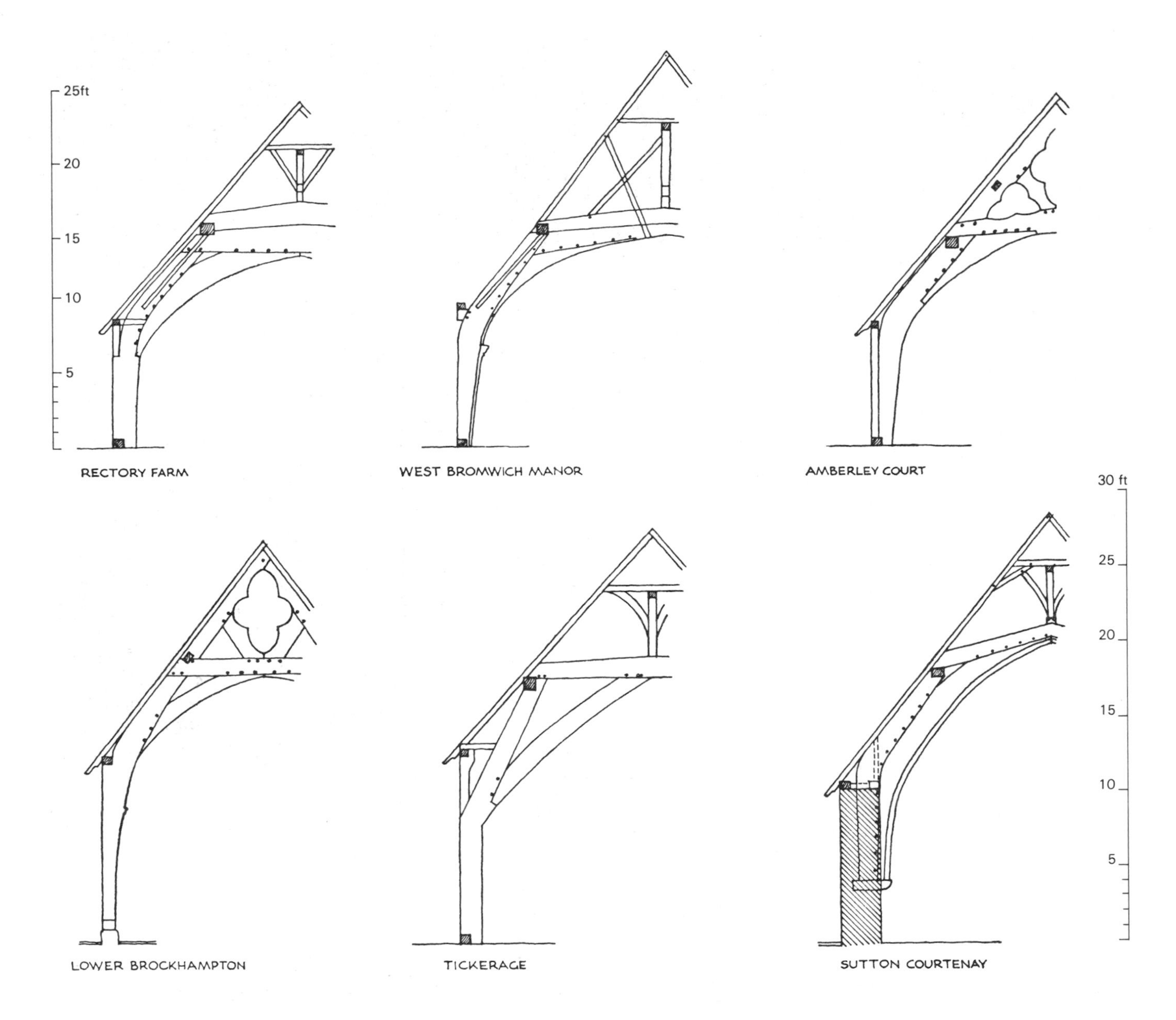

Figure 31 *Comparative base-crucks*

The means of recognition of rearing is to observe whether the cruck frame could have been completed with its blades, collar and braces on the floor, and reared *before* the plates were placed over or tenoned into them. In other words, was the frame 'closed' by means of the cross-beam connecting and braced to the blades? The most notable exception – an assembled base-cruck – is the barn at Frocester (Figure 33).[25] This, however, is an exceptional building, the original roof having been burnt down in the early sixteenth century and the present roof erected in 1546, utilizing the former cruck-blade *seatings* in the stone walls but entirely new cruck-blades. Rearing the new trusses would have been possible only by reducing the height of the standing walls by about 6 feet, and erecting a platform on which to frame and rear them. Instead, the blades had to be erected singly and held by scaffold until not only had the collars been tenoned over them, but also heavy knee-braces and struts 'offered up' to the collars

Figure 32 *Eastington Hall interior, as originally. By John Greaves Smith, courtesy Michael Dawes*

from underneath. The proof of this exceptionally awkward operation is in the wedging up of all the mortice-and-tenon joints, it having proved impossible to obtain tight joints between the blades, braces and collars by any other means.

Rearing was frequently the means of erecting aisled cross-frames as well as base-crucks, whether or not they were combined within the same building. The means of discovering the method here is again whether the frames are closed. For instance, those structures with a 'strainer'- or 'anchor'-beam set some distance below the post-heads and again braced (see Figure 14), as well as those in which the arcade-plate is set *on top* of the tie-beam instead of beneath it, were most probably reared. The latter arrangement is known as 'reversed assembly', and other explanations for it have been suggested (Figure 34).[26]

There are other sorts of 'crucks', one of the most notable groups being the 'jointed crucks' of the West Country,

Opposite: Figure 33 *Interior of Frocester Tithe Barn, Gloucestershire*

Below: Figure 34 *'Reversed assembly' of cross-beam and arcade-plate at Great Coxwell Barn*
(a) Normal assembly – a bredon barn
(b) Reversed assembly – Great Coxwell barn

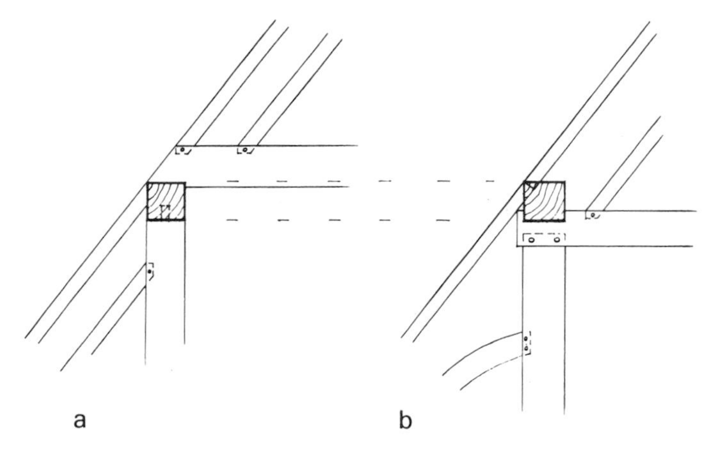

especially Devon and Dorset (Figure 35(a) and (b)).[27] These have haunched posts, which are sometimes quite short, set in stone walls near their head, or they are the full height of the wall-frame plus the haunch. Straight principals are scarfed on to the latter, and the purlins, of which there are generally two or three to each slope of the roof, are threaded through the principals. Lastly there are jointed crucks of many other forms, but it is likely that only those with blades of full height were reared.

Figure 35 *Jointed crucks, typical of the West Country*
(a) With continuous tenon
(b) With end tenon

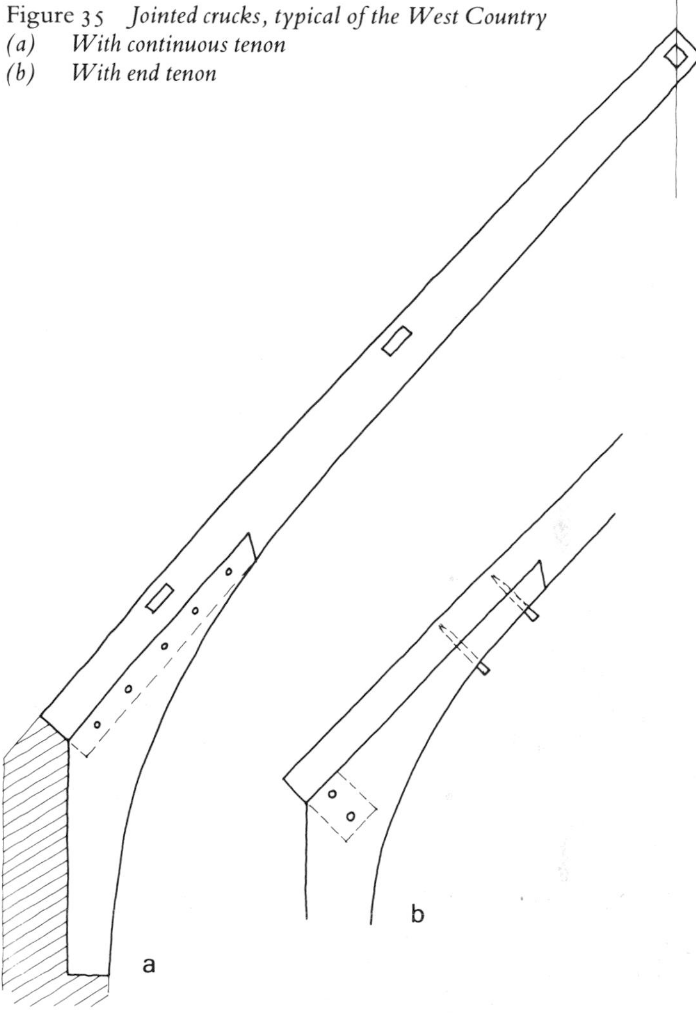

TIMBER

Properties and effects of environment

More must have been written about oak, especially English oak, than all other woods put together. As mere users of timber, for a somewhat special purpose, there is nothing we can add that has not already been said with far more knowledge and authority. Only the bare facts which the conservator should know are therefore put down, with a few observations from experience.

Oak is long grained, strong, durable and, not least important, workable. Though pliable for the first few months and even years after felling, it hardens with age and eventually becomes virtually impossible to saw or axe across the grain. Its medullary rays, flattened veins extending radially from the pith but seldom absolutely straight, provide natural cleaving surfaces. The rays also cause the characteristic 'clash' of cleft or quarter-sawn surfaces, distinguishing oak from all other timbers.

Its rate of growth is slow as compared with elm and other hardwoods, but may vary enormously even within the life of a single tree. In the lowlands with constant ground moisture growth is even and tends to be relatively fast, producing the 'complacent' oaks of dendrochronology, dateable only by the few 'signature' years of exceptional weather. But the rings of the slower-growing upland trees are as various as the series of summers through which they have been formed – to the dendrochronologists' delight (Figure 36(a) and (b)).

The shape of the tree also of course varies, from that of the parkland oak with spread of canopy perhaps 80 feet across, and gigantic branches thrusting from the trunk only a few feet above one's head, to slender forest or woodland poles – 'flitterns' – without a branch for 30 or 40 feet above ground (Figure 37(a) and (b)).

Shape and growth are determined far less by species than by environment, and man's persistent interference with the landscape alters the pattern of growth and predominance of a particular shape of tree from one age to the next. But since trees outlive humans several times over, the consequences of one generation's work are unpredictable and likely to be quite different from any intentions the planter may have had. Nor even can each generation easily see what is going on in its own time – except the more dramatic effects of felling.

Thus William Harrison[1] and others witnessing the great rebuilding of the last decades of the sixteenth and first few of the seventeenth century foretold imminent dearth of oak, seemingly confirmed by the practically universal changeover to brick that had taken place by 1700. But there is now considerable doubt that the shortage was more than relative, affecting only building timber, and even that is contradicted by the considerable number of

Figure 36 *Dendrochonological sections*
(a) Complacent oak

(b) Section from thirteenth-century structure with rings of extremely variable width. Courtesy Walter Horn

Figure 37 (*a*) *Typical parkland oak probably about 200 years old* (*b*) *Woodland oak poles or flitterns*

seventeenth-century yeoman farmhouses and town houses which surpass many earlier ones in the size and lavishness of their timbers. Perhaps it was not only the planting of the new parklands of the great Elizabethan mansion-houses that these writers had failed to note. A little later John Evelyn wrote his *Sylva* of 1644, the first-ever survey of the trees of England, of which the direct result was the replanting of 300 acres in the New Forest for shipbuilding on the orders of Charles II. Had this been 3000 or 30,000 acres, it would still not have made the slightest difference to the navy until decades after Trafalgar, and then it would have been too late!

It may be of interest as a comparison that, at the same time as Evelyn was conducting his survey on his own initiative, across the channel Colbert was officially surveying the French forests.[2] One result was that while in England parkland oaks continued to be planted right through the eighteenth century to the practical exclusion of forest trees, in France it was the forests that were favoured, the results of which were never more evident than they are today.[3]

Another diarist, Samuel Johnson's James Boswell,[4] recording their travels to the Highlands in 1773, portrayed a desert of trees 'from the bank of the Tweed to St Andrews' – the aftermath of centuries of border wars. He noted that he never saw a tree which could have been planted *before* the Union of 1705, and might today be surprised that the saplings he presumably did see have now resulted directly or indirectly in some of the richest woodlands in Scotland – except in that region's uplands, which are still as effectively denuded by sheep as the whole region was by war.

Later still, and in fact well into the nineteenth century, widely spaced parkland oaks for ever larger ships' ribs or 'compass' timbers were still being planted within reach of shipyards. If not already felled for wholly different use, they might still come in for reconstruction of the salvaged Mary Rose!

In the more domestic scene, a line of widely spaced mature oaks may still mark the hedge that has otherwise disappeared but which once defined an enclosure; or pollard oaks, overgrown and deformed, may still be seen in the fields surrounding a water-mill – perhaps only its site – where tanning, as well as corn-grinding, was carried on (Figure 38).

Thus every landscape has its memories in the form of oaks that have either missed out completely or outlived their purpose. Only conifers can hope to keep up with man's voracious appetite for timber.

Oaks and other trees for building

Of the two native species, pedunculate (*Quercus robur*) is the one known as English oak; but sessile (*Quercus petraea*) has the same properties, at any rate for building, and their wood is indistinguishable. The trees, often growing alongside each other in the wood or hedgerow, may be identified by the pedunculate's stalked acorns compared with the sessile's which are tight on the twig. Their leaves are the opposite – the sessile's with stalks, the pedunculate's without (Figure 39(a) and (b)). There are of course other differences for the botanist and forester, as well as several hundred species of oak in all, but these two almost exclusively provided the timber for all traditional uses, as they still do for the vastly different needs of today.

Brown oak, traditionally the heartwood of English oak and in greatest demand for cabinet work, was also used for building. Until recently the so-called beefsteak fungus peculiar to oak was believed to cause its colour. Its presence at the base of the tree is an indication that the wood will be brown, but the cause of both is ground conditions. A deep gravelly soil with a clay base colours the water drawn up the tree which, combining with tannin, darkens the cells. And the fungus is an effect of the same conditions, though indirectly. For the clay base resists downward growth of the roots so that they 'tiller out' and cause a rupture in the wood tissues at ground level. This is where the fungus gains its hold, to be similarly coloured.

In buildings, smoke from the hearth or atmosphere, if not paint and varnish, has long since disguised the natural hues, and it was only as a result of repairing the timbers at Severns in Nottingham that we found that brown oak had been used throughout. It may be somewhat weaker than normal oak as a result of the fungal attack, but this had had no effect on the Severns' structure.

Burr oak is another decorative disease, this time probably caused by injury in the young tree, suppressing its natural growth upwards and causing instead an outburst of small, epicormic growths clustering on the stem, like huge blisters. The resulting flowered pattern beneath the surface is still valued for veneers and occasionally found in buildings (Figure 40).

Figure 38 *Oak tree probably pollarded as a pole and allowed to grow into a deformed tree through the last 100/200 years*

Figure 39

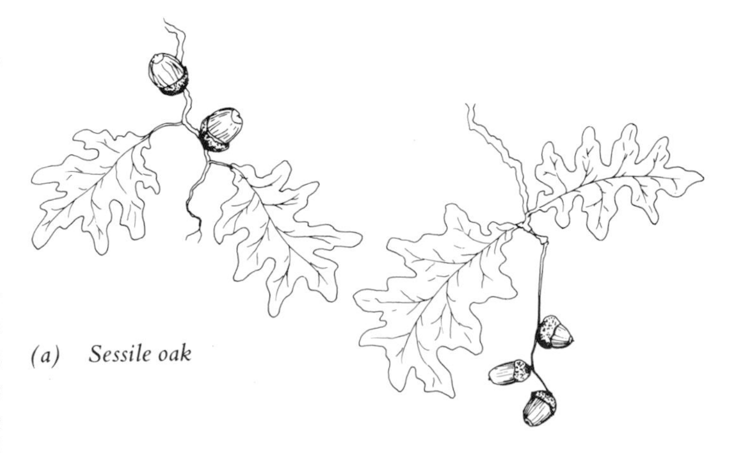

(a) *Sessile oak*

(b) *Pedunculate oak*

Figure 40 *Oak tree damaged by encircling when a pole, probably with metal ring, resulting in heavy burring*

Bog oak need only be mentioned in archaeological context. Having been preserved in peat bogs for centuries, the trees, mostly consisting only of heartwood, are valuable for dendrochronology, extending the annual ring formation much further back in time than is possible with structural timbers or, of course, standing trees. The size of oaks in the indigenous forest are also suggested by one unearthed at Stilton Fen in Yorkshire in 1948, estimated to have been 150 feet high.[5]

American and Japanese oaks are also used in construction. The former is faster growing and coarser grained than English oak, but the American white is a good joinery timber – and the red oak worth growing if only for its autumnal blaze. Japanese oak is smaller and milder and, with its silver grain, the most elegant of all for interior work. For small-scale external structures – though it is not used for the traditional Japanese house – it also presents few problems of twisting or warping, even in the English climate.

Only Spanish chestnut is a substitute for oak. In Kent it is said to be even more common in ancient structures; but it is extremely difficult to distinguish the one from the other after they have darkened and weathered. For many years it was believed that Westminster Hall was of chestnut; repairs, however, proved otherwise.[6] The rear hall at Leycester's Hospital in Warwick, of which the roof trusses have retained an unusual lightness of tone, were also said to be chestnut; more likely they have somehow escaped the effect of daylight, especially sunshine, and of smoke. The grain of chestnut is more stringy, and the absence of clash makes it less lively in appearance. Its general use today is still for fencing for which young coppiced poles are cleft, one of the few ancient traditions that still persists, mostly in Kent. The natural forms of the tree are also less varied and elbowed, though perhaps more graceful, than those of oak.

Elm, now tragically scarce in the landscape after the ravages of Dutch elm disease in the 1970s, is a huge rough tree growing at twice the rate of oak, sprouting suckers and shoots from the ground upwards and occasionally dropping great branches without warning. Its autumn leaves are of the clearest and most delicate yellow. In the sixteenth and seventeenth centuries the wood, recognizable by its coarser and wider rings, became as popular as oak (and scarcely less beautiful) for floorboards, purlins, beams and joists. It was little used outside except for weatherboarding or underground conduits and pumps, for it lasts better than other native timbers when permanently wet. Otherwise it is treacherous, and structural timbers, like the tree itself, are liable to rot quickly from inside to out. Many midland barns, also tragically diminishing in number, are weatherboarded in elm sawn through-and-through and recognizable by the waney edges – horribly overdone in later examples.

Ash, often closely related to the oak in place, as well as in the summer rhyme, is its complete contrast in form, but the sweetest of all timbers to cleave (Figure 41). Though its traditional use is for handles and shafts, historical documents[7] survive in which ash was specified at least for minor buildings or repairs, and young poles have also had a place in the restoration, as rafters, for a part of Chester House (page 53). But on the whole it is too pliant for load-bearing members and very susceptible to beetle.

Even poplars, now grown for matchsticks but capable of becoming large, have apparently produced crucks, though the only example is a Herefordshire barn reconstructed at Avoncroft, and this is disputed by some

Figure 41 *Young ash*

experts who have suggested the blades are of the hardly more appropriate willow![8] Poplar is good for floorboards (and clogs), and being 'non-slip' was used especially around dangerous machinery, but it has none of the structural qualities of oak or chestnut.

As for coniferous trees, in Kings Lynn[9] at least one medieval roof has been found which is entirely of fir, probably brought from one of the Hanseatic ports on whose trade the town was chiefly built. Conifers, particularly pine, may also last as long as oak, as this roof and, more convincingly, the stave churches of Norway demonstrate, but they hardly lend themselves to Gothic forms, despite the example at Kings Lynn.

Lastly greenheart, a true foreigner, has been given a use that no other wood could have fulfilled. That was at the White Hart in Newark where the client insisted that there should be no step from the pavement into the shop and the architect that the restored posts and studs must be tenoned into a sill-beam. Only greenheart, normally used for landing stages and locks, would last as long as the rest of the building, with three sides of the timber buried below the pavement and the top under constant pedestrian traffic. It is imported from Guyana as hewn logs about 10 inches square and 30 feet long, die-straight, but with a slight taper, representing the whole stem of a jungle tree. It is nearly twice as heavy as oak – about 80 pounds per cubic foot – and practically unworkable.

Thus many timbers besides oak have had, and still have, a use, but none, with the possible exception of chestnut, could be married to oak in repairs, their properties and behaviour being far too different.

Decay of oak

Much has also been written about the diseases of timber and the numerous chemical means of dealing with them. Again it is only possible to relate some of our own findings and refer the reader to more authoritative sources for the rest.

First, oak demands only the right conditions and it will outlast most kinds of stone. It is less vulnerable to atmospheric pollution and, correctly used, wears under wind and rain in such a way that its surface hardens instead of eroding or breaking down.

Second, the most common scourge of wood, the furniture beetle (*Anobius punctatum*), can only attack the sapwood of oak, unless it has been damp over a long period, by which time other more lethal forms of decay will also have set in. And since sapwood is of no structural value it matters little even if it has disintegrated into frass, as occasionally may be seen in outbuildings or buildings deserted over the years. Woodworm in oak therefore need only be controlled or eradicated for the sake of other woods and furniture in the same building. The best time to apply a preservative is in April and May when the brown beetle, about an eighth of an inch long, eats its way out from the pupal case, a few millimetres beneath the surface, and settles to lay its eggs to start the three or four years' life cycle all over again.

To preserve oak wholesale against this little pest by chemical means is a waste of time and money, and that 'the full treatment' is generally demanded by building societies, insurance companies and local authorities shows either innocent lack of discrimination as between oak and other timbers, or less innocent influence by those who stand to profit. Perhaps the most telling experience was in being briefed by a client, a notable entomologist, who had been presented as a gift with a full report by one of the larger wood preservation companies on the condition of the timber in a house he had just bought. He handed it to me with the comment, 'I know you won't take any notice of this and nor will I'.

The death-watch beetle (*Xestobium rufovillosum*) is a different matter, but its eradication by preservatives is even more pointless. For death-watch destroys heartwood, so that infected timbers *must* be replaced. Again however its presence is the result of conditions to which no timber should ever by subjected – constant dampness and absence of ventilation. An external four and a half inch brick skin and lath-and-plaster internally with layer upon layer of wallpaper are what the beetle best likes. It has been said that the larva must be present in the tree, the beetle having crawled (it cannot fly) through the ground into an already decaying heart, for any structural timber to be infected; hence its common presence in the larger members, most often wall-plates or arcade-plates, of churches and cathedrals. But this is not borne out by its inevitable presence in wholly encased timbers, regardless of size, that have not seen daylight for fifty or a hundred years. These may be reduced to dank yellow dust, but where the attack has just started, the passages of the larva may be seen in a cross-sawn section of the timber in the outer rings of the heartwood, and in a long section the larva itself, about half an inch long, may be found at the end of a 2 or 3 foot passage. The emergence holes are several times bigger than those of the furniture beetle, almost an eighth of an inch in diameter.

As for other insects, the wood-boring weevils help in the destruction of timber when it is already doomed by ground moisture and wet rot fungus, and the powder post and house longhorn beetles may also be found where oak is already damp and airless.

Further worsening of conditions may lead to more rapid decay caused by fungi, in particular wet rot and dry rot, now scientifically termed *Merulius serpula* instead of the more descriptive *Merulius lacrymans*. Dry rot is of course more serious than wet rot as it brings its moisture with it, sending water-laden strands or 'rhizomorphs' along the heart of the timber, and these can penetrate brickwork and so spread the fungus from room to room.

But it is also the rot to which oak is most resistant. In none of the timber-framed houses surveyed by us has the fungi of dry rot ever been found in the oak frame; it has in oak beams bedded in brickwork. In one house, mostly brick built, cellar beams and joists were infected a few years after we had carried out repairs and alterations. As the owner said, 'I fail to see why the house should have succumbed to dry rot after two hundred odd years.' The reason was simply that a watercourse on the adjoining farm had been diverted, so that the surrounding walls had became saturated in the space of the previous winter and water had seeped into the cellar to a depth of 12 inches.

The action taken was as follows:

1. We made sure the watercourse was reinstated.
2. We dug a trench all around the outer walls of the house to dry the brickwork below the other floors, before the rot could start on them.
3. We isolated all the cellar timbers by building blue-brick piers clear of the walls for the beams and bracketing steel angle-bearers from the walls for the joist ends (Figure 42).
4. We treated the timber chemically after burning the fungus off the surface and out of the shakes and joints.
5. We removed the skirtings and architraves in the room over the cellar, and
6. We dried the brickwork with blow-lamps – now looked on as old-fashioned, but still effective.
7. At each beam end we took out a one-inch thick cross-section, leaving the disconnected end still in the wall.

There was no rhizomorph in the heart of any of them, and, deprived of every source of moisture, both joists and beams (and the skirtings and architraves, replaced after six months) have shown no further sign of an outbreak over ten years.

The beginnings of wet rot, as a fine fungus on the surface of beams similarly supporting floors over damp cellars, are not uncommon, but are easily dealt with by ventilation and some means of getting the surface dry.

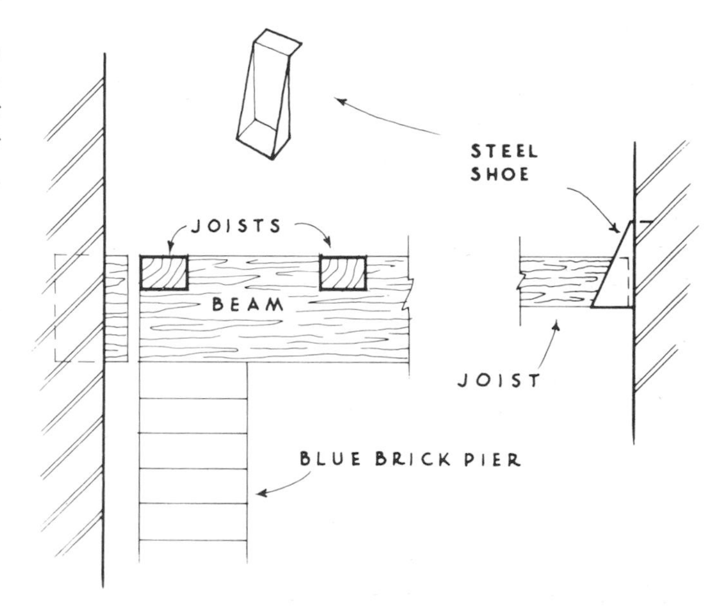

Figure 42 *Isolating infected sections from sound parts of beams and joists in cellar*

Thus dryness is all, normally best achieved by human occupation and use. Even the furniture beetle, needing starch produced by humidity in the sapwood, will starve in a warm, dry and airy house. Lastly, it is worth remarking that if chemical treatment had ever been necessary to preserve oak, there would not have been a single timber-framed structure left to treat.

Use of unseasoned oak

Oak was used green though preferably not directly after felling, as even a slight degree of drying out makes for stability. Felling was best done in the autumn or winter when the leaves had been shed and there was no rising sap. The timing of operations, as well as urgency, may be seen in a letter addressed to John Thoresby, Archbishop of York, in January 1356:[10]

> All the timber as yet obtained, which was supposed would be sufficient for quite a long time, is in the hands of the carpenter, prepared for setting up, if God wills, in the near future; and unless new timber is cut during the winter season, so that it may dry off 'exsiccari' during the summer, the carpenters and workmen employed on the building of the site work will, for lack of timber, stand absolutely idle throughout the next winter season. Be so kind, therefore, as to order the delivery of suitable timber, which could consist more of bent trees than those of greater price and value which grow straight up, to be cut during this present winter-time.

There is plenty of other documentary evidence. The king presents trees to a religious order for building a new monastery;[11] the manor lord provides the main timbers for the construction of a new house by one of his tenants;[12] there is a search for timbers of requisite size for a large new project, and so on.

Technically, the reasons are even more pressing. Unseasoned timber is the best to work. It will cleave, saw, hew and cut with ease, and even bend in the final structure to take up irregular loading or movement through drying. But this movement is extremely slight as each component is firmly secured within a completely interlocked frame – the best possible way to season timber. Distortion in the standing buildings of today is thus seldom, if ever, the result of warping, still less of using already warped timbers (even during construction, means of preventing the springing of rafters before they could be protected and held down by the roof covering were often adopted), and still less is distortion the result of using crooked timbers from the tree. Rather, the whimsy shapes and tipsy gables, beloved of artists, have been caused by alteration and

Figure 43 *Bailiff's House, Bewdley*

(a) Street front. Courtesy Wyre Forest DC

(b) Date plaque

mutilation exerting stresses that would have caused collapse of any other form of structure.

Finally there is the conclusive evidence of dendrochronology. The timbers of the Bailiff's House at Bewdley were proved by dendrochronology to have been felled in 1607. The date of the building's completion is 1610 (Figure 43(a) and (b)).[13]

Thus in repairs and reframing oak must again be used green, not only because of tradition but also for the very good reasons given at the end of this chapter. Shrinkage is inevitable and this presents problems in the combination of new with old, to be overcome as far as possible by the manner of jointing and pegging.

Size of trees

Though oak is still the most common of Britain's deciduous trees, few of them would even have been considered as suitable for converting into the structural components of a timber-framed building – least of all the ancient gnarled and twisted oaks of the 'traditional' village green. The open-grown trees of parkland and pasture would also have been of limited use. But an estate inventory of *c.*1500 describes hedge trees as 'great oaks', 'pantrees or purlins', 'joists', 'double-spars' (presumably studs), and 'double rafters'. Other hedgerow trees such as ash, poplar and willow are also described according to their purposes – as those, for example, of the wheelwright, shoemaker (clogs) and turner. Of greatest interest, the document is clearly referring to individual trees and poles (pantrees), and those described as double spars and double rafters must have been intended for halving.[14] Most of today's hedgerow oaks would probably have been classed as great oaks – too large to be specified as building timbers.

Today, the woodland and forest oaks, with straight stems of 3 to 4 feet diameter, are the trees most in demand by the merchant of home-grown timber. Anyone who has seen the huge delimbed butts disappear into the sawmill, and suffered despite blocked ears the scream of the powered band-saw ripping them in seconds into slabs and boards, must marvel that supplies of such trees have not long since been exhausted. Again, however, nothing emerges that would be of the slightest use to the medieval carpenter, except the boxed-heart beam in the middle or a pair of halved members from either side of the heart. But to obtain only one or two timbers out of every tree of this size would be extravagant to say the least.

The carpenter of the timber-frame tradition selected the *smallest* tree that would yield the required cross-section of the structural member; this was only economical. The timber also had to be straight, at least in one plane, and the heart had to be contained or exposed through its length. The trees that yield this kind of component are the young poles of woodland and forest. But what may have been the most prolific traditional source has now virtually gone. That was the coppice, or 'coppice with standards', in which poles were felled every sixty to a hundred years, the stools being allowed to shoot again. The standards, encouraged by the close-in poles to grow upwards rather than to spread, produced the larger beams and posts – some of which were enormous and today irreplaceable (Figure 44).

For instance, in 1970 six beams had to be obtained for the restoration of York Minster tower. The original members, which had decayed, were just over 50 feet long and about 2 feet square. The smallest tree that would yield such a beam would have a top diameter of about 3 feet. Its basal diameter (4 feet 6 inches above the ground) would be little short of 5 feet and the height to the crown about 100 feet (see Figure 44(b)). Only one such tree could be found, despite enquiries to every timber merchant in the country. Reluctantly the architect had to use steel lattice beams.[15]

The lantern over the Octagon at the crossing of Ely Cathedral has oak posts 62 feet high, pentagonal in section and 2 feet 6 inches across at the top. It is documented that the search in 1328 for such trees was 'far and wide' (see Figure 44(a)).[16] There was also a search for huge trees for the 65 feet span double-hammer-beam roof of Westminster Hall, begun in 1395 after the earlier hall of the eleventh-century Whitehall Palace, whose aisled roof it replaced, had burnt down. These were found near Farnham where the entire framing was done. The name of the place to which 'thirty strong wains' had to go was 'The Frame by Farnham', and from there 'twenty-six half-beams and twenty-six pendant postes' were carried in two carts with sixteen horses, making fifty-two journeys to Hamme on the Thames – one journey for every timber 'at 7s. 4d. a time'.

Lastly, the hall of the Bishop's Palace at Hereford has posts 4 feet 6 inches square at their base and 35 feet high;[18] these would require a basal diameter of about 7 feet (see Figure 44(c)). Built in the early twelfth century, this extraordinary hall probably does not belong to the timber-frame tradition at all, but rather to post construction. Recent archaeological excavations of other 'halls' of this or earlier date also suggest that their earthfast posts were mostly far bigger than the oak trees of today could yield.[19]

But it is not only the sheer size of the straight posts of these structures that stretches the imagination.[20] The

Figure 44 *Sizes of oak trees required for*
(a) Ely Cathedral Lantern
(b) York Minster tower
(c) Bishop's Palace, Hereford

timber vaulting below the Ely lantern consists of ribs out of timbers about 12 inches square, die-straight vertically from their springing to the lantern floor and forming a perfect segmental arch in elevation. The length of each of these measures on the circumference rather more than 30 feet, and there are no less than sixteen of them, perfectly matching each other (Figure 45). The secondary ribs of the vault are hardly less in length and there are forty-eight of them. Nor is this by any means the only Gothic timber vault. The transept towers of Exeter, the south choir of Rochester, the Presbytery of St Albans are other examples. And finally, the remarkable permanent scaffold of Salisbury spire has at its lowest stage eight identical long braces of *triple* curvature.[21]

These structures may represent architecture of a different status from that of the buildings and roofs with which we are concerned, but nevertheless their timbers all had to be obtained from the same forests and woodlands, continuously exploited for all needs since Roman times.

Cruck trees

Crucks represent yet another form of tree which, if not extinct, certainly takes some finding. Leigh Court tithe barn has eleven cruck trusses, spanning 34 feet (Figure 46(a)). Nine of these have blades about 35 feet in length and each represents a separate tree (see Figure 44(d)). But it is not their size that is so remarkable, even though the building is the largest cruck structure in existence, nor

(*d*) *Leigh Court Tithe Barn*
(*e*) *Typical Welsh cruck-built hall*
(*f*) *Typical 20 foot span tie-beam or floor-beam*
(*g*) *Typical rafters*

even their similarity of profile, but that they were probably all grown locally. This is strongly suggested by the dendrochronological samples taken to date the barn, all of which showed 'complacency' compatible with its lush surroundings. Today, the whole of England would doubtless have to be searched for a single replacement.

Stokesay Castle,[22] *c.*1300, also has cruck trusses of which each blade represents a whole tree, but instead of a gentle sweep there is a pronounced angle of about 35 degrees, and both the roof component and wall component are straight (Figure 46(b)). The latter in each truss has now become a vestige of the original, having gradually rotted through contact with the stonework of the walls, and progressively cut short. Such blades as these, at least as they used to be, would be even more difficult to find. Lastly, trees for the shorter, but immensely broad and sometimes even more acutely angled blades of a typical Welsh cruck-built hall – often the halls of the nobility – must have come from trees that also no longer exist in quantity, yet must have been among the most common of oaks in medieval Wales (Figures 44(e) and 46(c)).[23]

It used to be thought that a cruck-blade consisted of the main branch and stem of the oak tree, and that in the building this dual limb was up-ended so that the shorter branch corresponded with the wall and the longer stem with the roof.[24] This, however, does not stand up to examination, mainly because the point of branching would cause a serious weakness in the converted cruck precisely at its section where the outer fibres would be in greatest tension. Not less important, the labour and

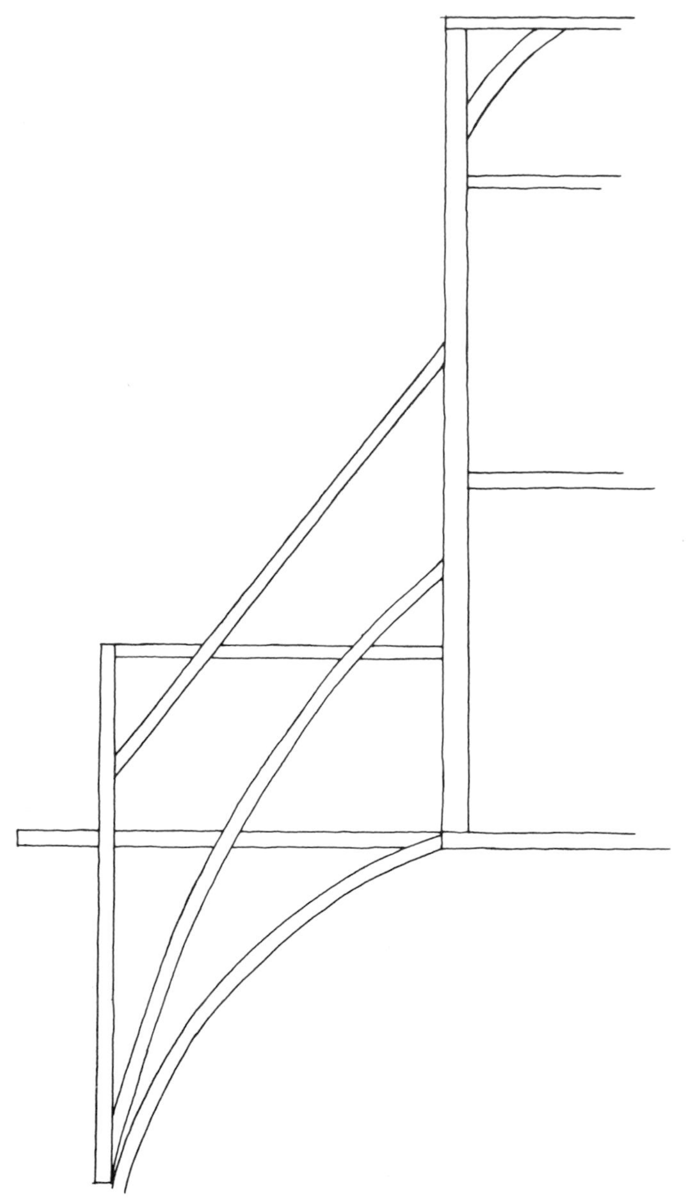

Figure 45 *Timber vaulting below Ely Lantern*

wastage of timber in paring down the stem so that it would taper towards the ridge of the roof, that is, towards the base of the tree, would have put the cruck carpenter out of business even before he started. The cruck thus represents the whole stem *below* the height of its major branching. No doubt the windswept sites of the Welsh coast and other similar regions were the source of such trees – now, like the Scottish lowlands, given over to sheep or coniferous plantation.

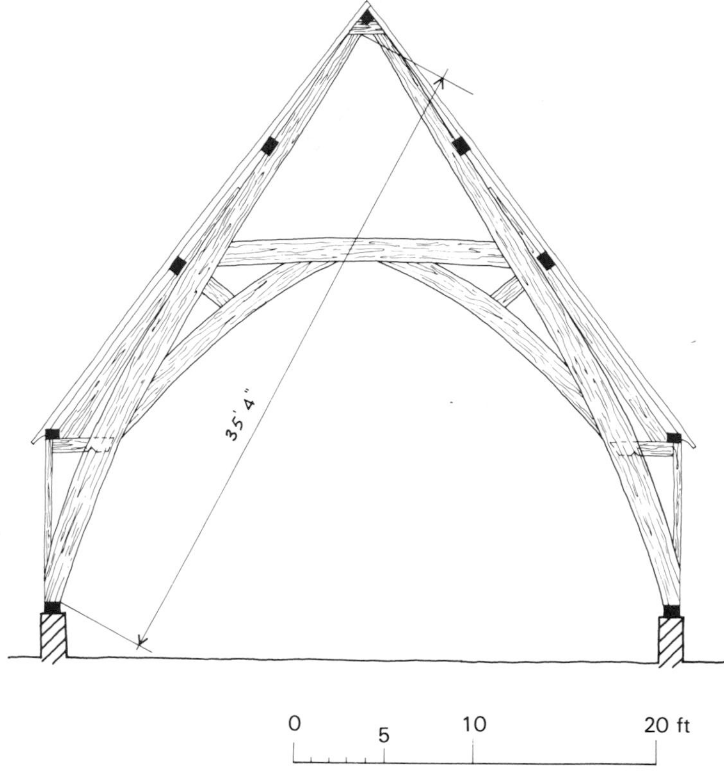

Figure 46 *Cruck trusses*
(a) Leigh Court Tithe Barn

Shoots from coppiced stools may also have produced crucks, starting at an angle and growing into cruck form without the aid of wind or slope. But these would be of the kind generally found in the medieval hall or barn (as Leigh Court) of the West Midlands, their shape only slightly swept rather than sharply elbowed.

Today's trees

Thus have oaks, not only for the better crucks and the largest beams and posts but also for the great majority of ordinary building timbers, become scarce or disappeared completely. And so for restoration one must make do with the trees of today, the open-grown relatively mature oaks with their great spread of crown or, as preferred by the timber merchant, the equally mature woodland or forest oaks with longer stems. Or one must persuade the merchant that the poles often left in the selectively felled woodland to grow to maturity, and even the bent branches of his larger trees, are worth bringing in and being given a corner in the yard. Unfortunately, since

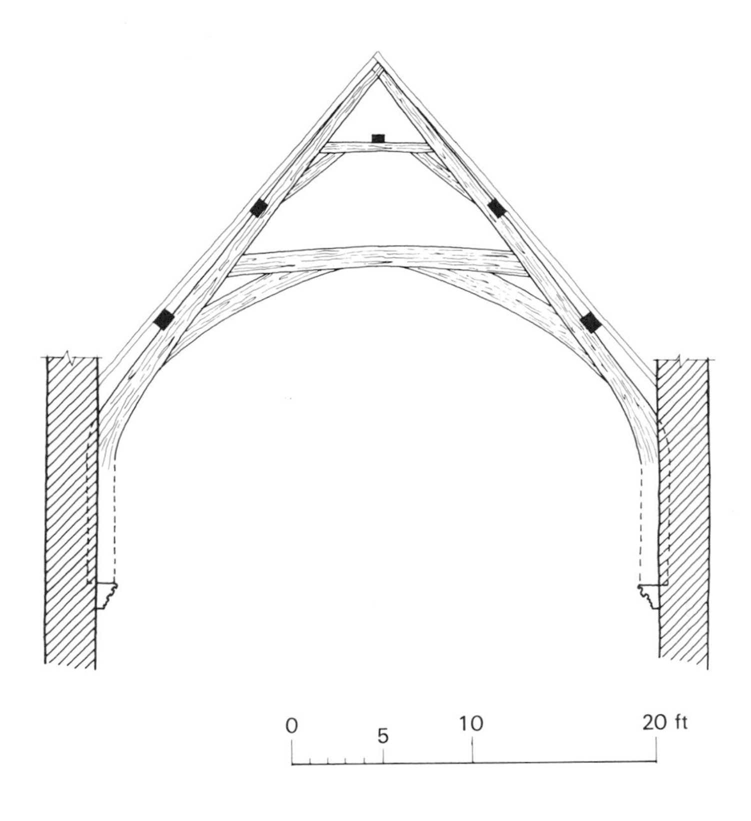

(b) Stokesay Castle

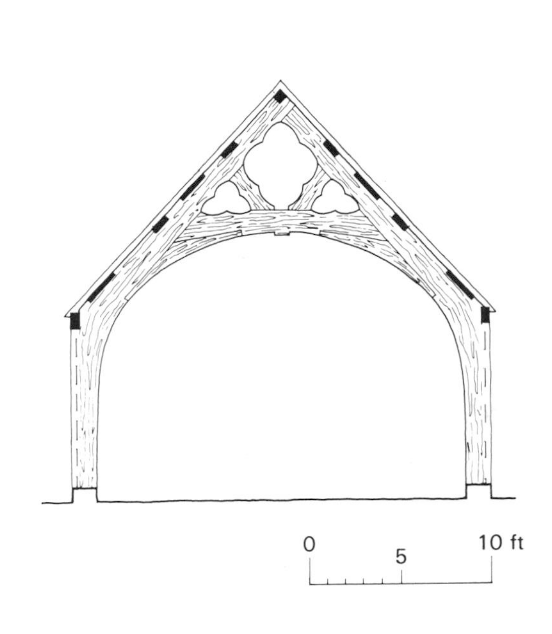

(c) Brynmelyn Llanwrin, Montgomeryshire. Courtesy Peter Smith

only about one in ten poles or limbs is likely to pass the test of straightness for timbers longer than about 10 feet, their careful selection and conversion would probably be even more wasteful than especially converting the over-large tree.

An example of such extravagance, even though it saved the day, was in the restoration of Middle Littleton Tithe Barn in 1975, when six new base-crucks were required (see Chapter 11). The originals had come from angled poles of perhaps a hundred years' growth, probably on the nearby Cotswold escarpment. The replacements had to be carved out of four-foot-diameter butts which, when felled and taken to the yard, must have been found to be too awkward to put through the band-saw; for they had been consigned to the nettles outside the storage area of the yard and there left for an unknown period. No doubt they would still be there but for the search on behalf of the barn.

Other instances of oaks too large for ordinary members, whether straight or shaped, are all too numerous. Probably the proper tree has never yet been selected for any restoration contract, except occasionally for such minor components as rafters, numerous enough to make their selection, felling and conversion economically worthwhile – if still not very profitable. The irony is the false pride in the completed job – because the new timbers are of so much 'better quality' than the originals!

Conversion methods

Hewing with the axe the four sides – or as many facets as the finished member might have – was the means of converting the log into boxed-heart members, from the giant beams, posts and crucks, already noted, down to the slenderest of members, such as rafters (Figure 44(g)). All were 'sized' for their particular functions, not by any sort of calculation but by the natural organic proportion of diameter to length of the stem of the tree. While this may result in some extravagance in terms of volume of material to actual loading and stress to be withstood in the structure – the components invariably being over-sized once they have been securely locked into the total frame – it was economical in terms of time and labour for converting and squaring. Nor was there any extravagance in that process, sapwood and even bark often being left on the finished member; chamfering the edges, though primarily decorative, could also be functional in getting rid of some of these layers.

Hewing was a relatively unspecialized skill, used in all kinds of construction from shipbuilding, post construction, bridges, scaffolds, cranes, engines of war and so on, down to temporary structures and fence-posts. Halving, on the other hand, with wedges and mallets, or with the saw, was the true characteristic of the half-timber tradition. This method of converting the tree determined architectural style and structural method as well as making best use of resources. The normal section of halved timbers is roughly in the proportion of two squares, with the heart exposed along the centre line of each half. This is the fair-face, always set to the outside of a wall-frame. For rain passing through the softer substance of wood between the annual rings cannot reach the inner face (Figure 47). Moreover, the rings at right angles to the outer surface are worn by weathering into hard ridges, channelling the run-off so that the timber dries quickly. Wall-frame members sawn through-and-through, on the other hand, will become wet to the inner face from heavy rain.

The halved face is also fair as regards appearance. It must be seen, not only outside, but internally from the end of the building that is socially the more important, the 'upper' end; it is the face most suitable for carving or moulding, the wood sculptor always preferring to 'cut into the teeth of the grain'; and it is the face into which the pegs are driven. Needless to say there are exceptions to these rules, but they remain the structural principles determined by practice.

For shaped members, halving is also the means of obtaining paired mirror-images – the Gothic element of the timber-frame tradition. Not only pairs of swept braces of all kinds but the most elegant of crucks are halved out of a single log. The effect of cross-frames composed of such members is best seen in large barns, of not only cruck or base-cruck but also aisled construction (Figure 48). Each cross-frame is perfectly symmetrical but is also subtly different from all the rest; for while the aim of the carpenter was to achieve unity throughout, the very nature of oak made that an impossible ideal. The arcade-braces and wind-braces in each bay lengthwise are also paired, but the slight variations between one bay and the next are more difficult to see. Nor is it always easy even to recognize a pair. The course of the heart-shake and positions of knots can often be seen to correspond to each other more readily than the symmetry of outline of the complete arch.

Even tie-beams were sometimes paired in adjacent trusses, but generally placed with halved face to halved face so that their profile when seen from the same position was identical (Figure 49(a) and (b)). Principal rafters, tapered or cambered, were also generally paired. Some gable-end elevations have not only paired braces, struts and principal rafters, but even corner posts. Cheylesmore (Chapter 8), Besford Church and the Bromsgrove House (Chapter 11) are examples of this.

Posts, whether halved or boxed-heart as most of them are, must be jowled at the head, sometimes also at the foot. The head jowl, at least in medieval structures, is of course formed by the grain, the tree having been felled close enough to ground level to incorporate the swelling of the buttress, placed uppermost in the structural frame. The swelling of the stem just below the lowest main branch could also provide the jowl, and there are instances of posts jowled top and bottom, obtained by selecting trees of the correct length so that both the buttress and branch swelling were made use of.

Tie-beams are also cambered in conformity with the grain, but since their camber is most often considerably steeper on top, the soffit being only slightly concave or sometimes perfectly flat, there has to be some compromise, generally by letting the grain run out along the lower surface. Crucks which taper towards both head and foot from the elbow may also have to be finished in the same way, the sliced fibres appearing on either edge according to the natural grain and required angle. There is so strong a resemblance in profile between heavily cam-

Figure 47 *Cross-section of stud from Shell Manor showing halved timber with heart-shake and heavily infected sapwood, also some penetration of beetle into heart-wood*

Figure 48 *Bredon Tithe Barn interior during reconstruction – note variation of profile of cross-braces as between trusses. Courtesy Martin Charles*

Figure 49 *(a), (b) Tie-beams halved from same log in opposite trusses at Chester House, Knowle, Warwickshire*

bered tie-beams and crucks that rebuilt structures, incorporating timbers from the earlier one, may have an original cruck making do for a tie-beam.

As for the method of halving, it is not always easy to decide whether timbers were cleft or sawn. Cleaving has the advantage of speed. Two halves of a log 20 feet long may be produced in a matter of minutes, compared with an hour's back-breaking and blinding toil with the saw. And the finished member is in every way superior. The heart is exposed through the whole length, no fibres are broken, the surface is clear and uniform throughout (Figure 50). But the tree must be exceptional. In fact few trees are of equal radius at any given section; more often they are lozenge-shaped with wider rings on one side than the other – perhaps reflecting aspect. Nor does the grain always grow straight even in a die-straight butt; more often it spirals, and the bark does not reveal what is happening inside (Figure 51(a) and (b)). Even the highest skill in selection cannot forestall these problems, and since the carpenter or builder would probably have to buy a parcel of trees, as the merchant today, he would not want to add to the inevitable number of rejects by cleaving them.

The saw overcomes all these problems. The sawn surface may be rough, and the heart present an uneven line, disappearing and re-emerging further along the timber. Strength may be lessened through ripping some of the fibres, but never to any critical degree, and the sawn timber is doubtless less stable than one that was cleft. But the effects are otherwise the same – notably, the heart-shake must open, as a result of releasing the tension of the annual rings, producing a slightly convex surface. The larger the timber, the more pronounced the angle. At Baguley Hall at Wythenshawe,[25] Manchester, where the intermediate posts of the hall are nearly 2 feet across, it was long believed that the hall had been built to boat-shaped plan – the result of Viking influence!

It seems that the frame-saw was preferred to the pit-saw until a very late date – seventeenth century or even eighteenth.[26] This is certainly the impression conveyed by medieval illustrations. It could be set up anywhere, and so would have the advantage of avoiding constant moving of heavy timbers to and from the pit. Digging and walling the pit would be additional labour; it might also be found to be in the wrong place before first framing of the structure had got very far. Thus the pit is more likely to have been a feature of the sawmill rather than carpenter's yard, but it remains a question for the archaeologist. Wind- and water-powered sawmills existed in the Middle Ages and are referred to in seventeenth-century documents as both being set up and pulled down because the sawyers refused to work them,[27] though they seem to have flourished on the Continent. But the most rapid development of water-mills for sawing was in seventeenth-century New England, with its untapped resources of oak and plentiful fast-flowing rivers. By contrast, the hand-worked pit-saw remained a feature of the English village right through the last century.[28]

Figure 50 *Ham Wood, Hereford and Worcester – timber suitable for stud after cleaving. Note heart-shake in centre of log*

Timbers are worth examining for evidence of mechanical sawing, with its regular widths between strokes, and for hand-sawing with its irregularity. The slight curve of the circular saw is easy to distinguish and, even easier, the wide straight cuts of the band-saw.

The development of saws is not matched by an improvement in the quality of finished timbers. On the contrary, their quality was never again so high as in medieval

Figure 51 *Ham Wood*
(*a*) *Long pole ready for cleaving*

(*b*) *After cleaving. Note spiral of log and ill-aligned heart-shake. Neither of these defects could be discerned in the growing pole*

structures – provided, of course, that comparison is between buildings of the same social and economic standing. Middle Littleton Tithe Barn is an exception. There must have been some reason, such as the exploitation of the fertile Vale of Evesham for other crops – place names suggest even wine-growing – for the poor quality of its structural timbers; many have rounded edges of sapwood and even bark, and the four aisle-posts are each of two separate timbers scarfed together. There is also evidence that the timbers came from two different yards. All this suggests that resources were limited, even for the wealthy Abbey of Evesham (Figure 52).

In general, however, the changes as time goes on are towards diminution in size of components and, in particular, the more sparing use of broad tie-beams and posts. Boxed-heart members are now sawn on all four sides; the quality of shaped braces and struts declines, first becoming smaller and then giving place to straight-sawn members; crucks, except for the lower-quality buildings, go out altogether, though cranked principals, often sawn to shape, persist right through the eighteenth century and beyond.

These changes reflect not only increasing demand for timber-framed buildings, at its height from *c.*1550–1650, but also technical changes towards greater production, with increasing dependence on the saw, the quality of which had no doubt already been enormously improved. The decreasing use of the axe and adze not only reduced the dominance of the natural grain in shaping the timbers but even led to its disregard in such members as jowled

Figure 52 *Middle Littleton Tithe Barn – interior after restoration showing poor quality of original timbers. Posts have scissor-scarfed extensions contemporary with original structure. Left-hand post has new piece 8 feet long scarfed on at bottom. Courtesy Martin Charles*

posts, the jowl becoming a sawn protruberance of various shapes, more or less ornate, and much more prone to split off than the natural jowls of medieval posts. But at least the correct tree was still selected, sometimes in the lesser buildings being rather too small for the job. And halving remained the rule, at least for wall-frames.

Today with virtually only two tools – the chain-saw for felling and trimming, and band-saw for converting – the finished timbers, instead of being the 'natural' and most economical product of the tree and tool, are expensively produced simulations. If the appearance of hewn timbers is needed, they must be sawn on all four sides, then finished with the adze, and so it is better to forgo appearances. Swept braces are likely to have come from large bent branches rather than poles, their convex back pared down by sawing instead of axing. Only the concave soffits still have to be finished with the adze. The wandering edges of rafters or joists can be obtained with the chain-saw, after careful setting out along the grain, and then again the chisel or adze must remove the sawmarks.

Lastly sapwood, inevitably present where it could do no harm in traditional timbers, can now only be retained if a special visit is made to the forest, the merchant persuaded to fell a particular tree and convert it in a way wholly inappropriate for his equipment. Thus has economy been turned on its head. Fortunately the further operations – jointing, chamfering, moulding and carving – are still for the man on the job, and he is still the master of the hand tool, though perhaps a little out of practice with the axe and adze.

The specification for timbers

The following specification clauses are a guide rather than the law, for if they were applied to the letter hardly a timber could be accepted from the merchant. Nevertheless, each one as it comes on the job must be carefully examined and passed if the heart-shake is in approximately the correct position at each end. If not, the member must be rejected. Sometimes a new halved timber will match the old to which it is to be jointed as a repair, not only in the alignment of the heart, but also in the widths of its annual rings. Then it is time for a celebration (Figure 53)!

Figure 54 shows stages in timber preparation.

General

1 Oak shall be unseasoned native oak from selected timbers or standing trees. All required members shall be obtained from logs of the least diameter that will yield the correct finished cross-section of the member. Small areas of sapwood are permitted but no sapwood shall be exposed externally in the finished building or covered up internally.

2 Members 8 inches by 8 inches or larger in cross-section shall be obtained from boxed-heart logs squared with the axe or adze on cambered surfaces (top and bottom in the case of beams) and sawn on the sides.

3 Members of 6 inches or less in their lesser cross-sectional dimension shall in all cases be heart-sawn. Straight members may be sawn or axed on the other three sides, the heart surface being used for the upper or outer surface in the completed building. Curved members may be sawn on the face opposite the heart surface but must be axed or adzed on the curved top and bottom edges so that their final shape exactly conforms with the natural grain. Shaped members such as jowled heads of posts, tie-beams etc. shall also be axed or adzed to shape in accordance with the natural grain, sawn surfaces being permitted on the sides and other face.

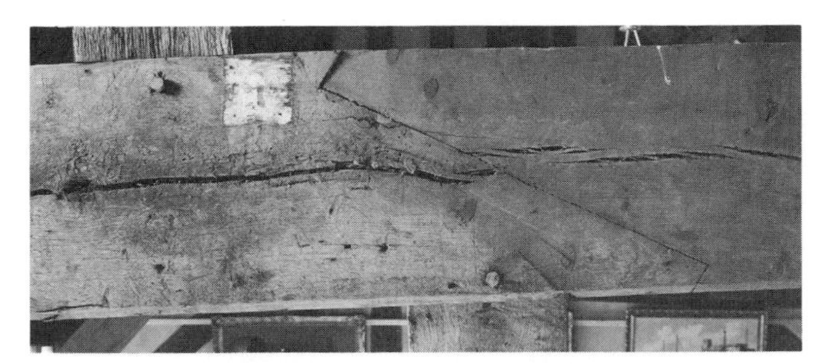

Figure 53 *163–5 Spon Street, Coventry. Tie-beam repaired with splayed-scarf, new timber on right with heart-shake continuing alignment of original. Tree-ring widths also similar – note the line of the scarf has had to be drawn in on the photograph to make it visible! Courtesy Martin Charles*

Figure 54

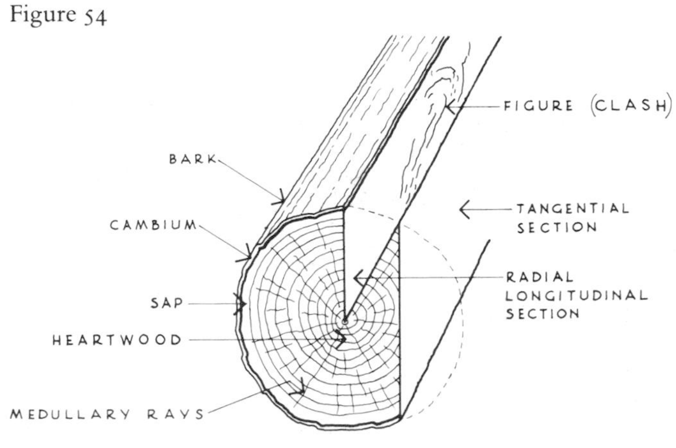

(a) The log

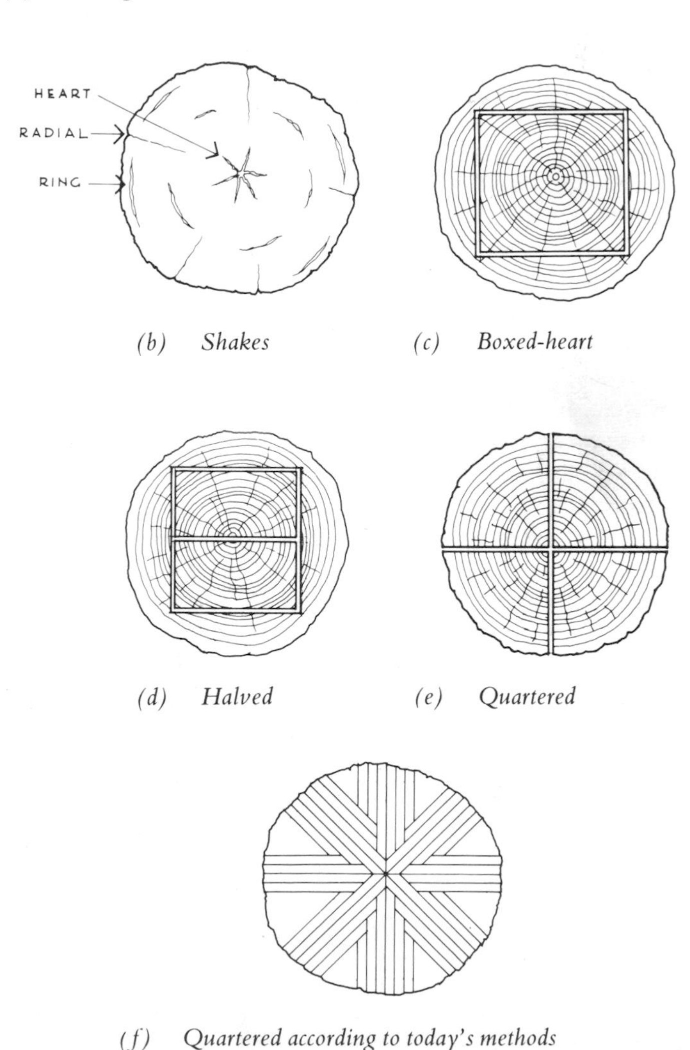

(b) Shakes

(c) Boxed-heart

(d) Halved

(e) Quartered

(f) Quartered according to today's methods

4 On finished surfaces sawmarks may be left visible but axed or adzed surfaces shall be smooth. Visible adze marks are not permissible. Chamfers shall be axed, chiselled or planed.

Conversion

5 The following list summarizes the normal method of conversion for the various members, but the repair schedule shall be referred to for each member individually:

Member	Conversion
Beams (floor-, sill-, cross-)	Boxed-heart
Purlins	Boxed-heart
Plates (wall-, arcade-, collar-, sole-)	Boxed-heart or quartered
Ridge	Boxed-heart or quartered
Posts	Boxed-heart or halved
Tie-beams	Halved
Girdings	Halved
Bressummers	Halved or quartered
Studs	Halved
Braces	Halved
Brackets	Halved
Principals	Halved
Collars	Halved
Struts	Halved
Joists	Halved or quartered
Rafters	Boxed-heart, halved or quartered
Bargeboards	Quartered

Old timber

6 The reuse of timbers in any position or for any purpose other than that for which it was first designed will not be permitted. On no account shall an old timber or part of it that is fit for reuse be rejected without the architect's approval.

Reused timbers

A not unimportant footnote might be added: 'No old timber of another building shall be used for repair or replacement.'

Because such timbers will not warp or shrink, it is often believed that they would solve the problems of marrying new to old, as well as look better. It has even been suggested that a bank for old timbers should be set up for restorers, as for bricks, stone, slates, tiles and a whole number of joinery features salvaged from demolitions. The reuse of old timbers might also seem to be justified by sound precedent in the many reused timbers found in buildings of the sixteenth and seventeenth centuries especially, and increasingly with subsequent alterations to them or their rebuilding in the eighteenth or nineteenth century. Clearly it was only economical to take still serviceable timbers from the old house and lay them aside for possible reuse in its new or altered successor. Thus beams, wall-plates, purlins and even the occasional post or cruck, as already noted, whose mortices or other signs or shapes give away their origin, may be found. But they are seldom, if ever, framed in. They are generally roof members or beams that can be simply laid on or built into solid walls.

In the repair of framed structures the problems are quite different. Since every structural timber is a 'special', the likelihood of an old one exactly fitting is infinitesimal. Even for adding a new end, say, to a partially rotted wall-plate, the chances of picking up the exact section are hardly better. But, presuming success, then not only must the splice be formed but also mortices, bird's mouths and probably a dovetail cut for components that are quite different from those for which it was first made. Thus the finished article will be so peppered with its new and original joints as to be useless. By selecting a member of larger section, such as a floor beam, to fulfil the same purpose, its paring down to lesser section, thus eliminating its original mortices, would demand such labour and expense as to put it out of court long before the jointing stage had been reached. Yet one hears of the enthusiastic restorer searching for timbers in the local authority or demolition contractors' yards, among piles of mouldering beams impossible to lift or move, or acquiring a barn no longer wanted by the farmer (as with most timber farm buildings today) which must then be totally demolished for the few undamaged members it may contain. But soon, with his saws, planes and chisels broken by the rock-hard wood and buried nails, he must see that this is not the way.

Nor are we building timber-framed houses from scratch, but instead repairing highly complicated structural frames to preserve or reinstate their original form and efficiency. This makes much heavier demands on the carpenter, working within fixed limits of length and space, to achieve exact fit and shape, then does the open-ended erection of new framing. And lastly, the moral aspect of pirating one building for the sake of another, thus perpetrating the deceptions of nineteenth-century restoration, must surely be taken into account.

ORGANIZATION AND FRAMING

History

There are many similarities between today and the Middle Ages in the way building contracts were organized and carried out. The main difference in our field is the central role of the carpenter, whose skills also had to be those of architect, forester, feller, joiner and builder all rolled into one. He was also wood-carver for the embellishments of all but the most humble of the dwellings he designed and built.

Though in larger contracts he often took second place to the mason, the mason still depended on him for model-making, templates, scaffolding, centering and most of the temporary equipment, and of course for floors and roofs. On some projects they were equals. Henry III sent both his master mason, Henry de Reyns, and master carpenter, Simon de Norhampton, to consult and advise on York Castle in 1244–5; and again, in 1256–7, he appointed two others, Master John of Gloucester and Master Alexander, 'to be chief master of the royal works touching their respective crafts this side the Trent and Humber'.[1] And such men as Hugh Herland, the carpenter of Westminster Hall roof, and William Hurley of Ely lantern, were unsurpassed as architects by any standards.

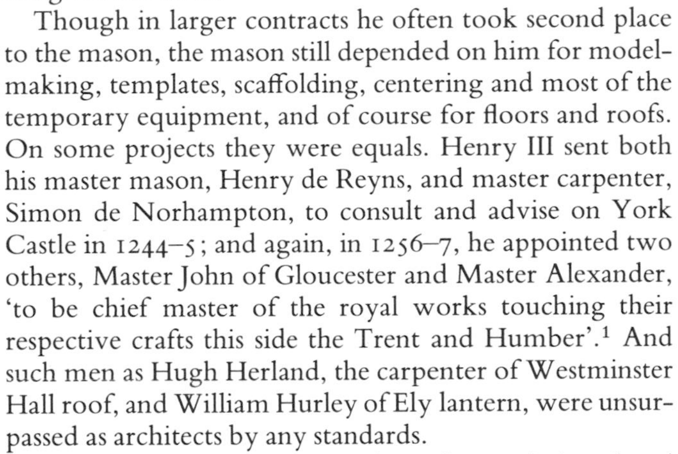

A chronicle of 1194–8 describes a house designed and built by the carpenter, Louis de Bourbourg, at the castle of Ardres. He was 'not much inferior to Daedulus in his skill in this art'. The house was certainly remarkable and not only for its date. It had three storeys, with a 'labyrinthine entrance hardly to be penetrated', and the living accommodation, as most commonly on the Continent, was on the first floor. Here were the great chamber, buttery and pantry, the dormitory of the attendants and pages, and in the 'inner part of the great chamber there was a private room partitioned off where they could light a fire at dawn or dusk, or in case of sickness or at times of blood-letting . . .'. The kitchen was at the same floor level within a separate structure. On its lower floor were pigs, geese, capons and other birds for killing, 'always ready for eating'. On the top floor 'were made sets of rooms in which the boys and girls separately, as was fitting, and the master of the house slept . . . . Staircases and passages led from floor to floor from house to kitchen and room to room, as well as from the house to the lodge.' There was also a chapel, 'like the tabernacle of Solomon in its carvings and paintings'.[2]

Also as today, the client took greater or less part in the design of his building, according to his own capacity and interest in such things. A house at St Pancras Street, Winchester, to be built between Easter, 8 April 1436, and Lammas, 1 August of the same year, was to be 'as the tracing showeth drawn in a parchment skin between them made'. Thus the client, the Warden of New College, Winchester, seems to have designed the project with his carpenter, John Berewick of Romsey. It also seems that framing in the carpenter's yard must already have been completed, for the document reads: 'The same John to begin the work the Monday next after Easter week in that the same John have no default of timber neither of carriage for the same timber when it is wrought to carry it to Winchester to the ground aforesaid.' Whether he was to be responsible for its carriage and had already been the carpenter for the timbers being 'wrought' is not so clear. The Warden, however, undertook that John would 'have no let (hindrance) in rearing of the house'. Payment was to be in stages; the first payment a week after he started; the next, five weeks or so later, and the last on completion ten weeks later, the entire contract amounting to $11\frac{1}{2}$ marks, about £7 13s. 4d.[3]

As for building in the village and countryside, the manor court rolls give many examples of tenants having to build a new house of so many bays under threat of various penalties or fines (sometimes a capon) for failure or delay. The lord normally provided the main timbers, and in the fourteenth and fifteenth centuries in Worcestershire these most often seem to have been crucks.[4] The village carpenter, though not generally mentioned until it comes to the accounts, would be the builder.

There was also spec-building even in the Middle Ages. Spon Street, Coventry, was built up at the same time as, and following, the construction of the city walls, in the latter half of the fourteenth century. Terraces and semis, with halls unbelievably small, but by no means jerry-built, were put up by the landlord, St Mary's Priory, for sale or letting to craftsmen and artisans.[5]

It is, however, in the sphere of drawings, both as regards their purposes and technique, that there has been least change. Though few have survived, there are enough to show that they ranged from working sketches and more finished 'client's' drawings to working drawings and full-sizes, the last still to be seen on the tracing floors of York Minster and Wells Cathedral. Drawings of the greater buildings, especially cathedrals, are of utmost intricacy; there is no fudging of detail even to the accurate drawing-in of hundreds of croquettes. Plans of buildings are the rarest, the most famous and longest to have survived being the plan of St Gall, of the ninth century.[6] Designed for a Benedictine monastery as a 'type-plan' or paradigm, it nevertheless goes into every detail of the site layout and function and size of buildings, all of which have now been 'reconstructed' both graphically and as a model.

A drawing of a town house recently discovered had been used as a binding for the papers of the Register of Jerome de Ghinucci, Bishop of Worcester 1522–35.[7] Binding seems to have been a not unusual fate for large drawings on parchment. This one had lain in Worcester Cathedral library ever since the bishop's term of office. The building it shows was typical of the larger town houses of that date, of which the ground floor was shops,

Figure 55 *Sixteenth-century drawing of timber-framed house*
(a) *The drawing. Courtesy Diocesan Registrar and Worcester Record Office*

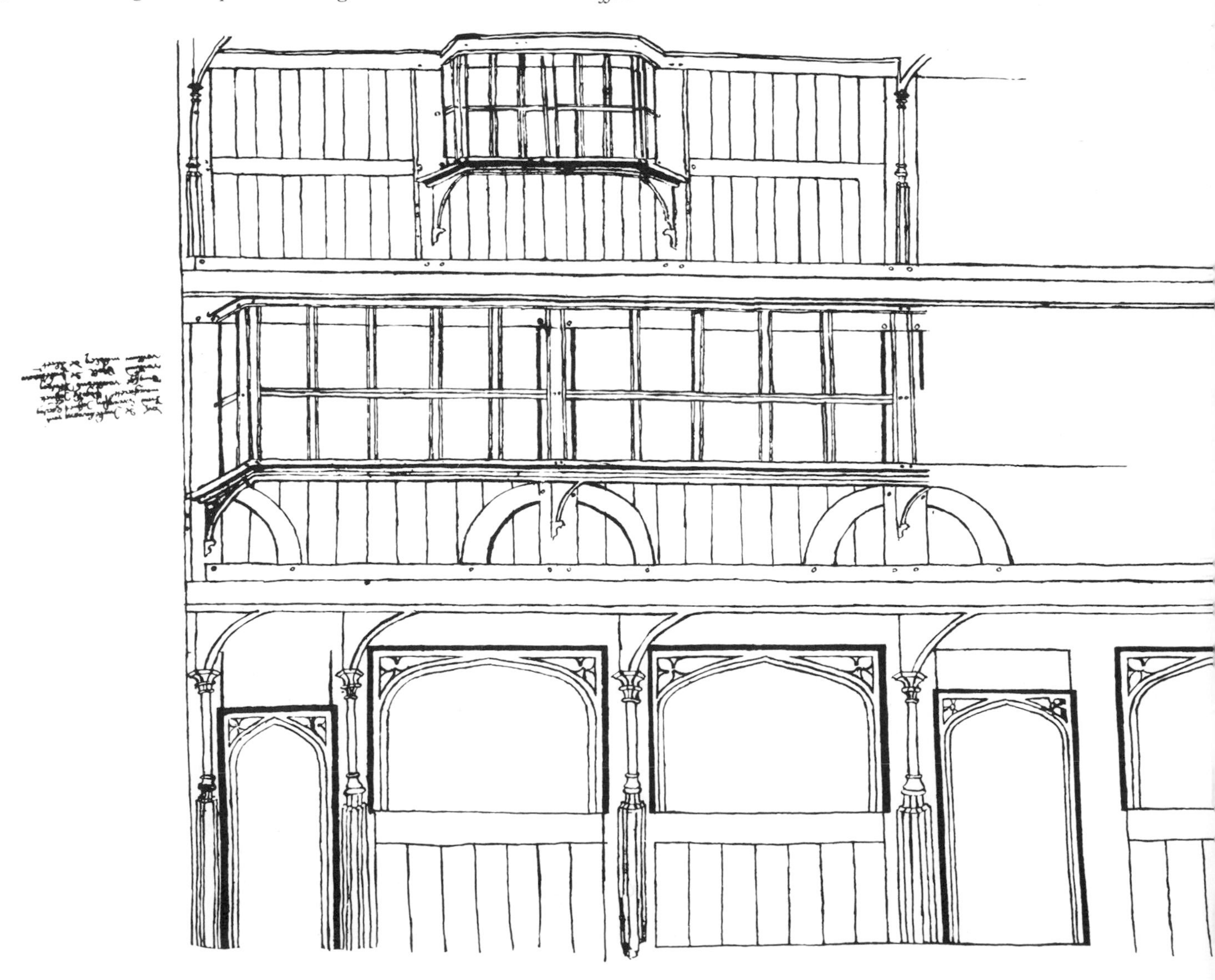

the first floor a workshop, and the top floor storage or a further workshop. The living accommodation must have been at the back, possibly grouped round a court-yard. Several comparable buildings survive, its most striking resemblance being with the Abbot's House in Butcher Row, Shrewsbury. On the back of the parchment are roughing-out details for a fine arch-brace roof-truss and jetty of the same or another building. Even the pegs indicating the mortice-and-tenons are drawn – as they always should be (Figure 55(a) and (b)).

Drawings would probably not have been necessary only for the peasant houses and barns mentioned in the

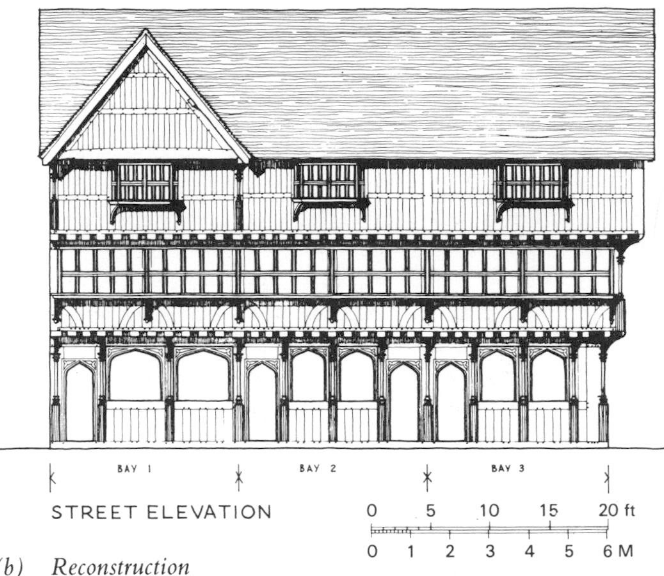

(b) Reconstruction

Figure 56 *Typical medieval town house*
(*a*) *Plan*
(*b*) *Elevation from passage*
(*c*) *Opposite elevation*
(*d*) *Street front – T I*
(*e*) *T II*
(*f*) *T III*
(*g*) *T IV*
(*h*) *T V*

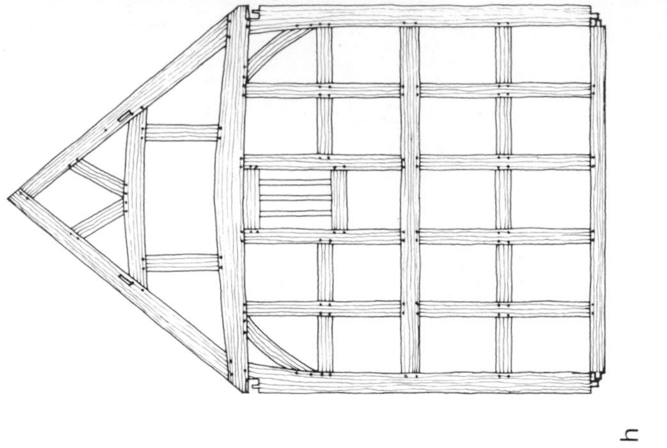

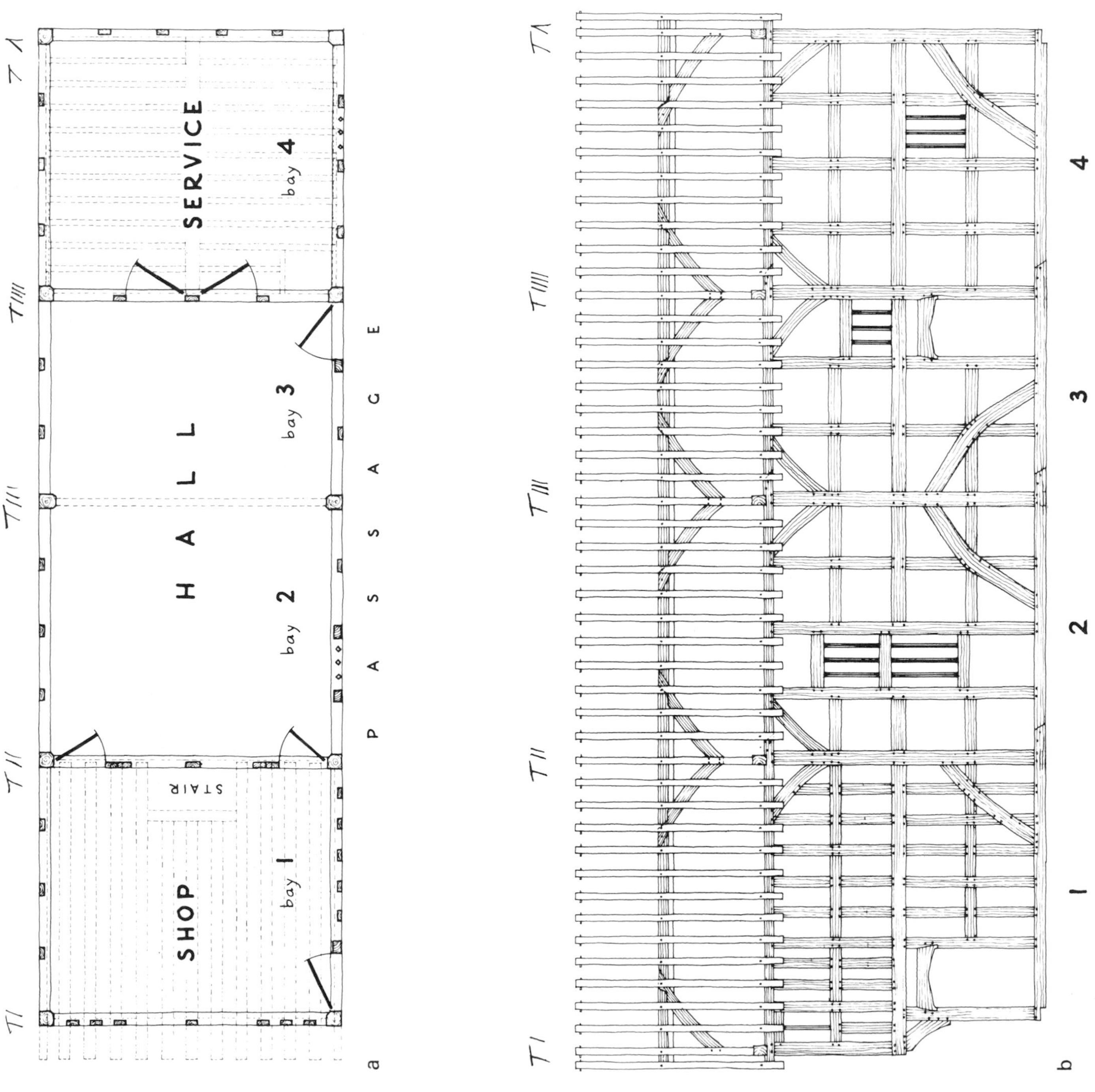

a
T I
T II
T III
T IIII
T V
SHOP
STAIR
bay 1
H A L L
bay 2
bay 3
SERVICE
bay 4
P A S S A G E

b
T I
T II
T III
T IIII
T V
1
2
3
4

h

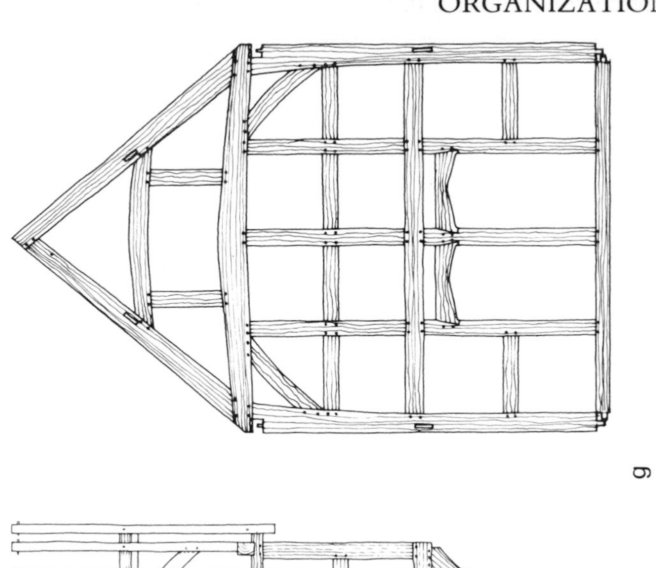
g

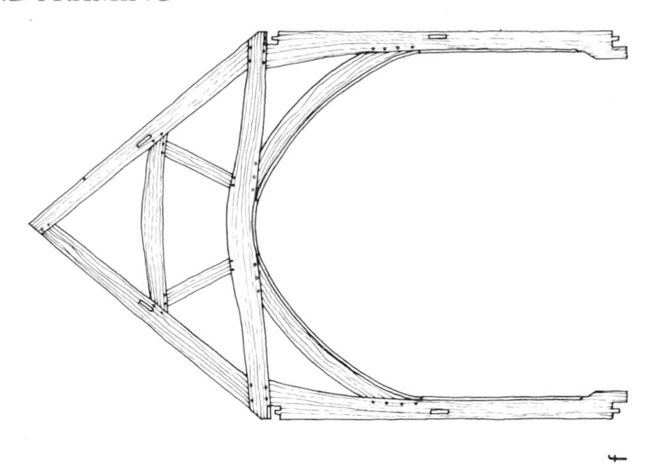
f

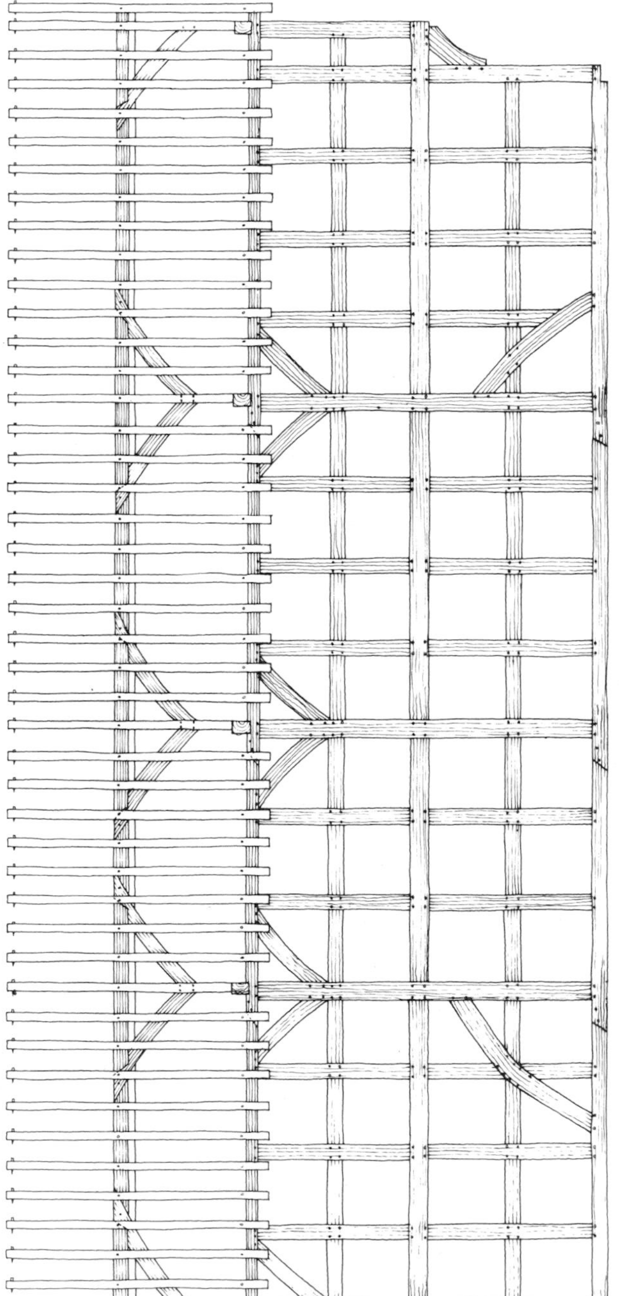
c

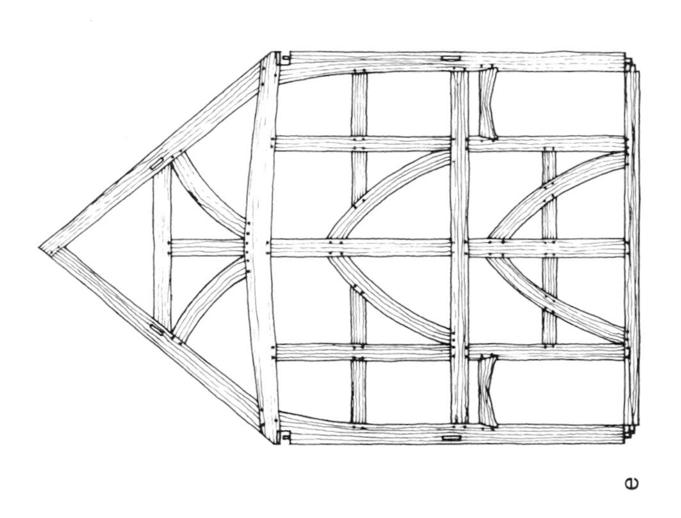
e

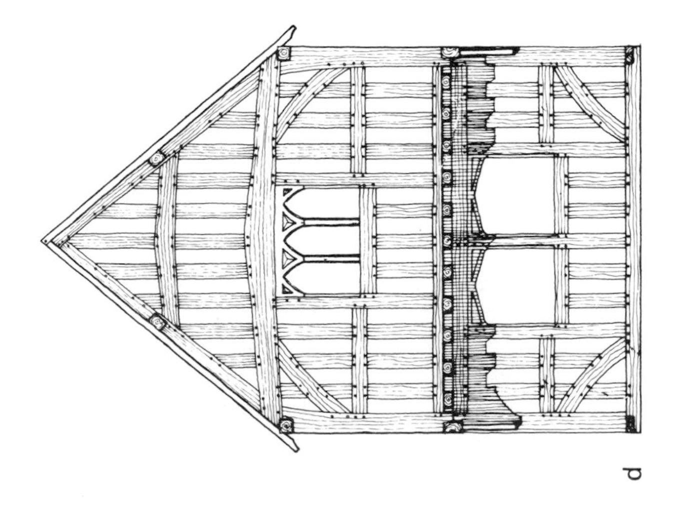
d

court rolls. The village carpenter would know what was required and be able to set out the frames directly on the framing floor, deciding its construction when he saw what main timbers the manor would provide.

The carpenter's yard

The yard was the common factor for the carpenters of whatever standing. Their location and organization is a subject hardly yet explored. The yard at Farnham has already been noted (Chapter 2), and such pub names as 'Beetle and Wedge' (especially when found at *Moul*sford, yet another term for mallet, and situated by the Thames) give some idea of their functions as manufactory and entrepôt between forest and building site. Medieval paintings show carpenters at work, but generally with more regard for the tools and their manner of use than the structure, which is usually highly conventionalized and technically inaccurate, perhaps because it did not yet exist for the artist to copy. One illustration, however, shows carpenters performing every operation, from cleaving and frame-sawing to drilling peg-holes, just outside the moat of a besieged castle, and quite oblivious of the arrows flying between the tower they have just built and the castle ramparts.[8] No doubt for the larger works, apart from those for warfare, it would often have been more convenient and economical to set up the yard at the building site than to double-transport every timber, depending of course on the geographical relationship of forest, yard and site. This might also have been the case in the village or farm, but in general the permanent yard was the place where the timber-framed building had to come into being.

Its organization and the highly skilled crafts that went on in it have no parallel today. In common with the present yards of timber merchant and builder it no doubt had offices, workshops, stores and temporary shelters. In addition there would be the carpenter's house and probably accommodation for workers and apprentices. There might even have been a farm as still survived in many carpenters' establishments in the last century. But the real difference was in the work that went on and the way it was done.

Here were brought the butts and branches of all shapes and sizes directly from the forest. The timbers required for purposes other than a specific building would be separated, the largest butts probably reserved for planks and boards, panelling and shingles – products that were in constant demand; the longest and straightest stems for rafters; the branches for furniture and tool-making, carts, wheels and equipment; the shorter for fences, gates, joinery and furniture; and the near-brash timbers for staves, laths and the thousands of pegs needed for any and every structure. The finished products of these departments would also have their separate storage areas and method of stacking before the next operation, the planks and boards 'in stick' to season for a matter of years, during which no freshly converted timbers could be placed on top of them.

One well-known timber merchant relates that in his

> childhood days when we lived in a house within the timber yard there were piles of crooked trees and limbs specially put aside for boat and roof timbers, from which architects and boat builders came to select what was suitable for the particular building or boat they had to build or restore.[9]

As for the main structural timbers already selected for each contract even before felling, such as posts, tie-beams and principals, these had to keep their identity and position right through the complicated and lengthy process of converting them into frames. On the principle that what went into a stack first had to come out last, the whole campaign had to be preplanned for the final, strictly ordered site erection; this might be months or even a year or two into the future, but once under way could not be delayed by having to search for a misplaced member.

Sample building and numbering system

We shall illustrate the process by taking as a sample a fifteenth-century town house of more or less standard medieval plan (Figure 56). Such a house might be found anywhere, its exact replica nowhere. It is end-on to the street, with shop and jettied upper chamber in the first bay, the hall, open to the roof, in the next two bays, and buttery and pantry in the fourth, rear bay. The front has a characteristic pair of shop windows with let-down external shutters as a counter, and the main door is in the side wall, opening off a passage or wynd. This also leads to the door into the hall, and beyond that is the service bay. The yard with its well and kitchen within a range of outbuildings might have completed the accommodation. Beyond that again would have been the garden and an orchard. There would also have been a cellar under the front bay 1, this and the upper floor being approached by a ladder stair at the back of the shop. All of this, however, has been omitted from our sample.

Before describing the framing of the four bays of the house only, and discovering the number of its components, joints and pegs and the number of trees that were needed to construct it, there are a few preliminaries to be noted. The first is the numbering system applicable in principle to all medieval buildings: all components of each separate truss and cross-frame bear the same Roman numeral, always on the 'upper' face. The first truss is normally at the upper end of the building and the first to be erected. Thus in a multibay building, all the components of the first truss are / or ⋌, of the second // or /⋌ and so on, the fleck on the final digit indicating that it is to the right of the centre line as seen on the upper face. The whole series would appear as:

/ // /// //// Λ Λ/ Λ// Λ/// ⩚ X

⋌ /⋌ //⋌ ///⋌ Λ Λ⋌ Λ/⋌ Λ//⋌ ⩚ X

The reason for the inverted V is that a slip of the scribing knife, drawn towards the body, could convert it into X. Similarly, the use of four digits for IV prevents its being read as VI when, as delivered to the site, the top and bottom of the member cannot be distinguished. In multistorey buildings the timbers for each wall-frame at different heights may have a stroke on top to indicate the first upper floor, two strokes for the second storey and so on:

/̄ /̿ /̿̄ //̄ //̿ //̿̄ .... X̄ X̿ X̿̄

Such is the system as researched in France in the last century by Viollet-le-Duc,[10] but which is equally applicable to England and the rest of Europe. There are, however, far more buildings, no doubt in all countries, that do not wholly conform with it than those which do. Bredon Barn is the only 'perfect' example so far found, even to the numbering of the longitudinal components – the arcade-plates, purlins and arcade-braces. These are numbered I and II in bay 1, III and IV in bay 2 and so on, thus doubling up on the cross-frames, down to the last bay 9, where their numbers are XVII and XVIII.

The buildings that do not conform may differ in the sequence of numbers, their style and the tools used to scribe them, even their positions on the timbers – sometimes they are on their inner edge, sometimes on the mortice-and-tenon shoulders and therefore impossible to find without opening the joint.

The system underwent considerable change in the sixteenth and seventeenth centuries, in response to changes in the design of buildings. Increasing standardization meant that each component, instead of being numbered according to its relevant frame, could now be interchanged with its opposite number in any identical frame. There was thus less emphasis on upper and lower end. Many barns of the period were erected from the middle, or threshing bay, outwards, so that all except the two end frames faced inwards. In more complicated, multistorey structures, every timber of each elevation had to be numbered, in contrast with medieval buildings whose side-wall timbers are seldom numbered at all. But it is impossible to discover a general system. At the Ancient High House in Stafford, probably the largest and certainly most complicated town house in England, the architect's ingenuity in devising Roman numerals gave out and he had to resort to Arabics as well (Figure 57).

Figure 57 *Samples of numbering from various buildings*

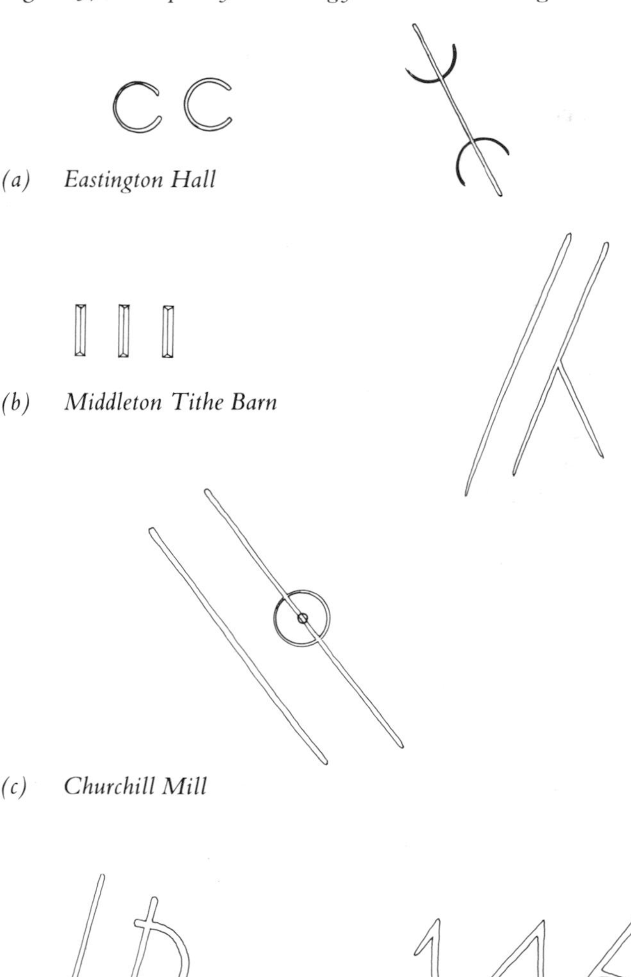

(*a*) *Eastington Hall*

(*b*) *Middleton Tithe Barn*

(*c*) *Churchill Mill*

(*d*) *Ancient High House*

The organization of the carpenter's yard for framing a structure such as this was the climax of centuries of experience and development, and could never have been thought up for just one contract.

Another of the 'mysteries' of structural carpentry is that it is a straight and rectangular art, despite all its swept, shaped or simply rough timbers. Splices, scarfs and tenons must all be cut along the grain and mortices must be at right angles to the drilled face. It is not possible to 'angle' longitudinal joints. Only the cock-tenon of wind-braces in base-cruck structures may be skewed vertically to engage the mortice of an arcade-plate (Figure 58). But town sites, especially, are seldom rectangular. The wall of a building, if it is to conform with an irregular line or curve, can only do so in straight facets of which each is a complete frame, unjointed to its neighbour. Posts must thus be juxtaposed and their contiguous faces shaped to the radius of the angle.

The non-rectangular plan, generally a parallelogram, is often the result in town buildings of earlier ridge-and-furrow strips and the first street or lane having been driven across them obliquely. The plots might have a further kink in their lateral boundary as a result of the 'heading'. All the buildings in Spon Street are examples of this. Their corner posts are to a greater or less degree trapezoidal in cross-section, their two outer faces being angled to the site, their two inner ones at right angles to their respective wall-frames, for the panels must have square edges (Figure 59(a), (b) and (c)). Jettied floor beams or joists are skewed to the frontage; roof trusses and rafters are also set skew to the wall-plates. But these, and in fact all the skew 'joints', are simply seatings of trenches for laid-on members; all the mortice-and-tenons remain rectangular.

Sites which narrow from front to back present the same kind of problem. If the building is end-on to the street, either the wall-plates must be inclined upwards toward the rear and top tenons of the studs and posts also inclined to conform with them; or the ridge (and purlins) may slope downwards; or, again, the pitch of each successive truss may be increased to compensate for their reduced span. But in all these variations, it is only seatings that have to be skewed.

The setting out of such buildings on the framing floor, on the other hand, is considerably more complicated. First, the shape of the plot in plan must be exactly reproduced. This could either be done by accurate measurement of the site, transposed on to the framing floor, or by laying out the actual sill-beams on the plot, jointing them there, and then returning them to the yard, where they could be laid out again as a full-scale plan. By either method,

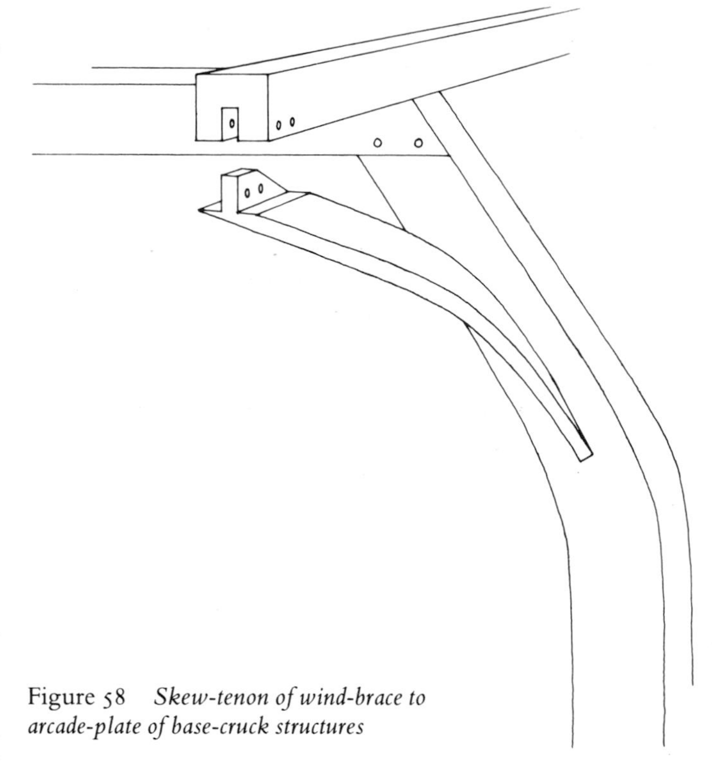

Figure 58 *Skew-tenon of wind-brace to arcade-plate of base-cruck structures*

Figure 59 *Non-rectangular structures*

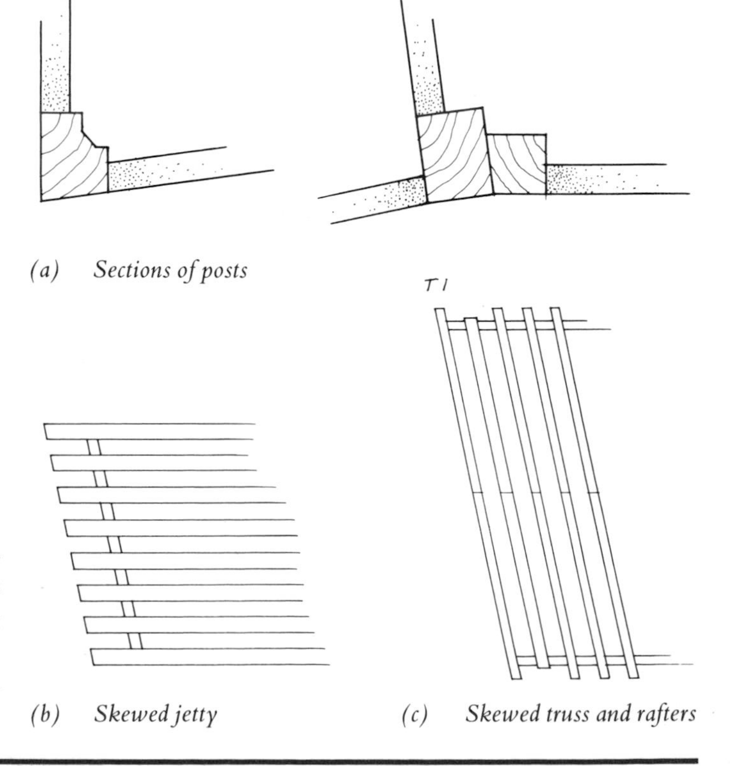

(a) *Sections of posts*

(b) *Skewed jetty*

(c) *Skewed truss and rafters*

the non-standardization of the frames would of course demand more work in their setting out than if, as in our example, all are of the same span and profile on a rectangular plan.

Framing

So, to revert to our relatively simple building, it is still necessary to emphasize that, in practice, actual circumstances would alter every step described. Yet every step had to be taken somehow or somewhere for the final structure to emerge.

The first conversion was done as soon as the timbers had arrived at the yard. They were sawn to length, making allowance for top and bottom tenons and some extra, and then squared or halved and worked into their final shape. Boxed-heart members might be chamfered and the chamfer-stops carved even at this stage, but window-heads, mullions, door-heads and brackets would probably only be blocked out. Drawings and schedules were the guide, and in this state they awaited first assemblage on the framing floor (Figure 60).

First the sill-beams were laid out, perhaps already fully converted and jointed from their earlier laying out on the site. They must be jointed over each other at the corners, the transverse sill-beams being halved under the lateral ones to allow the mortices for the corner posts to lie longitudinally. Alternatively, the transverse beams could be tenoned into the lateral beams. The latter would then project a short distance beyond the face of the cross-member or the tenon-and-mortice might be set back an inch or two from the corner. Yet another method is to butt the cross-beams against the sides of the lateral ones and secure them by means of jowled posts with a tenon lying in the direction of each sill (Figure 61(a), (b) and (c)).

Pegs and string lines gave the outline and key points

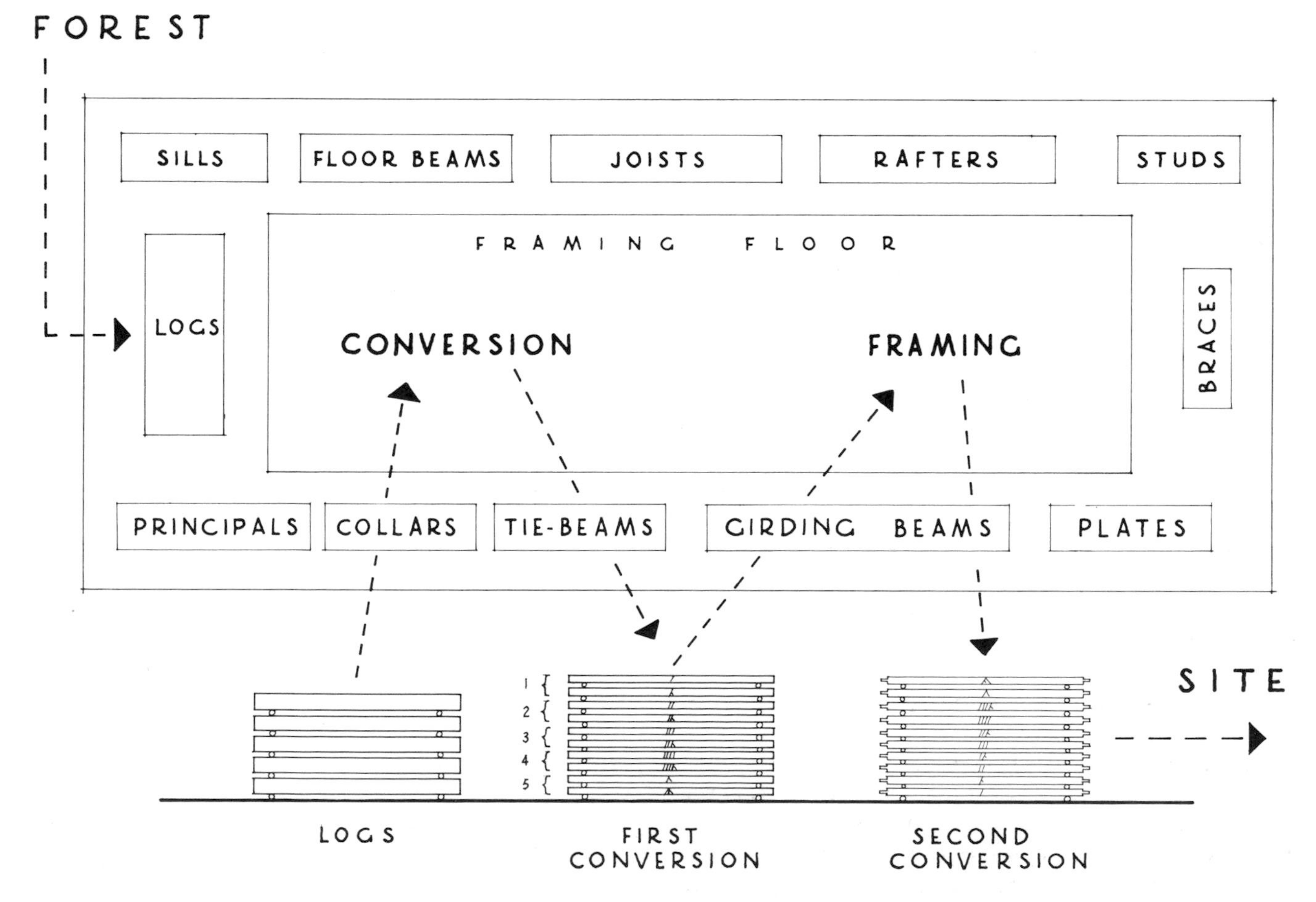

Figure 60 *Diagram showing organization of yard and progress of timbers from forest to site*

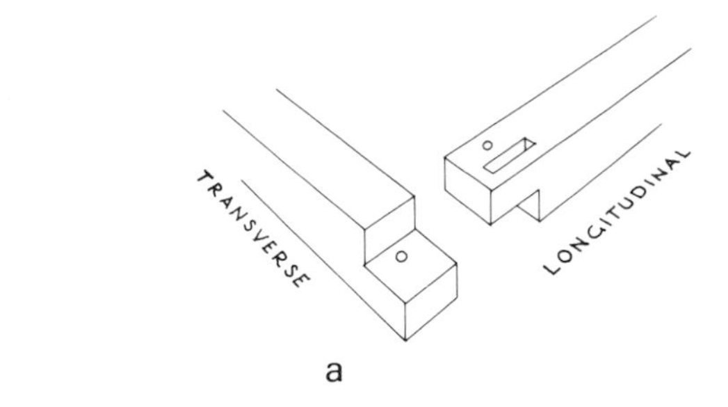

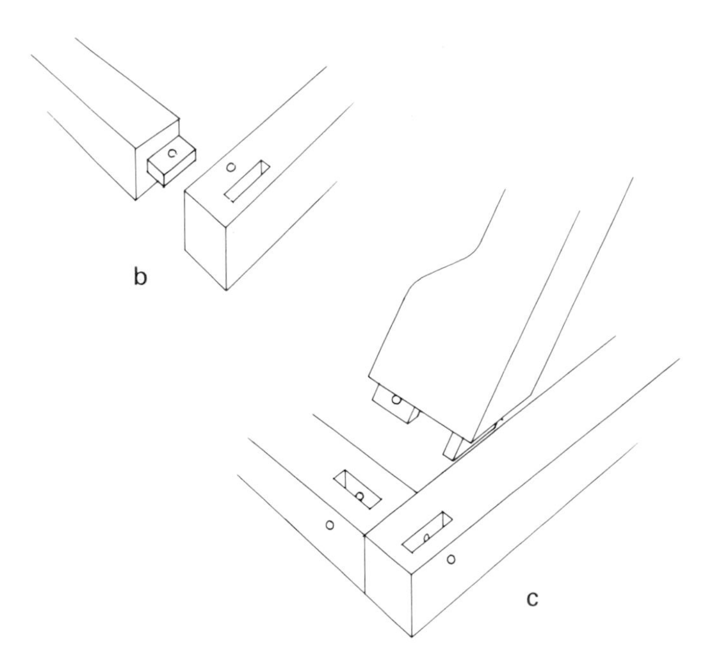

Figure 61 *Common joints for sill-beams*
(a) *Half-lapped*
(b) *Tenoned*
(c) *Bottom jowled post*

of the building's cross-section or end elevation – the bottom of the sill-beam, the top of the wall-plate and slope of the roof. Intermediate levels – plate of ground-floor framing, girding-beam and first-floor joists – would also be necessary for the jettied front frame, probably the first to be laid out (Figure 62(a)).

On such full-size diagram, all the cross-frame timbers could be set out frame by frame, from T I to T V. The fair-face was always upwards – hence 'upper' face both technically and, in the finished building, socially. Only the lowest end frame, truss V, being external, would have to be reversed – though there are exceptions – when the building was erected. The outline of the joints was marked on every timber with a straight edge and scribing knife and, after at least the main members were numbered, their even more painstaking second conversion was begun. The secondary timbers, so long as they were always in the right stack, could be picked out and assembled on site, generally without the aid of numbers. They may even have been delivered to the site from standard stacks and there have been finally converted and jointed just before erection.

We now look at the first frame in a little more detail. It has four posts, two of them jowled, but all of them, we have presumed for this illustration, halved out of a single tree, though it is perhaps more usual that each post should be boxed-heart and so out of four separate logs. Their joints are shown in Figure 62(b) and (c), but again those of the jetty must not be taken as standard. Alternatives are shown in Figures 80, 81 and 82.

Next, the lower part of the frame may be completed by laying the sill-beam and top-plate over the two lower posts and marking out the joints. The swept braces would probably also be included. Finally, the mortices for the secondary members – the studs and rails – must either be marked by means of measurement or by placing them on the frame. At this stage there are already four main timbers to be worked on the bench, and twenty-three secondary members, with altogether fifty-two mortices to be drilled and chiselled and fifty-six tenons to be cut, and this represents barely one-third of the work for T I only. The same labours then had to be executed for the upper part of the frame and the roof truss.

The same laying out of truss II, a closed frame, follows, and for this we have adopted a design with features characteristic of the upper wall of medieval halls (Figure 56(e)). The members are sill-beam, full-height posts, girding-beam, tie-beam, collar and principals; and the secondary timbers are three pairs of matching braces, six studs, six rails, two shaped door-heads and a central strut.

Truss III is the open-arch truss demarcating the upper from the lower bay of the hall (Figure 56(f)). This is architecturally the most important of all the frames. Many of them had cusped members above tie-beam level, as well as chamfered and moulded posts, tie-beams and braces. Many also had a mantle-beam about 6 feet above sill-beam level. But we have kept this frame to the simplest. The posts are chamfered on both sides and the chamfers are carried round the big arch-braces and the shouldered soffit of the tie-beam to form a continuous arch. These open-arch trusses do not of course have transverse sill-beams, but are tenoned into the lateral sill-beams.

Truss IV is the more ordinary lower wall-frame of the

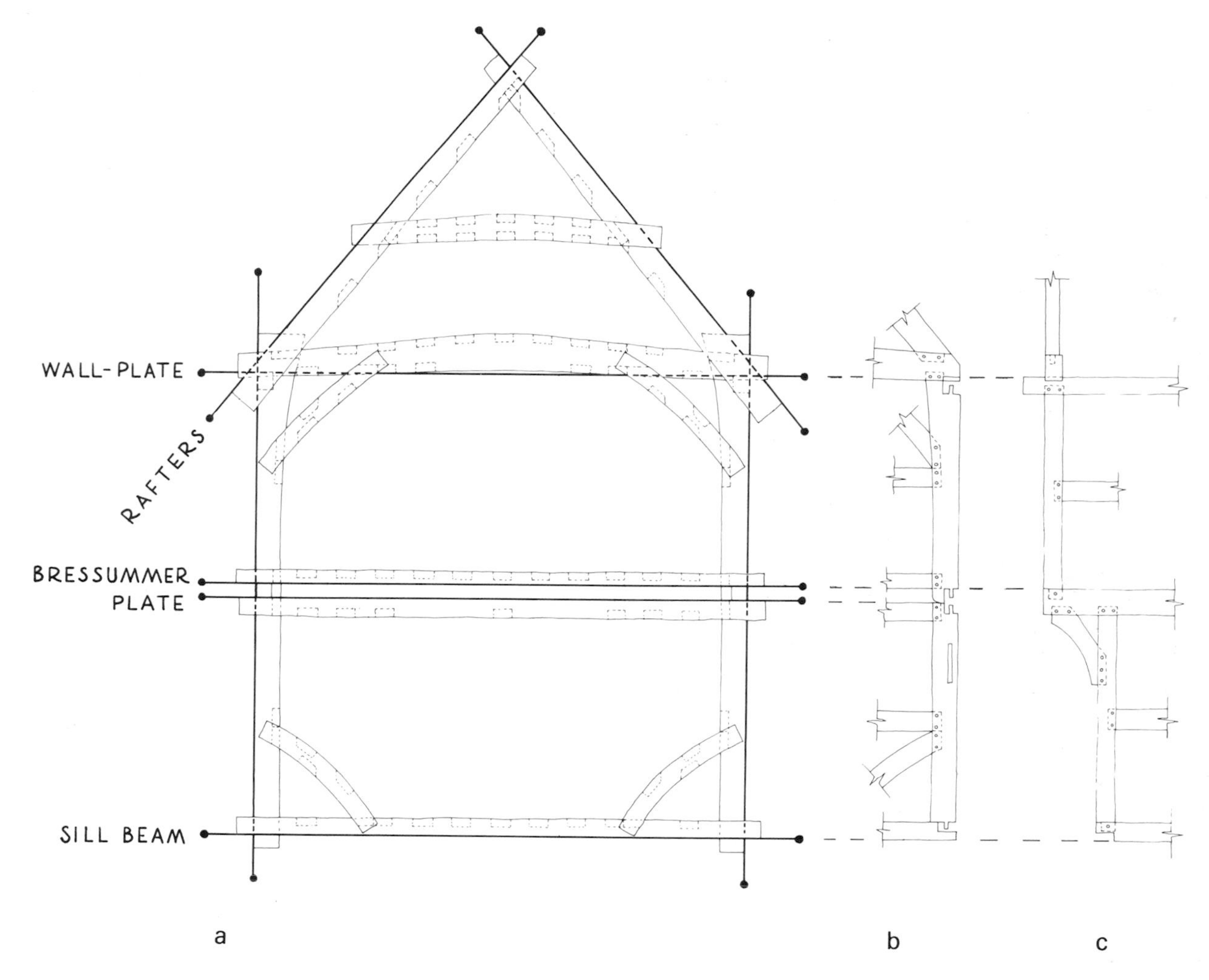

Figure 62 *(a) Laying-out of front frame, T I. The string lines indicate levels of components named – thus upper face of rafters, top of wall-plate, bottom of bressummer but top of girding-beam, bottom of sill-beam*

(b) Detail of setting out joints for right-hand posts
(c) Side elevation of posts showing jetty construction and joints

hall, entirely of straight members except for a small pair of braces and two door-heads (Figure 56(g)). And lastly, truss V is also uncomplicated except for the central window. The service bay has an upper floor, as bay 1, but without a jetty (Figure 56(h)).

So each cross-frame came into being. Then the framing floor would be repegged for the side wall-frames and roof, each side being set out separately (Figure 63).

The joints would include scarfs for the sill-beams and wall-plates, and dovetails and bird's mouths in the latter, as well as the mortice-and-tenons for the studs and rails. There are also mullion-mortices for windows, and shouldered mortice-and-tenons for door-heads. The rafters, also bird's mouthed, and scalloped at their feet, would complete this operation.

Lastly, the upper floors of bays 1 and 4 require the squaring and cutting to length of the joists for bay 1, and the longitudinal beam for bay 4 tenoned vertically at one end and horizontally at the other, with joists jointed into each side of it, their outer ends resting on the girdings. All of this could be done on the site.

By the end of these operations the carpenters will have produced nearly 500 components out of about 180 oak trees and poles.

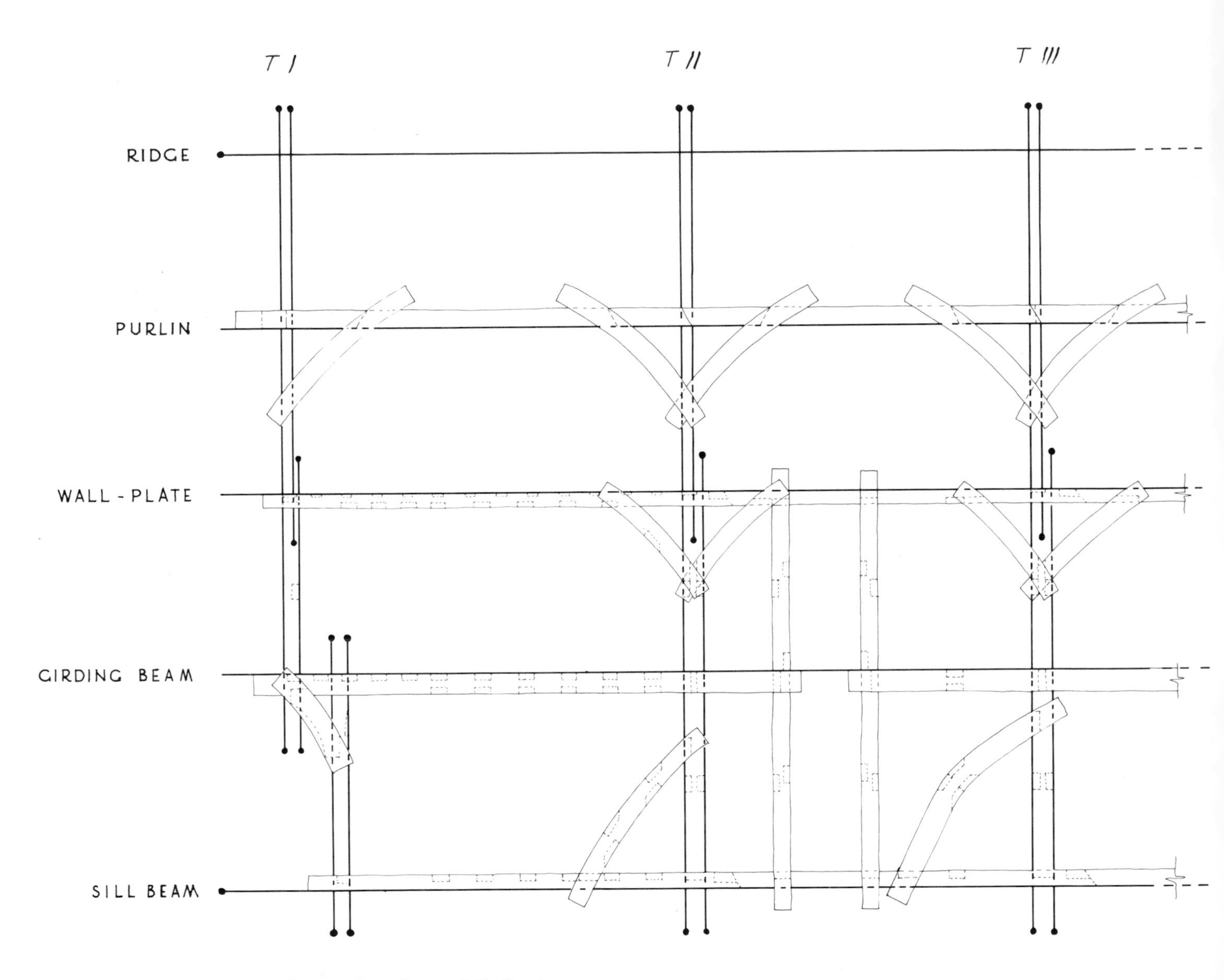

Figure 63 *Setting out on framing floor of part of side framing*

Site erection (Figure 64)

Delivery of the first frame to the site must have been a huge relief after the sheer slog of conversion and framing in the yard. First, of course, the sill-beams were placed on their stone plinths. It is more than a tradition that the first post to be erected was that on the left of the front elevation, as seen from the outside (Figure 56(d)), for the bottom corner of the roof corresponding with this post is also the starting point for the thatcher. This may be coincidence, but if it assisted in the follow-on of trades there is reason for it. The post is braced in both directions and the carpenter works across the front and down the side elevation (Figure 56(c)), completing the ground storey of bay I with erection of posts / ʎ and // //ʎ. Then the upper floor can be laid, and the left-hand upper post starts the upper frame, its first member being the bres-summer, or sole-plate, laid along the outer ends of the joists. Thus can the whole of bay I be completed while the 'lower' bays of the building are being framed up to

Figure 64 *Re-erection of the Bromsgrove House at Avoncroft*

(a) Erecting the frame of the solar wing – the first post to be erected was that of the ground floor at the further left-hand corner
(b) Solar wing frame completed, and hall framing with reinstated chimney on left. Courtesy John Aulton

wall-plate. Whether the wall-frames are completed all round the building before truss I is erected, or whether progress follows a bay-by-bay sequence, including the roof structure, is a matter of choice, depending as everything on the decision at the initial design stage. There was no doubt a way of doing it typical of a particular yard or region, but while regional patterns of details and joints can easily be discerned, we can no longer watch buildings being erected, only occasionally *re*-erected.

Table 1 summarizes the whole process according to the members. First their scantling is given and then the top diameter of the tree or pole, including sapwood and bark, that would most economically yield them. The total length of each set of timbers is next given, then the method of conversion and consequent length of the butts. This gives finally the approximate number of trees, assuming that on average each tree felled would yield a 20 foot butt. Such trees would cover about two acres if all of them were suitable. More likely selection would reduce their number to about one-fifth or less. Therefore not less than 10 acres would have to be searched for the construction of a building even as minor as this one.[11]

Table 1

Member	*Scantling (inches)*	*Top diameter of tree (inches)*	*Total length of timber (feet)*	*Conversion*	*Length in full butts (feet)*	*Approximate number of 20 foot butts*
Sill-beams	8 × 8	12	204	BH	204	10
Posts I	10 × 6*	18	40	H	20	1
Posts II IV V	9 × 9*	17	105	BH	105	5
Posts III	12 × 9*	20	35	BH	35	2
T I plate gd fl.	8 × 6	12	18	BH	18	1
T I bressummer	6 × 5	9	18	BH	18	1
Girding-beams	11 × 5	16	130	H	65	4
Cross-beams	11 × 5	16	60	H	30	2
Wall-plate	7 × 7	10	140	BH	140	7
Purlins	8 × 6	12	140	BH	140	7
Tie-beams	15 × 7	22	100	H	50	3
Principals	10 × 5	15	150	H	75	4
Collars	10 × 5	15	50	H	25	2
Struts (straight)	8 × 4	12	30	H	15	1
Braces (swept)	9 × 4	14	130	H	65	4
Braces (swept)	10 × 5	15	120	H	60	3
Wind-braces	9 × 3	14	130	H	65	3
Studs	8 × 4	12	760	H	380	19
Rails	8 × 4	12	340	H	170	9
Rafters	5 × 3	8	1344	H	672	84**
Upper floors						
Joists bay 1	8 × 7	12	200	BH	200	10
bay 4	5 × 4	16	240	Q	60	3
Beam bay 4	10 × 10	16	16	BH	16	1**
Door-heads, windows etc.	15 × 4	22	20	T	20	1
						187

BH = boxed-heart, H = halved, Q = quarter-sawn, T = through-and-through.

* Jowled posts, jowls obtained from buttresses, but extra minimum diameter as posts used upside down and sapwood not permissible at foot.

** 16 foot butt, halved.

Floorboards omitted as obtainable from stock.

As for the labours, in addition to the final conversion of the timbers, Table 2 shows the jointing required frame by frame. Assuming, purely for the sake of illustration, that each joint requires two hour's labour on the bench, including the time required for handling the timbers, and that the craftsmen work a forty-hour week and a fifty-week year, it would need a year and a half of craftsman time to make the joints – excluding their pegs!

Table 2

Building part	*Mortices*	*Tenons*	*Halves*	*Dovetails*	*Bridles**	*Pegs*
T I	155	147	2	2	—	311
T II	82	59	2	2	—	156
T III	41	19	—	2	—	84
T IIII	70	49	2	2	—	136
T V	76	59	2	2	—	144
W elev.	118	139	5	5	8	254
E elev.	163	179	5	5	8	331
W roof slope	—	26	8	—	21	142
E roof slope	—	26	8	—	21	100
Totals	705	703	34	20	58	1658

* Given as complete joints, bridle and mortice.

Details

The subject of framing cannot be left without a note on the finer details of framing methods. First, draw-boring is of fundamental importance in obtaining a tight frame. The outer wall of the mortice, after the tenon has been inserted for the first time, is drilled, and the tenon just marked with the bit. It is then withdrawn and the tenon drilled an eighth, perhaps as much as a quarter of an inch nearer to the tenon's shoulder; sometimes the entry into the hole for the peg is eased off, making a slightly spooned-out ellipse. Meanwhile the mortice member is drilled right through, before the tenon is replaced. When the frame is finally erected, the peg is driven in for the first time, tightening the joint as it goes in and biting at the interface of the two components (Figure 65). The pegs are left projecting, to be finally hammered in as the last job. They were never sawn off flush either at the upper or lower side, though the points might be cut back, particularly where they could be dangerous. It has been noticed where sometimes joints have been taken apart that tenons may have been drilled more than once as if the first time was a mistake and the carpenter incompetent. Of course there are other reasons, as drilling and pegging requires the same skill as does the rest of framing. It is more likely that where the components concerned are 'standard' their position in the first or second framing, at whichever stage they were drilled, was in a different frame from the one in which they were finally fitted. Or again, it is impossible to say how often frames have been taken apart in their long history, especially in repeat structures such as rafter roofs, and put back in different order, or with parts interchanged. This is certainly not the first age in which whole buildings have been dismantled and re-erected, but it may be the first in which every effort is made to put it back *exactly* as it was before.

There is another apparent anomaly that researchers have discovered in many roofs and in different regions. These are holes drilled 2 or 3 inches deep into the lower face of rafters and a few inches above their feet, so that

Figure 65 *Bredon Tithe Barn – brace showing evidence of draw-boring in the slight gouging out of the hole as the peg was driven in*

Figure 66 *Rafter-hole – always on 'lower' face. Courtesy Martin Charles*

they are more or less directly above the wall-plate (Figure 66).[12] Various theories, for instance that they indicate sprockets pegged to the side of the rafters, or that each rafter couple was framed to a jig either on the ground or during erection, have been advanced. Since no pegs have ever been left in these holes, they probably contained iron dowels removed when they had fulfilled their purpose and no doubt reused next time. This suggests that prefabricated rafter-couples, between first framing and final erection, had to be restrained against warping and springing until final pegging and the weight of the roof covering could hold them down permanently. Certainly no member has greater tendency to spring than a long slender pole, especially when sawn on its top and lower face to obtain the straightness necessary in that plane. It seems therefore than when stacked in the yard, probably vertically as many prefabricated trusses are today, the rafter couples could be held at the apex by their permanent joint, but would have to be prevented from spreading at their feet by temporary light-weight planks or jigs notched over the dowels.

Figure 67 *Scotch*

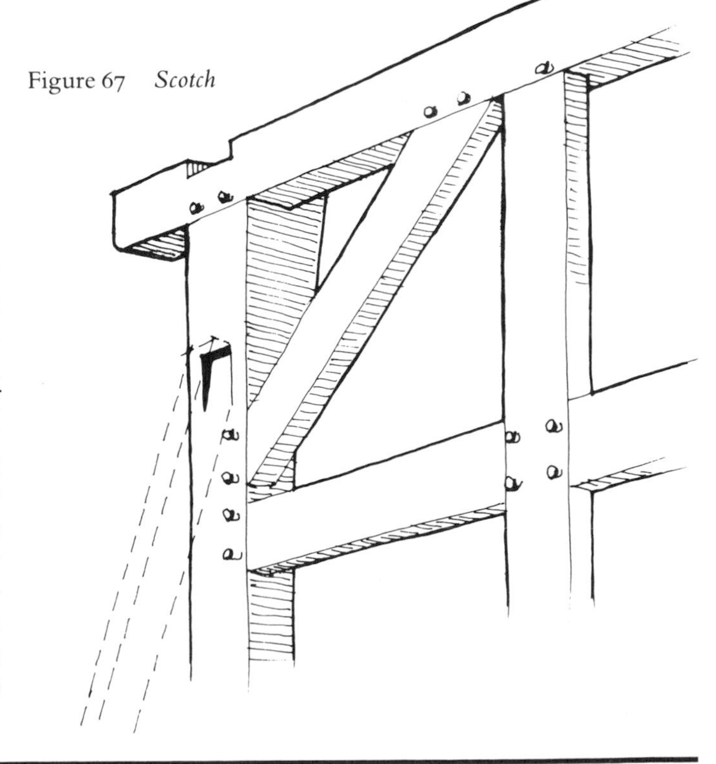

'Scotches', those splayed recesses cut into the outer face of wall-posts one or two feet below the wall-plate, are hardly a mystery (Figure 67). It is only perhaps worth saying that they have nothing to do with rearing; they are simply the means of scotching the top of a raking shore to prop the wall-frame before it had been permanently secured by the wall-plate and tie-beam. They are much more likely to have been cut on site than in the yard. Wattle grooves and stave holes, on the other hand, were certainly part of prefabrication.

Lastly, it is worth remarking that a special tool, the bizaigue, is still used in France for restoration, but may never have been introduced into England. It may be described as a large twybill with chisel edge at one end and gouge at the other (Figure 68). The mortices are drilled with (again in France) the old-fashioned auger, which their carpenters say are more efficient, at least for larger-scale carpentry, than modern power drills, and then the man stands on the timber and drops the chisel end of the bizaigue accurately on to the mortice walls, shaving off the wood between the drillings, finally finishing them with the gouge. There are bizaigues of different sizes for larger or smaller mortices.

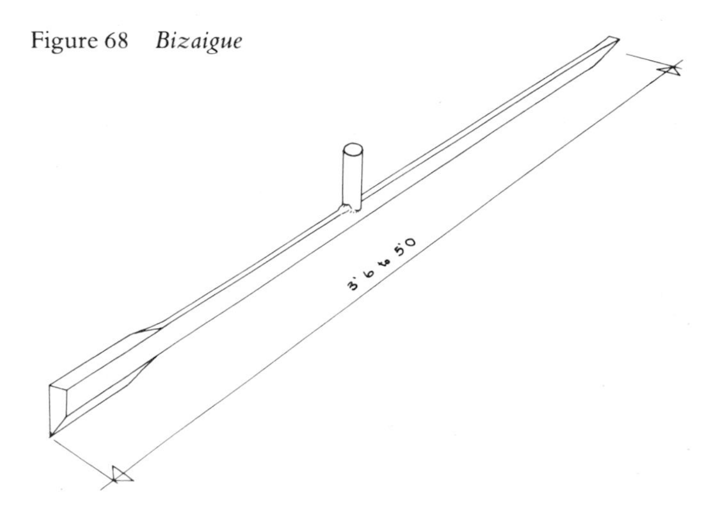

Figure 68 *Bizaigue*

HISTORICAL CHANGE

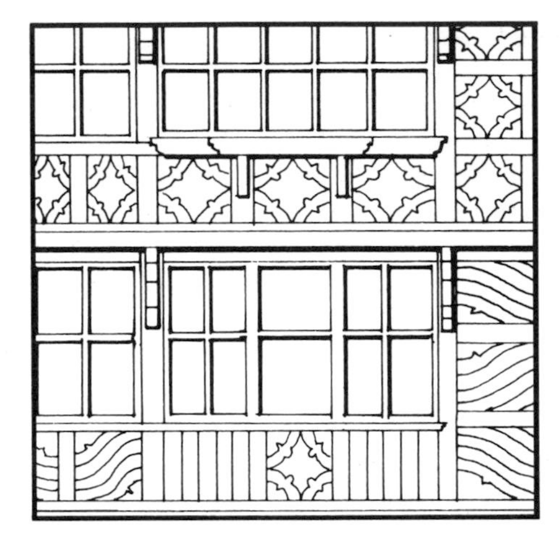

The history of the lesser house shows not only the way most of the population of England lived, but also – particularly in the case of timber-framed houses – those architectural and technical changes which are still the most reliable key to dating. For architectural style, proportion and decoration override regional variations. The latter might be called the constants. They are such things as the pitch of the roof – shallowest for the huge stone slates of the north, steepest for thatch or shingles – and the design of trusses and size of timbers – again largest where the weight of the roof demands strength, and lightest where the climate is softer and unlikely to blow the house down. There are also secondary regional influences – the relative exploitation of timber, accessibility to other regions, even from overseas, special economic circumstances of the community and so on. But these are for the specialist historian, and it is only necessary for us to note that it is dangerous to draw general conclusions from his findings.

Architectural style, on the other hand, disregards regions and changes with fashion. Occasionally the source of a new style can be pin-pointed. It has been suggested that the revolution that created Romanesque and Gothic owed *something* to the form of great timber structures.[1] There can, however, be no single cause of change but rather a gradual build-up of influences from many cultural sources until there is a more or less dramatic breakthrough. Then the new fashion spreads, albeit unevenly, from region to region until all but the most humble or utilitarian of buildings reflect it.

Technical advance follows, as the builder – the engineer in our own day – must solve the problems posed by the new architectural ideas, and in doing so opens up new possibilities for the designer. Historically, from the time when each region was a relatively isolated cultural unit, the differences between them have tended to close up through the centuries. But so long as local materials were in use, uniformity was unattainable. Even today local tradition is in some regions, such as the Cotswolds, a conscious element in the design of buildings and so helps to preserve their local character.

In the historical study of the timber-frame tradition, because its structural system is so direct and undisguised, each change and trend comes out more clearly than in any other building system. Indeed, many stone buildings can be dated only by their timber roofs or floor-frames. On the other hand, decorative stonework is stylistic *par excellence*, and when combined with timber – corbel-stones supporting posts, for instance – the stone may date the timber.

The first period convenient for this review is 1100–1300, which saw the full development of the hall-house and possibly more by way of architectural style than has so far been noticed. There is evidence that the hall and solar were originally not only structurally separate but separated by some considerable distance. Boothby Pagnell in Lincolnshire is the standard example.[2] There are several later ones, the distance becoming successively less. There was a five-foot separation between the hall and solar at Martley Rectory (Chapter 11), but at Shell Manor (Chapter 7) and many other late medieval buildings the distance is enough only to plaster the panels; in yet other instances this is difficult or impossible. Later on the two elements become not only functionally but also structurally unified, the side wall of the solar cross-wing providing the support for the wall-plates and roof members of the hall, and also for the longitudinal floor beam if the hall were floored.

The next period is 1300–1550, through which the hall-house persisted but with every possible combination and permutation of its arrangement with other elements – the solar, service and other ancillary accommodation.

Then in the period 1550–1650 the hall was superseded (though not quite finally) and the beginnings of the 'modern' house appeared, with fireplaces, glazed windows, staircases and separate rooms for sleeping and waking.

Finally from 1650 timber framing was progressively taken over by brick and stone until only the meanest cottages were still timber built.

1100–1300

There is in this earliest period a wide gap in the buildings' record between the great halls of kings and bishops at one extreme and the lowliest peasant houses at the other. The former are known both by archaeology – the Cheddar Palaces, Yeavering and other sites – and by standing buildings or vestiges of them – Hereford, Farnham and Leicester Castle. The peasant houses, all of post construction, as at the medieval villages of Goltho and Barton Blount in Lincolnshire, are known only by archaeology.[3] When the middle-range houses eventually appear towards the end of the thirteenth century they are the fully developed hall-house of aisled or base-cruck construction or the occasional Wealden. All are of some architectural pretension and superbly built, as they would have to be to last seven or eight hundred years. They represent the next rung down from the great halls.

There are also the stone-built upper halls of the Jews Houses in Lincoln, and Boothby Pagnell, also in Lincolnshire, and Norbury Hall in Derbyshire. The only evidence of a timber upper hall is one shown in the Bayeux tapestry, being set on fire by the Norman invaders, and although on the English side of the Channel it is, like all upper halls, Norman. They are clearly not a precedent for the peculiarly English ground-floor hall derived directly from barns or other simple forms of shelter.

So what of the rest of the houses of this period – the lesser halls of probably villein status up to that of priest and squire? That archaeology draws a blank virtually proves that they were timber framed. For when one of these is demolished and its plinth stones 'robbed', all signs of it would disappear at the first ploughing, and its traditional mud floor at the first shower of rain. Thus it is likely that in archaeological excavations of deserted medieval villages such houses are missing. Mostly they would have been of no great size, perhaps only of one bay with diminutive solar or parlour and no service accommodation; cooking and the rest would have been done outside in a post structure that might well have been mistaken for the house on the archaeologist's plan. Shell Cottage (Chapter 11) may give some idea of the smallness of such halls. Another of similar size is known from a seventeenth-century inventory which notes 'a hall and chamber beyond the hall' plus a number of other rooms, including parlour, best chamber and so on.[4] This house was one of very few inventoried that could be located; the hall and its chamber turned out to be a tiny separate medieval building, in use as a workshop and toolshed, the rest of the inventoried accommodation comprising a separate two-storey-plus-attic house of much later date and standing some distance away. But neither Shell Cottage nor this – at Lower Grinsty Farm, near Redditch – are of course of this earliest period. Those that were would clearly have been rebuilt, probably several times over, and in towns the chance of survival would be even less. There are indeed very few sites anywhere which had not already been built on before their present occupants were erected. The only building on a virgin site, so far as I know, is Eastington Hall, and that, significantly, is one of the oldest and most splendid of the region's base-crucks (see Chapter 11).

Nevertheless there may be a few first-builds still standing. A very unusual cruck-hall was found by Stanley Jones at the back of a shop in Tewkesbury.[5] It is peculiarly lofty in its interior proportions and has intermediate compound principal rafter trusses, something otherwise unknown in the cruck tradition. But its outstanding character results from the combination of gigantic shaped timbers in the main trusses and the comparatively flimsy straight members elsewhere (Figure 69). This characteristic, as already noted, may be exceptionally early.

Another stylistic misfit is Martley Rectory (see Chapter 11), which also has massively shaped timbers for its substructure (post-and-truss construction) and the remains of typically slender members in the roof (Figure 70). No other such complete examples are known,

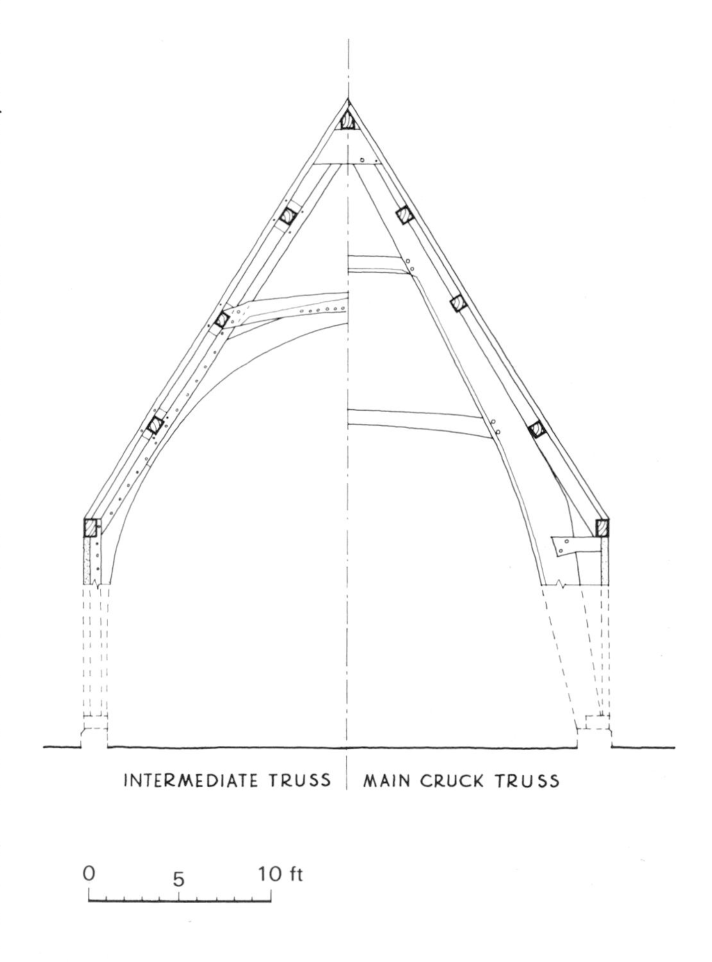

Figure 69 *81–82 Barton Street, Tewkesbury. Section of hall showing intermediate and main truss*

Figure 70 *Martley Rectory*

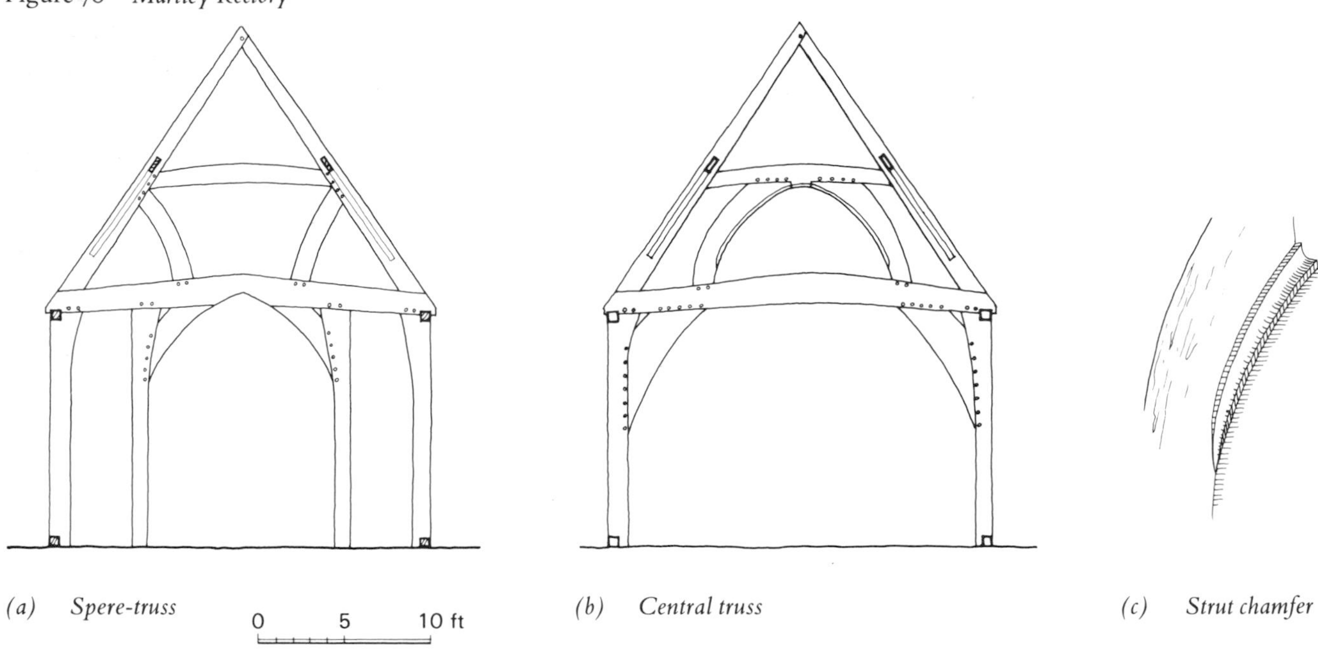

(a) Spere-truss *(b) Central truss* *(c) Strut chamfer*

but there are vestigial remains – a section of a massive cruck bedded in later stone or brick walls, the sockets of arbitrary half-lap joints, reused timbers with over-large mortices. Dendrochronology has hardly yet scratched the surface of dating but, to judge by results so far, many buildings may be found to be much older than is at present believed. Thus the base-crucks and aisled halls are unlikely to remain for very long the only acknowledged survivors from this period.

1300–1550

This was the supreme age of oak. In town houses and guildhalls, manor houses, barns and roofs, especially church roofs, there is nothing to compare in any other period, nor in most of the rest of north-western Europe. For our purposes there could be nothing better or more readily accessible to study than the many timber church porches, for these show in miniature most of the structural details of domestic roofs, as well as the style in each period (Figure 71(a), (b) and (c)).

The period begins with the heavy cusping of secondary members, the struts and braces (Figure 72), sometimes even in outside frames, and the 'scallop' arrangement of timbers in each bay of the wall-frame (Figure 73), all of course in mirrored pairs – at least in the superior buildings, as most of them are of this age. In the lowland counties

Figure 71 *Church porches showing architectural styles*

(a) Dormston Church, near Pershore – probably fourteenth century

the scallop pattern is less in evidence, but long curved braces from sill-beam to posts go the full height of the storey wall-frame (Figure 74). Crossover braces are also more prominent, especially in Lincoln and York (Figure 75). Rectangular panels universally exist, but they are larger and more irregular than the later ones, and often intersected by big swept braces.

The general style is Gothic. Apart from the 'natural' pointed arches of paired braces, its forms are also reproduced in pointed-arch door-heads, made from a pair of quadrants out of the buttress of a fairly large tree (Figure 76(a)). Ogee lintels are no less common, but differ from the stone version in that, carved out of a straight log, their profile is M-shaped rather than a pair of shallow Ss (Figure 76(b)).

Traceried windows are, or at least were, also common, but few originals have survived. A trick to be noted is that the tenons by which a traceried window-head is

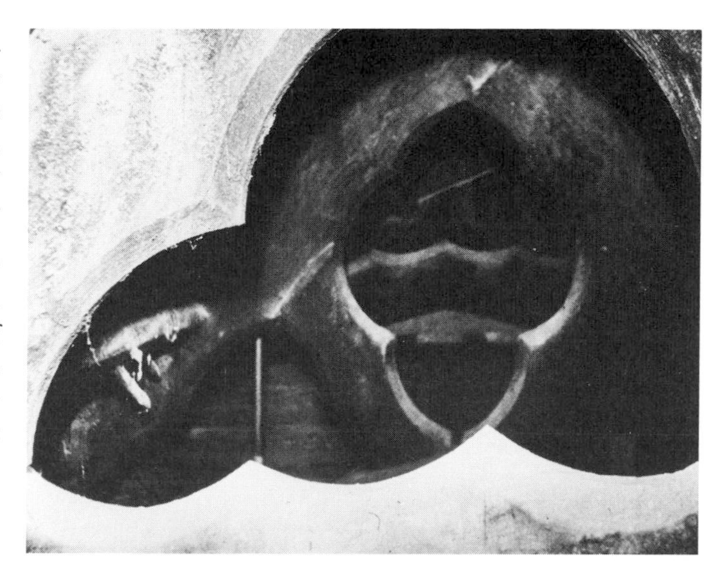

Figure 72 *Cusping of roof members at Amberley Court*

(b) Himbleton Church – fifteenth century

(c) St Kenelm's – late sixteenth century

Figure 73 *Typical wall-frame patterns*

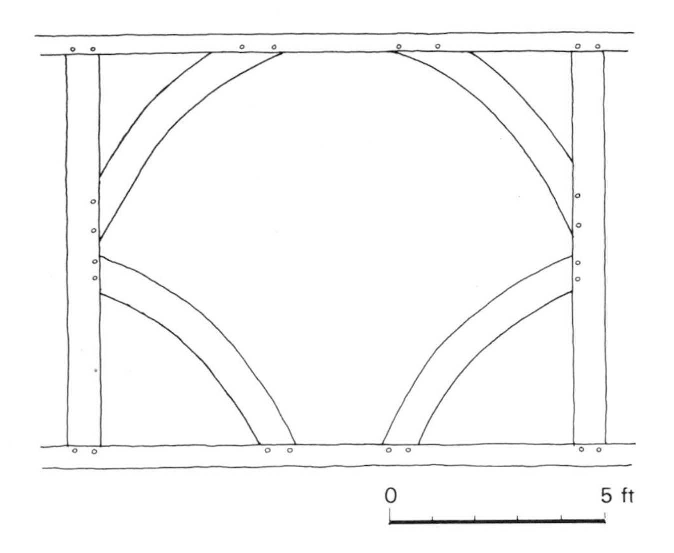

(a) Scallop pattern, as at Chorley House, Droitwich

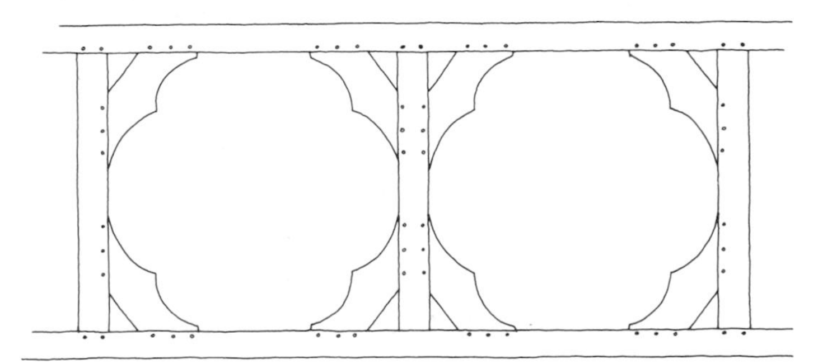

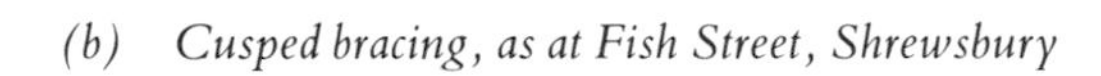

(b) Cusped bracing, as at Fish Street, Shrewsbury

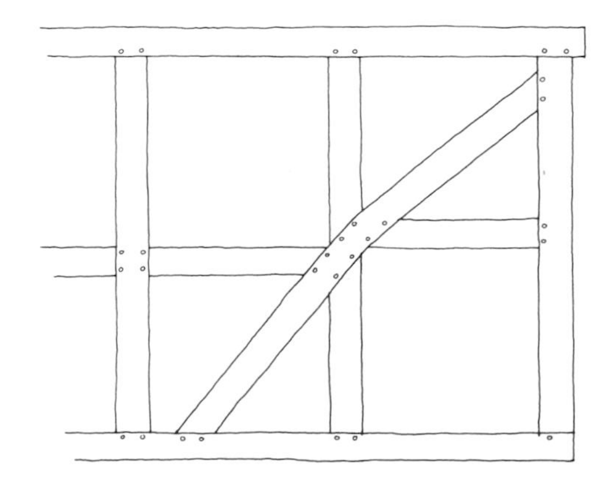

(c) Full-height swept brace, as at Lower Norchard, Peopleton

Figure 74 *House at Lindfield, Sussex – probably fourteenth century*

Figure 75 *Merchant Adventurers' Hall, York; fourteenth century. Example of scallop pattern and crossover braces*

Figure 76 *Typical medieval door-heads*

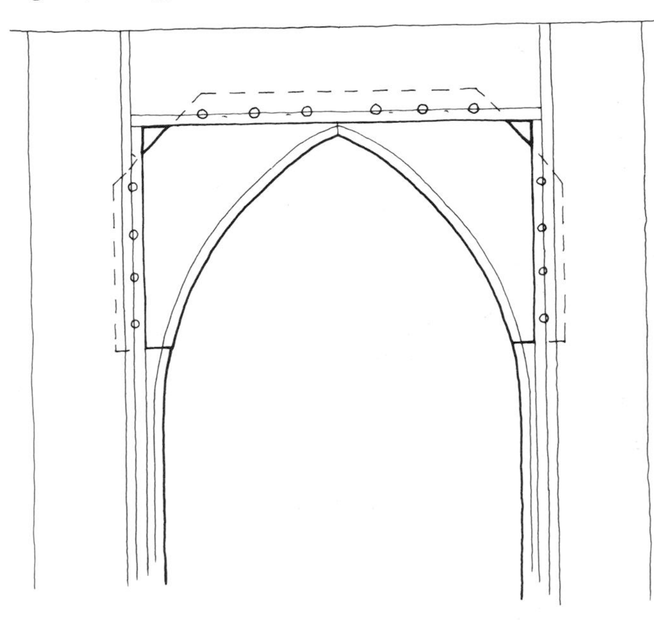

(a) *Pointed*

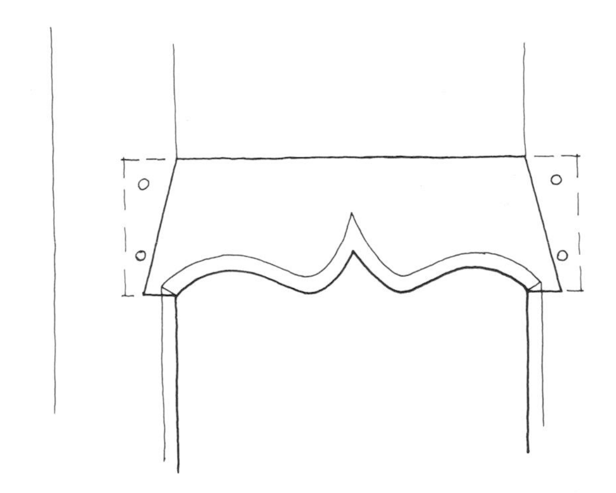

(b) *Ogee*

Figure 77 *Typical traceried window-head showing construction and method of tenoning the head to the stud*

secured must be some way down the receiving post or stud, for otherwise the top tenon of the latter engaging the plate must be cut back and so weakened (Figure 77). Long mortices with this detail are a sure indication of former traceried heads and must not be mistaken for a plain head with a gap between it and the top rail or plate.

The tall window of the upper bay of the hall has entirely disappeared, except for a few restorations (Figure 78). Traces of them, however, may still be found in the form of butt-ends of the mullions, wherever the wall-plate,

Figure 78 *Bromsgrove House, Avoncroft – restored hall window*

Figure 79 *Typical mullion sections*

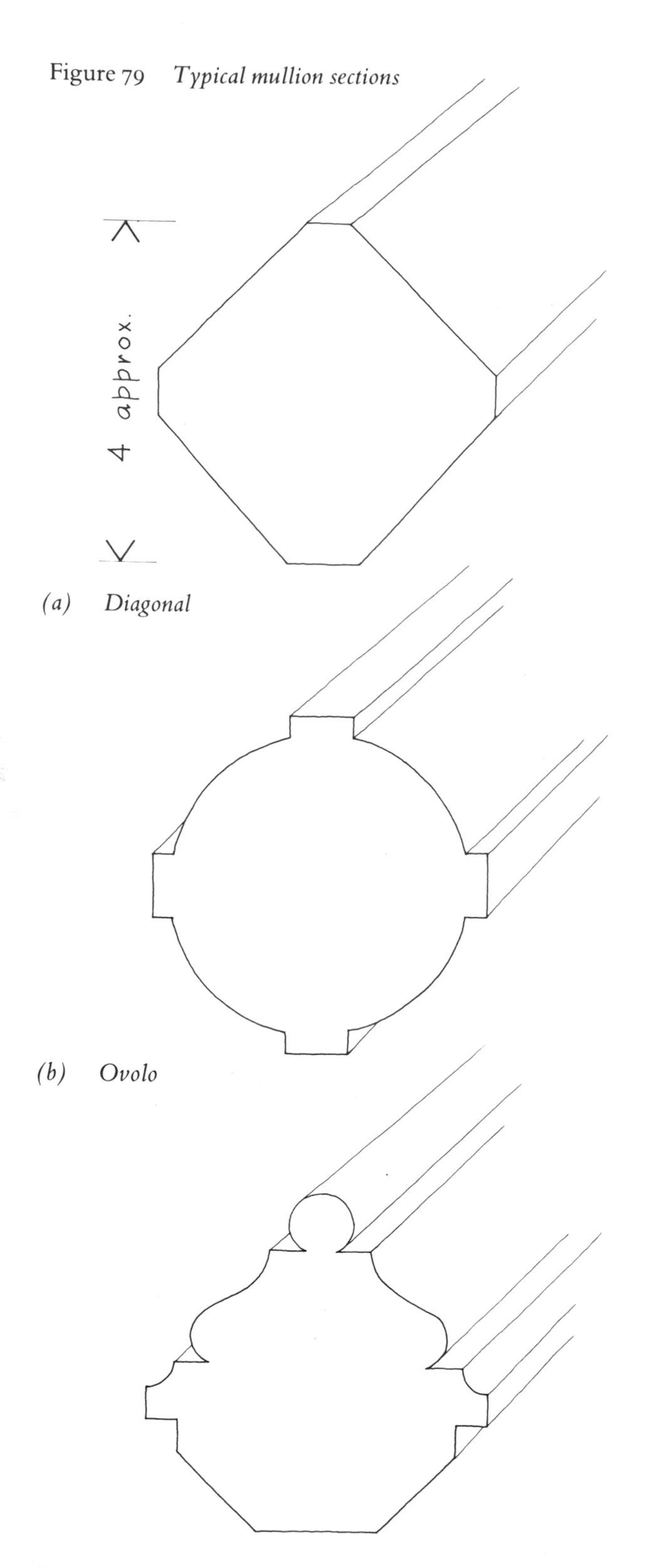

(*a*) *Diagonal*

(*b*) *Ovolo*

(*c*) *Bromsgrove House parlour, decorative moulding being to the outside*

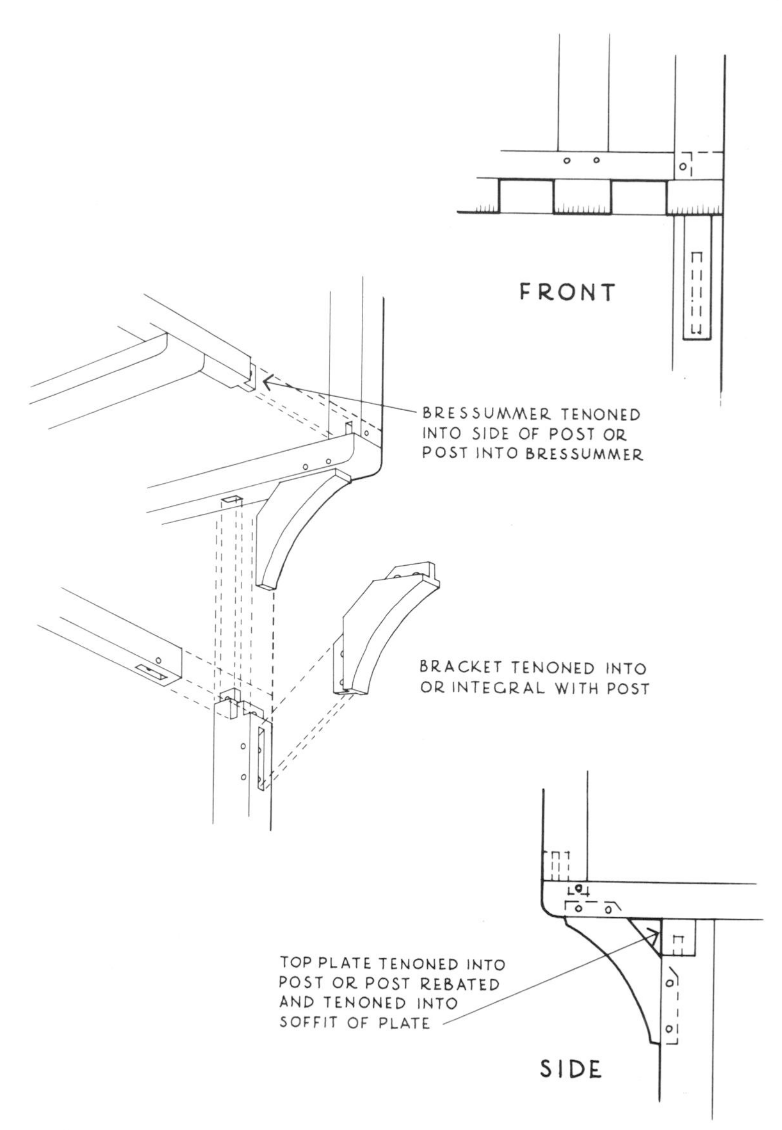

Figure 80 *Typical front jetty in which floor joists carry bressummer and are of same section as girding- and cross-beams*

lower rail and studs are chamfered. There may also be mullion mortices, much smaller than those for studs, and not pegged. Sometimes the 'shadow' of the mullion's cross-section may still be visible in the soffit of the lintel. They were generally square in section, set square or diagonally, but many were more decorative; the later ovolo-mould became universal for at least a hundred years, even persisting, in very reduced size, as glazing bars for the earliest sash windows of the end of the seventeenth century (Figure 79).

In town houses, the literally outstanding feature is the jetty. There is considerable variety of construction, par-

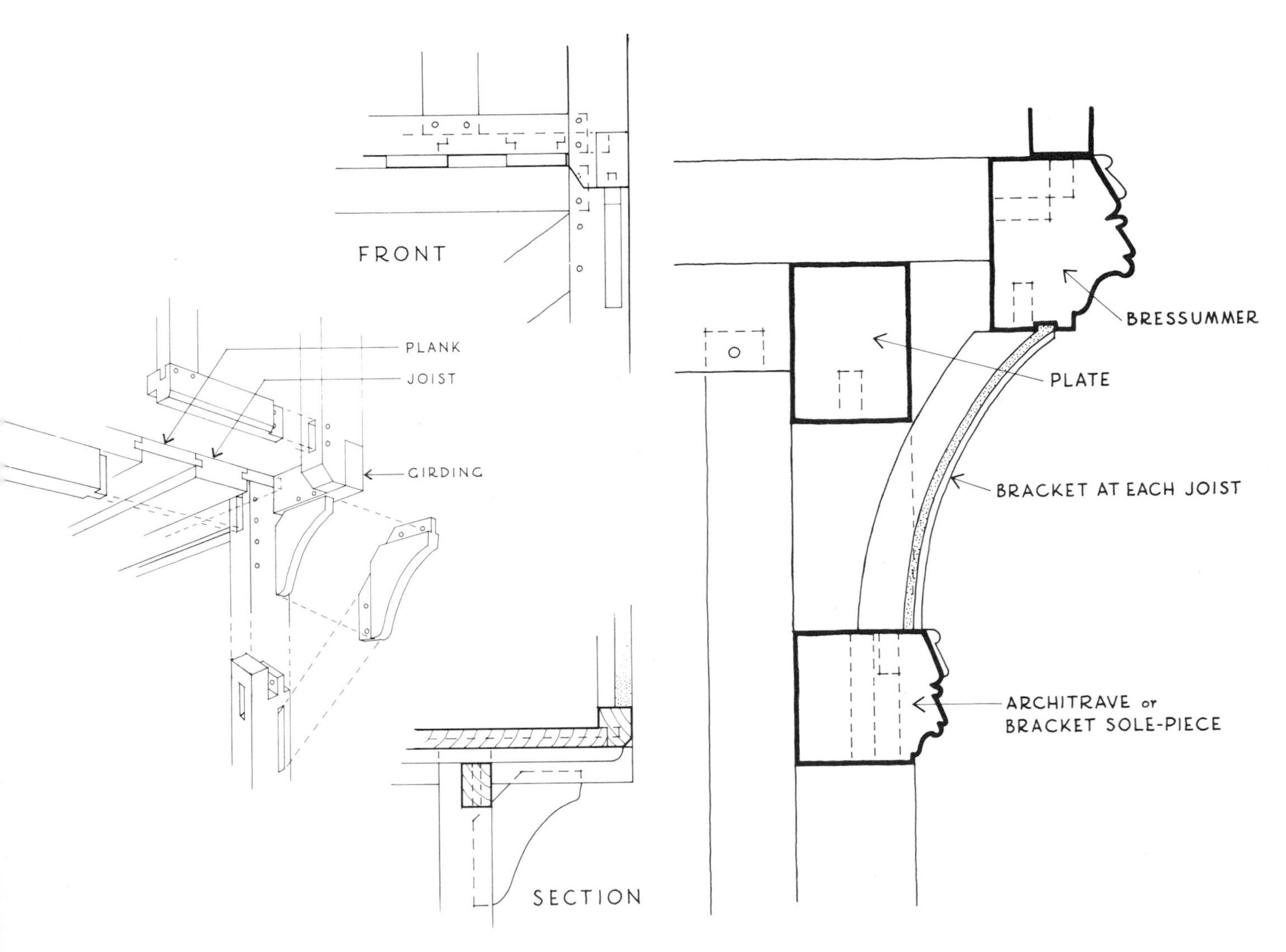

Figure 81 *Variation of Figure 80 in which girding-beams are of different section from joists, and bressummer is rebated to protect joist ends; example from Bromsgrove House, Avoncroft. Note floor of grooved joists and tongued planks*

Figure 82 *Variation in which joists are tenoned into bressummer; example from White Hart, Newark. Brackets are integral with posts and profile as that of brackets. See also page 224*

ticularly at the corner posts, but the principle is uniform. The floor joists, laid flat, are cantilevered over the plate of the front lower frame. Of heavy sections, generally 9 to 11 inches by 5 to 7 inches, the projecting ends appear as giant dentils carrying a relatively slender bressummer. Bay-dividing beams (the cross-beams of the intermediate frames) are of the same section as the joists, but bracketed to the post, brackets and posts in some early examples being integral (Figures 80 and 81). Later, the brackets may be at every joist and shaped to form a plastered cove (Figure 82).

The corner jetty is the carpenter's supreme achievement, as at the famous New Inn in Gloucester. The corner post, or dragon-post, could possibly be made from the buttress of a large tree and inverted. More probably it came from the stem and swelling of the first branch (Figure 83). The skill in *seeing* it in the growing tree and knowing how to extract it is perhaps comparable with Michelangelo's recognition of David in the block of marble.

Considerably simpler, though complicated enough in its jointing, as Figure 84 shows, is the dragon-post with attached instead of integral bracket. This is the general method from the beginning of the fifteenth century. At

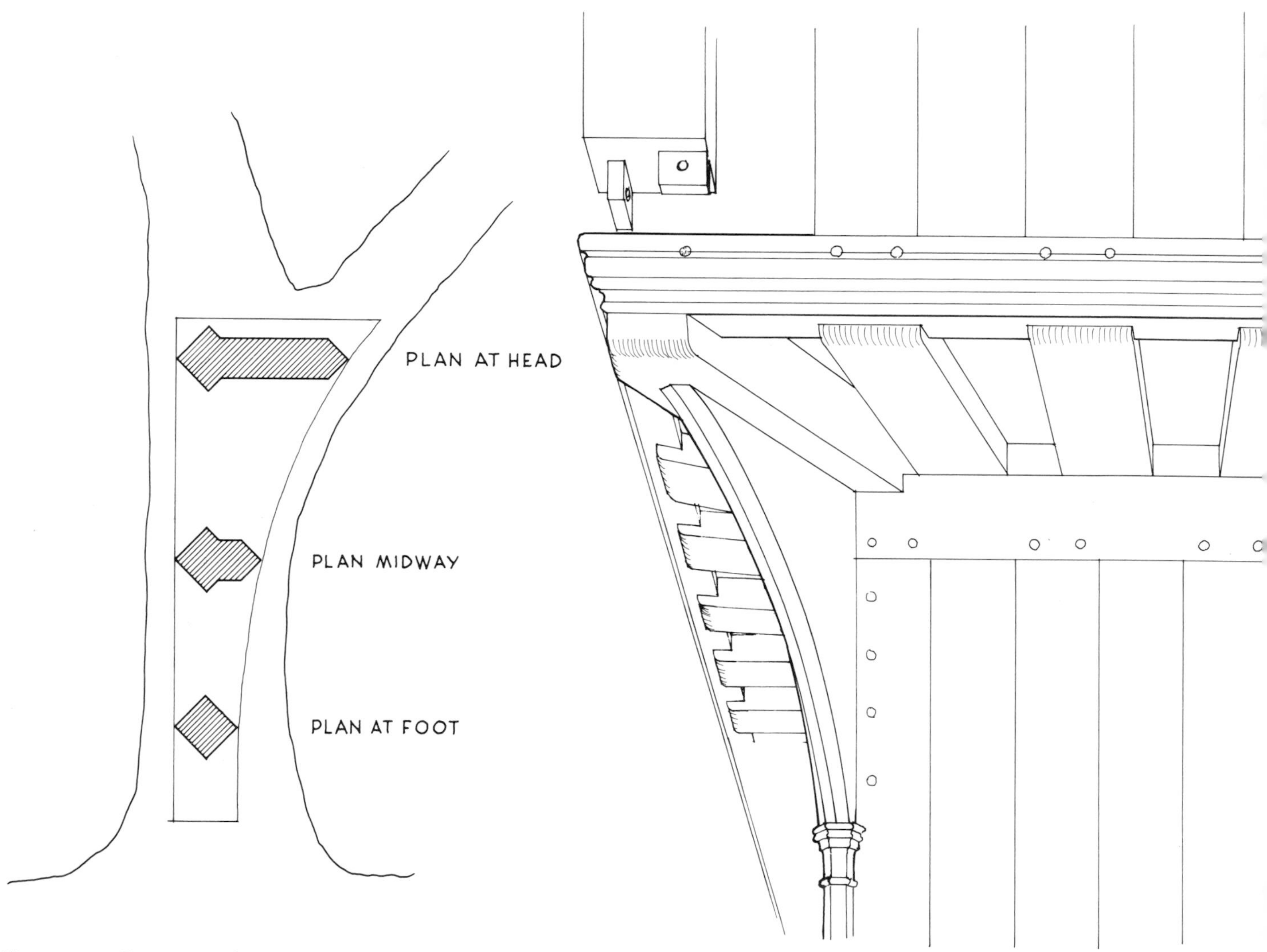

Figure 83 *Conversion of stem for dragon-post – profile of 'brace', integral with post, varies according to individual design*

Figure 84 *Corner jetty from Yardley Old Hall, Birmingham*

the same time the deep L-section bressummer appears, giving scope for richer mouldings and soffit coving, as well as concealing the joist ends.

The question, 'Why the jetty?', is so common and the answers so various that to hazard another might be only to increase the confusion. Nevertheless, the first reason, as so often, lies in the erection process. When building over a street there is virtually no other way to get the structure up. For by means of the jetty no timber is more than storey height, and each floor provides a platform on which the next frame can be erected (Figure 85). So it is possible to handle every timber and there is no need for scaffolding. By contrast, handling full-height posts on the street front and holding them upright while the girding-beams were jointed in before any floor beams could be laid, would be dangerous and probably impracticable. Of course it is possible to joint posts end-to-end, but when beams also have to be secured at the same level the result is one which any wise carpenter would try to avoid (Figure 86).

This is the best reason for the jetty. Some secondary reasons for it are: more space is gained on the upper floors, shelter is given to the one below, and the floor load can be considerably increased for the same depth of joist, the cantilever setting up opposite bending moments which are far less than one in the middle of the span.

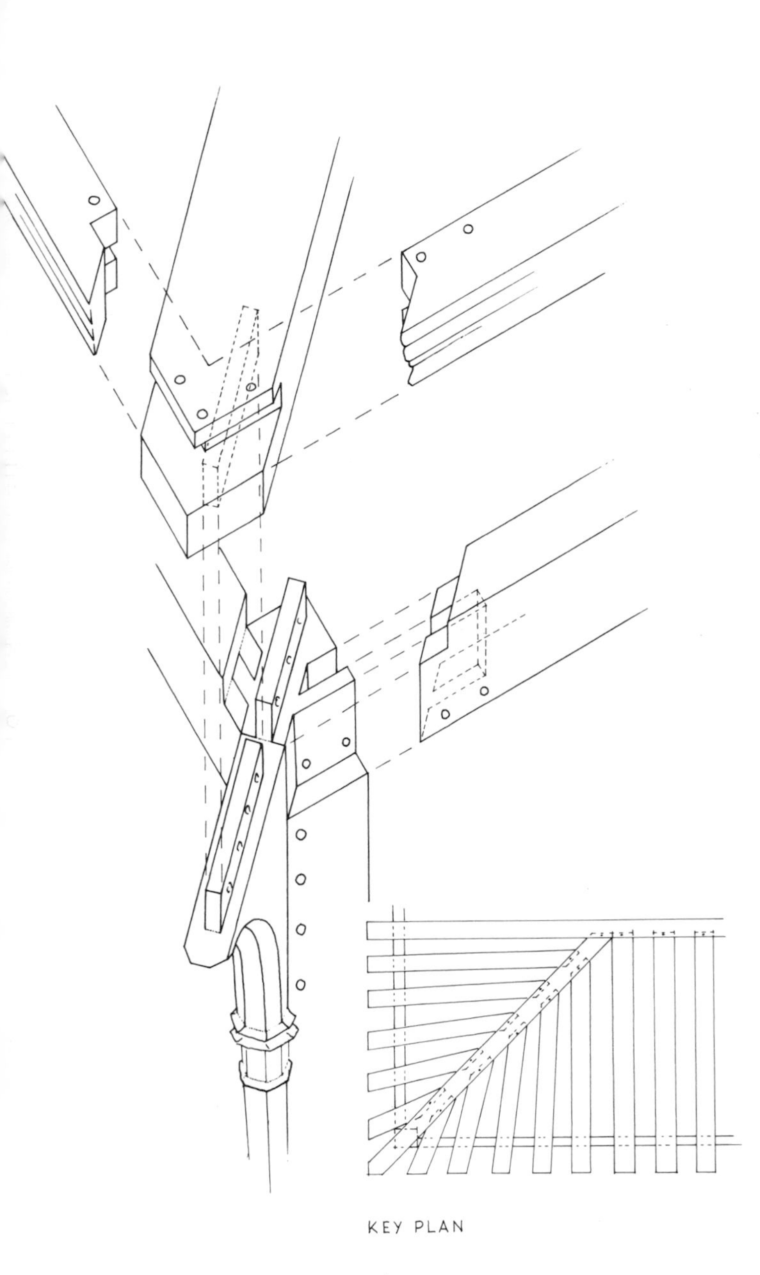

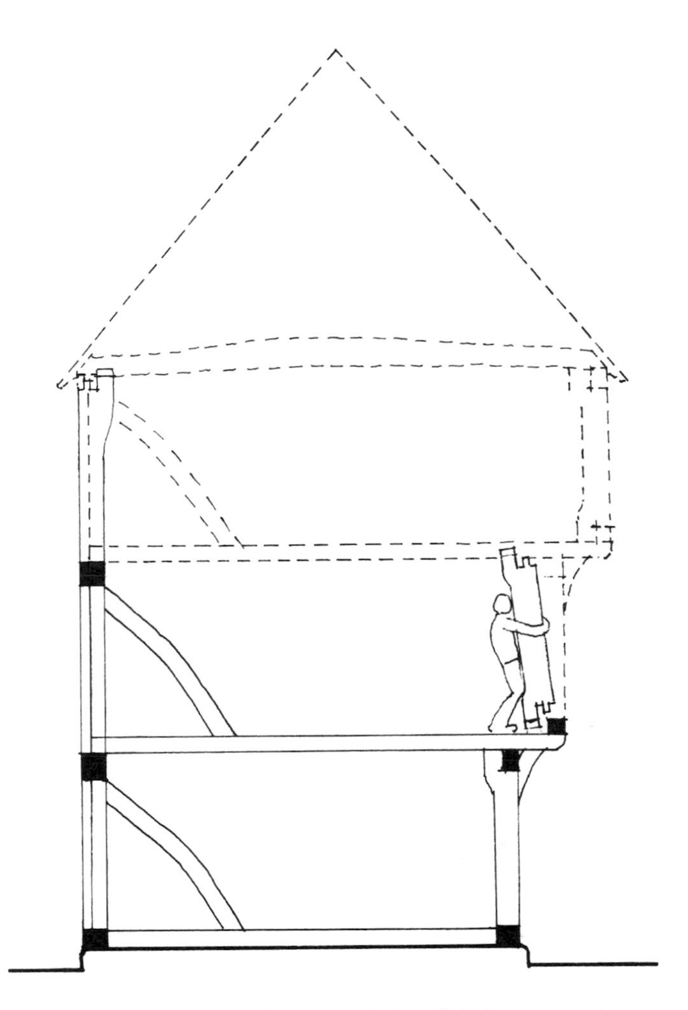

Figure 85 *Use of upper floor instead of scaffold for constructing jetty and usual bracing of rear posts*

Through the whole period, style is changing. From the earliest robust construction, emphasized by heavy braces and relatively sparse studs and rails, everything becomes lighter and more refined. By 1400 the scallop pattern was on its way out, to be replaced by close timbering, very literally reflecting Perpendicular. Braces were still curved but smaller and lighter, and soon only the top-brace, from post to wall-plate, was retained.

There is a slight difference in close-timbering as between the lowland and other counties. In the former the uprights are usually uninterrupted through each storey, probably deriving from earthfast palisade construction. In the latter, the usual presence of the mid-rail

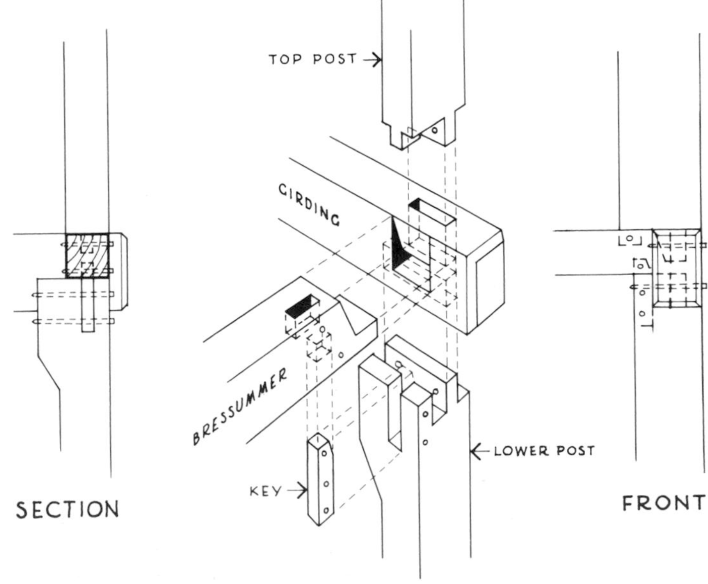

Figure 86 *Jointing of posts and girding-beams without the jetty – from Ancient High House, Stafford*

Figure 87 *Close studding*

(a) House in Robertsbridge, Sussex, typical of lowlands

(b) Friar Street, Worcester, typical of midlands

suggests that close-timbering is merely an enrichment of square panelling (Figure 87(a) and (b)).

Mouldings also become more refined, while four-centred arches take over from pointed and ogee (Figure 88). Where the latter persist, which is surprisingly often, they are shallower; ultimately, beyond this period, they become virtually straight with only a central mitre and quadrant at the shoulders (Lane House, Chapter 11).

The painting of timbers also comes in during the fifteenth century.[6] Vine and simple floral patterns seem to be the earliest, on timbers and panels alike (Figure 89). Later, these have often been overpainted with false close-timbering in complete disregard of existing widely spaced timbers and curved braces (Shell Manor, Chapter 7). Externally more buildings must have been brightly painted in reds, greens, yellows and golds, than the single extant example of the White Hart at Newark might suggest. For there are many historical references to the painting of buildings, and even blacking is mentioned,[7] but reds and ochres are the most common. At the very end of our period, John Leland, the official antiquarian to Henry VIII wrote (1539–40):

> The towne of Bewdeley is set on the syd of an hill soe coningly that a man cannot wish to set a towne bettar. It riseth from Severne banke by est upon the hill by west; so that a man standinge on the hill 'trans pontem' by est may descrive almost every house in the towne, and at the rysynge of the sonne from este the hole towne gliterithe, being all of new buyldinge, as it wer of gold.[8]

Figure 88 *163–5 Spon Street, Coventry – four-centred arched door-head*

Figure 89 *Astwood Court, Feckenham, Redditch – painting of timbers in solar roof*

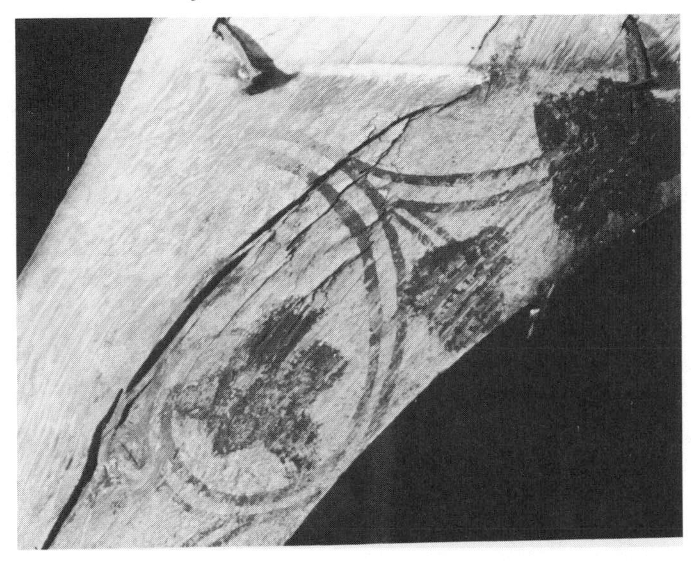

Needless to say every house was of timber, even though the simile might seem more appropriate for the mainly brick and tile town of today.

A medieval street of timber buildings may be difficult to imagine with their overhanging gables and barge-boards, bressummers, brackets, windows, door-frames and so on, not only carved and moulded but picked out in colour. Not even those continental towns where the tradition of painting has survived can give a true impression, for the richness, some might say, gaudiness of medieval decoration has gone. Even roofs, it seems, were coloured, if the discovery in Coventry of some medieval crested green-glazed ridge tiles is anything to go by.[9]

1550–1650

The house, in both town and country, was completely transformed by the effect on living habits of the Tudor economic and social revolution and by England's tardy architectural renaissance. By the time of the Civil War all semblance of the hall-house had gone, except in some farmhouses and cottages where something of the atmosphere of the communal hall perhaps survived in the kitchen.

William Harrison wrote the best of the contemporary descriptions:

> It is a world to see. No oak can grow so crooked but that it falleth out to some use. How divers men being bent to building do daily imagine new devices of their own. In the proceeding also of their works, how they enlarge, how they restrain, how they add to, how they take from, whereby their heads are never idle, their purses never shut, nor their books of account never made perfect. *Destruunt, aedificant, mutant quadrate rotundis*. . . . There are old men yet dwelling in the village where I remain which have noted three things to be marvellously altered in England within their sound remembrance and other three things too, too much increased. One is 'the multitude of chimneys lately erected, whereas in their young days there were not above two or three, if so many in the most uplandish towns of the realm (the religious houses and manor places of their lords always excepted) but each one made his fire against the reredos in the hall where he dined and dressed his meat'. The second is flock beds and bolsters substituted for straw pallets and a log of wood, and the change 'from platters into pewter, wood spoons into silver or tin'; and the third, '. . . as for stoves, we have not hitherto used them greatly, yet do they now begin to be made in diverse houses of the gentry and worthy citizens who build them not to work and feed in, as in Germany and elsewhere, but now and then to sweat in as occasion and need shall require it'.[10]

To trace the story, the first change was in the medieval hall itself, with the insertion of a floor and enclosed fireplace, the latter generally backing on to the passage. The two are interdependent both functionally and structurally.

In the typical two-bay hall, the inserted longitudinal beam was tenoned into the cross-beam of the upper-end truss and its other end supported on the new fireplace jamb or the chimney breast. The joists then, of course, spanned crosswise, tenoned into the beam and resting on a ledge

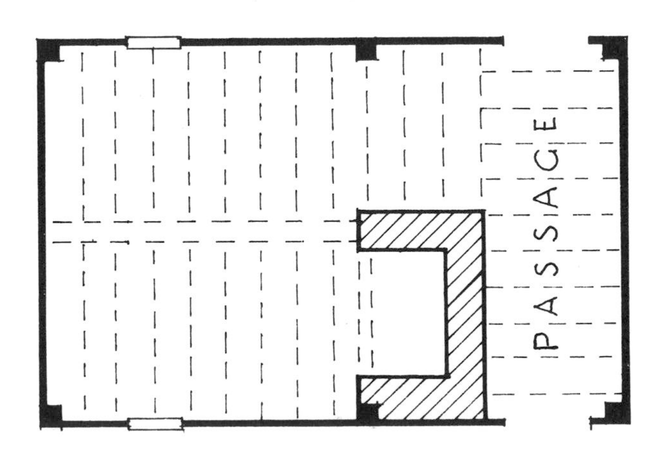

Figure 90 *Typical two-bay open hall showing normal method and position for inserting fireplace and floor beam in bay 1 – construction of floor in bay 2 varies*

Figure 91 *House with half-floored hall – from Bromsgrove House, Avoncroft*

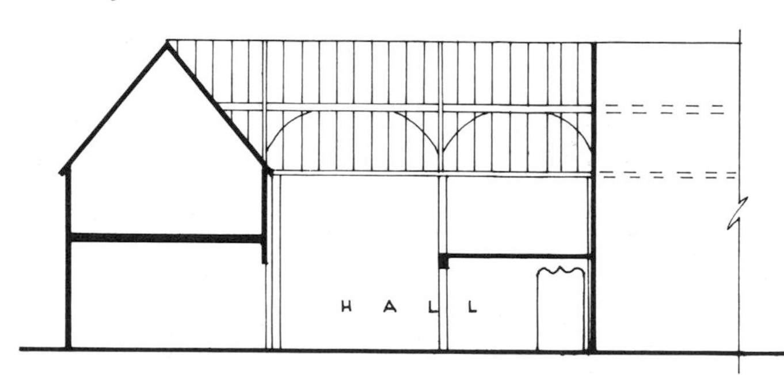

(a) Plan

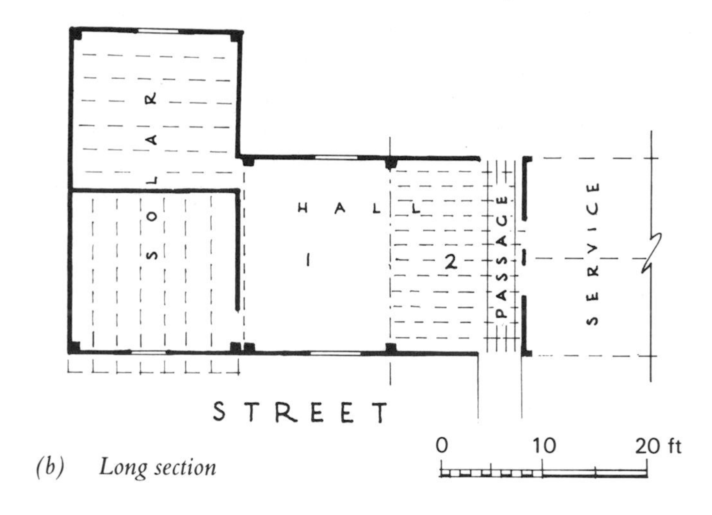

(b) Long section

pegged to the girding-beam of each side wall. This became the normal method for houses built with an upper floor from the start (Figure 90).

In the half-floored hall the floor was simply extended across the open bay and the fireplace built under the existing floor which was then partially removed for the smoke-stack, generally timber framed (Figures 91 and 92). The new floor joists could span either way. In Coventry they were generally from front to back, presenting a continuation of the jetty of the floored bay; the upper wall-frame then had to be brought forward, which generally meant taking out the original and framing anew to match that of the already jettied wall (Figure 93).

Attic floors could also be inserted simply by spanning joists between the tie-beams, either tenoned at one end and dropped into notches at the other, or housed into notches at both ends.

It is worth noting that crucks were peculiarly unsuitable for the insertion of floors because of the obstruction of the blades and, especially, knee-braces at the upper level. That many medieval cross-wings are now combined with sixteenth- or seventeenth-century 'hall-parts' may suggest that their original halls were crucks. In the few extant cruck halls now floored the offending members have been severely mutilated, if not removed altogether.

Figure 92 *Bromsgrove House, in course of dismantling – showing fireplace and timber-framed chimney-stack; also longitudinal floor-beam inserted when the hall became fully floored*

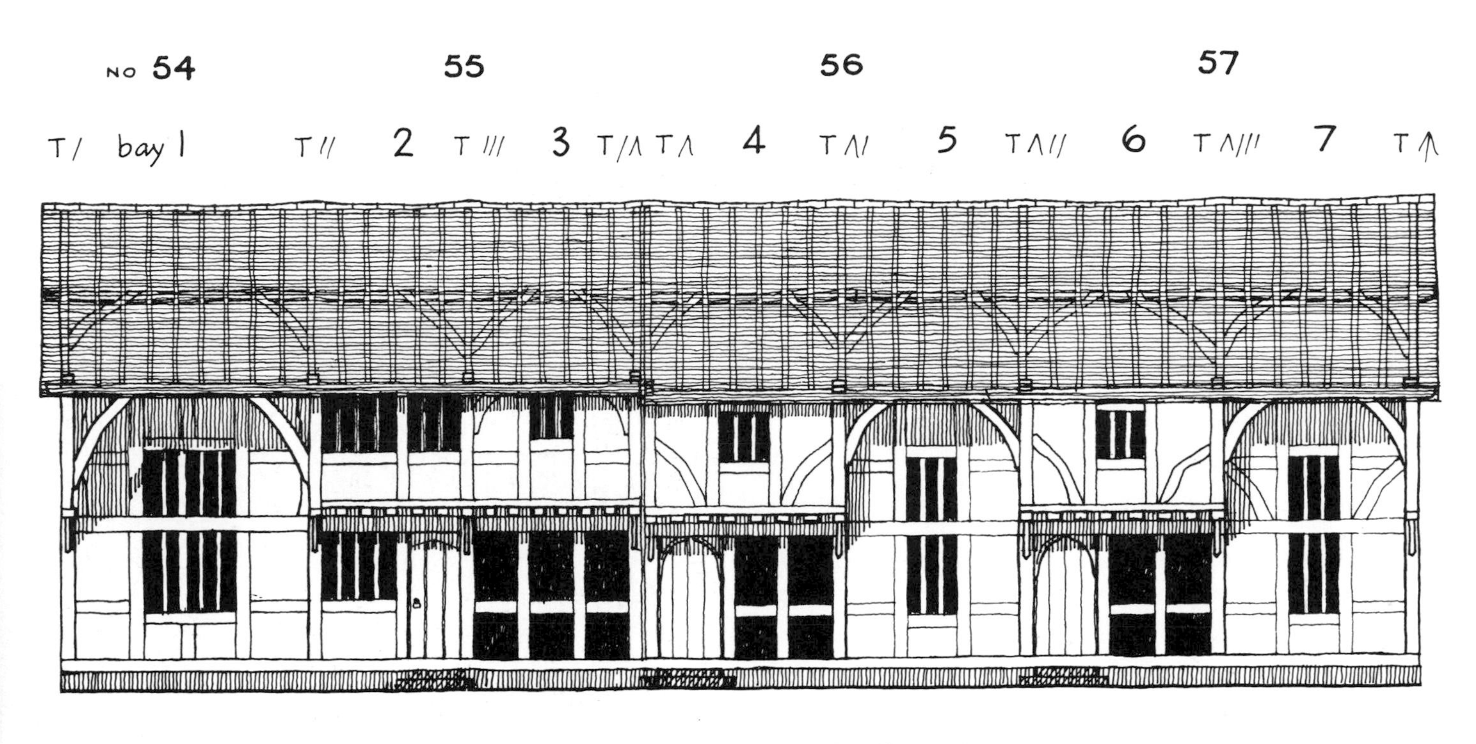

Figure 93 *Row of half-floored halls from 54 to 57 Spon Street, Coventry; the recessed bays 5 and 7, which contained the hearth, are internally open to the roof. Number 55 shows the result of flooring one such bay (bay 3), thus presenting a continuous jetty through its two bays. Number 54 is a single-bay hall considerably larger than the others with further accommodation at the back*

The town house

As for new buildings, town houses of *c.*1600 exemplify all the changes and architectural character of the period. First, to complete the story of the attic, in the earlier post-1550 buildings constructed end-on to the street the only changes were that the formerly cambered tie-beams gave place to level floor beams, and the roof structure, having been the pride of the carpenter-designer for centuries, became utilitarian. No more wind-braces were used than were necessary, and the purlin-to-principal joint could be trenched without causing aesthetic offence. Truss design also changed. By setting the collar higher, headroom for a doorway between the bays became possible, but usually there was also an interrupted lower collar, each half being trimmed into a stud, or queen-strut, to form the frame of the doorway. The roof structure now had two purlins to each slope to conform with the collars (Figure 94).

The next development was to change the alignment of the long axis so that it was now parallel with the street. The front gable, which was still fashionable, had become a dormer window, generally the full width of the struc-

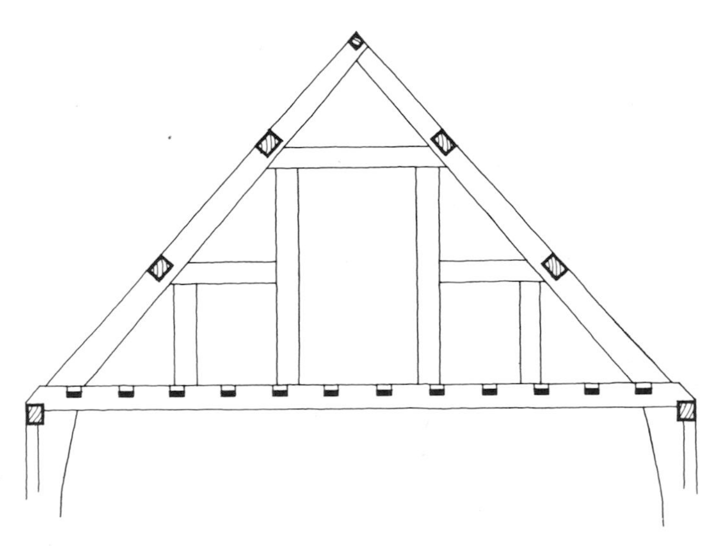

Figure 94 *Typical c.1600 truss permitting access between bays*

Figure 95 *Nash House, Worcester*
(a) Cross-section of attic *(b) Roof plan*

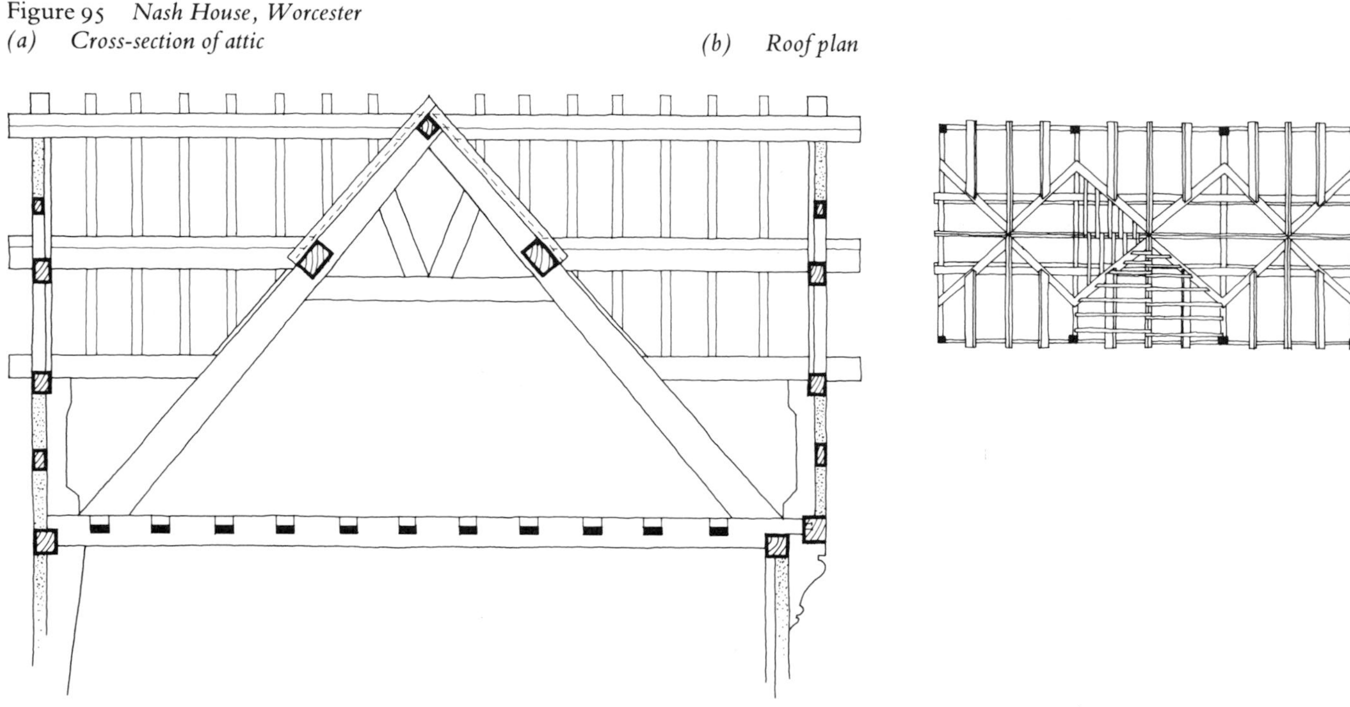

tural bay and as high as the former gables. A series of three such dormers was usual but the number was unlimited (Figure 95). The attic floor was jettied and to gain headroom the wall-plate was raised on short posts. The tie-beams, now across the building from front to back, had to be trimmed, like the collars already noted. At the back, the main posts might be high enough to reach the tie-beam, or short posts could be used here as well as at the front, or again, the trusses might be lopsided as at the Ancient High House, the rear principals having their feet on the uppermost floor beam (Figure 96).

The medieval style of jetty had also gone. Now the joists, hardly more than 4 inches square in section, were tenoned into the bressummer which, spanning between bracketed posts, was designed to take their load, as well as that of the wall-frame above. Thus the structural principle of the medieval jetty had been reversed. In practice, however, the joists still rested on the plate and so could also take some of the load. The brackets, invariably 'console' brackets, were integral with the posts. The corner jetty had no dragon timbers. Instead the corner post is bracketed both on its front and side; this demands a pretty sizeable tree, the bulk of which is then cut away below the brackets. Thus a 12 inch square post has to come out of a butt of about 3 feet diameter. The bressummers carry over the brackets to a mitred-and-tenoned corner joint. The side joists are short and trimmed into the adjacent front-to-back joist (Figure 97).

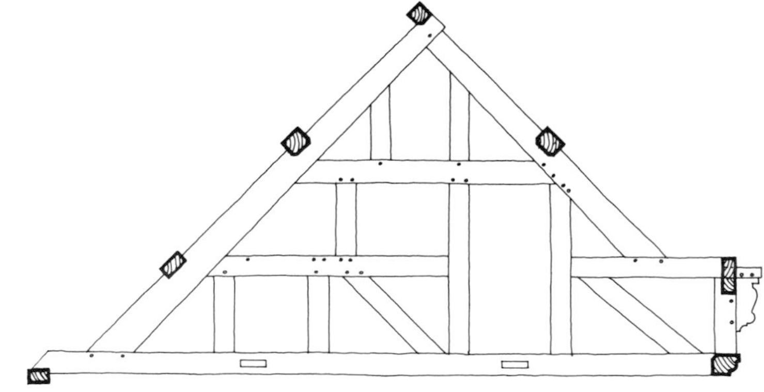

Figure 96 *Ancient High House, Stafford – cross-section of typical truss*

The floor framing usually consists of the primary cross-beams and two bridging beams, all morticed-and-tenoned, the central row of joists also being tenoned at both ends into the bridging beams (Figure 98). Putting such floors together on the framing floor is relatively easy; however, when it is remembered that two heavy beams with about eight joists spanning between them had to be tenoned into a still heavier cross-beam which, in its turn, had to be tenoned into the bressummer at one end and

Figure 97 *Ancient High House, Stafford*

(a) Plan at corner showing floor construction and corner post with integral brackets

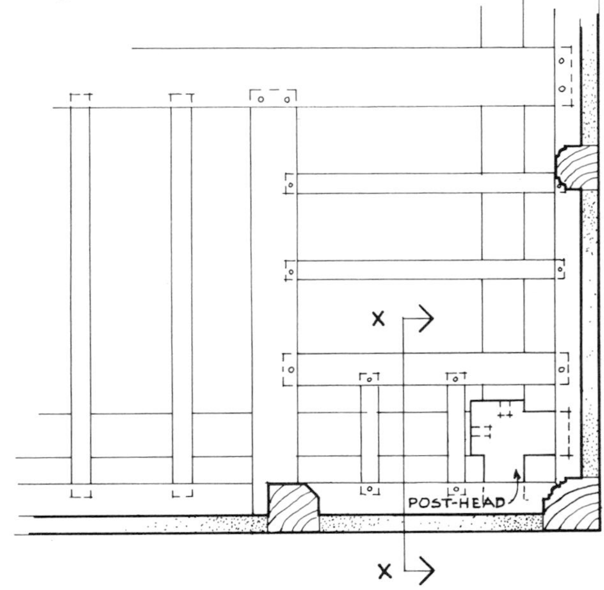

(b) Plan of head of post and required size of tree

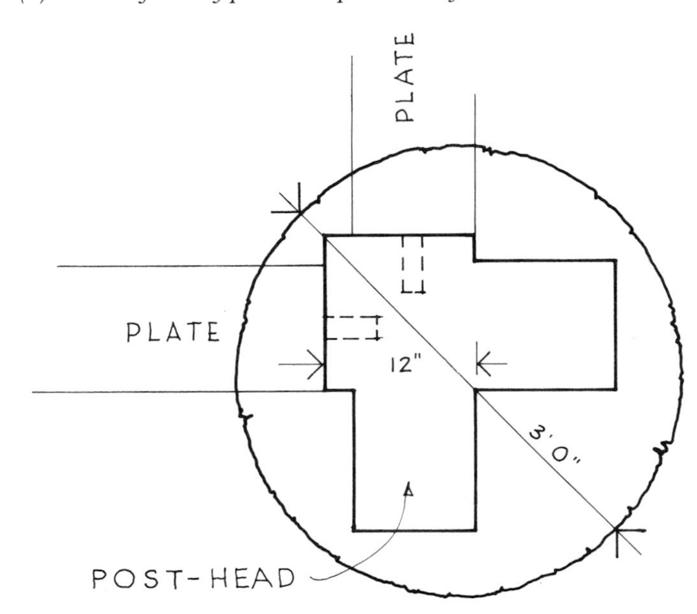

(c) Section X–X

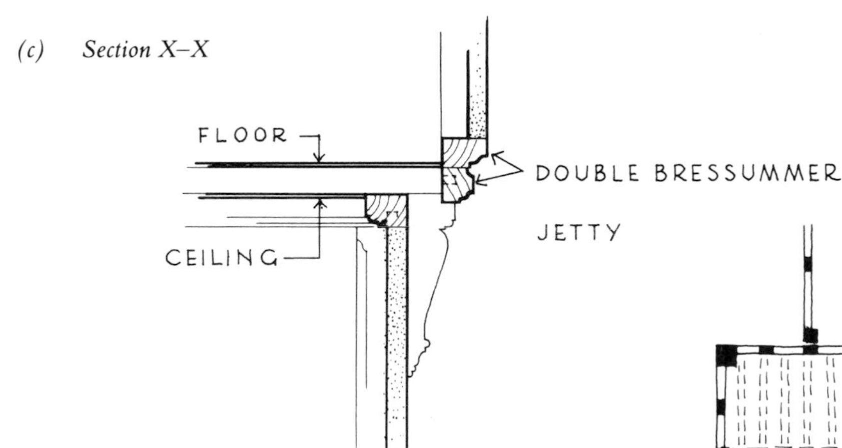

Figure 98 *Nash House, Worcester – plan of typical floor framing with mortice-and-tenoned cross-beams, bridging-beams, and transverse joists. The joists in the outer rows are tenoned at one end and rest on the girding-beam at the back and are cantilevered over the plate at the front to form the jetty*

into a post at the other, or dropped on to the post tenon, it is plain that the craftsman's ability to handle timbers was no less than his skill in working them.

Floors were also framed in ways that would make an impressive ceiling. Two examples are shown; both provide puzzles, as loved by Elizabethans, in how they should or could be erected (Figure 99(a) and (b)).

The type of floor in which the boards lie parallel with the joists may be earlier than Elizabethan, but the finest example partially rescued from a house in Bromsgrove High Street demolished in 1966 is of that period (Figure

Figure 99 *Typical Elizabethan floors*

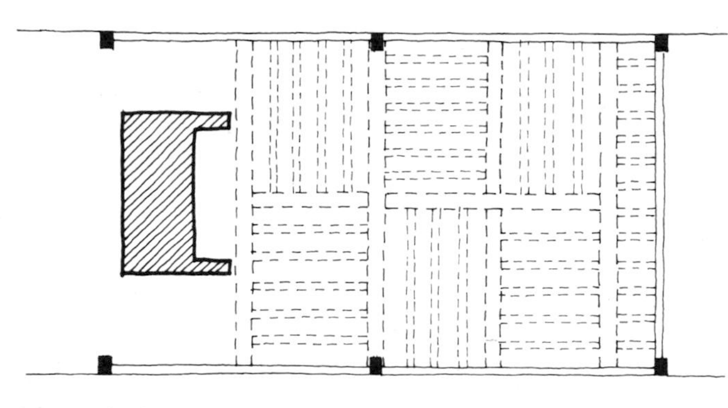

(a) Chorley House, Droitwich

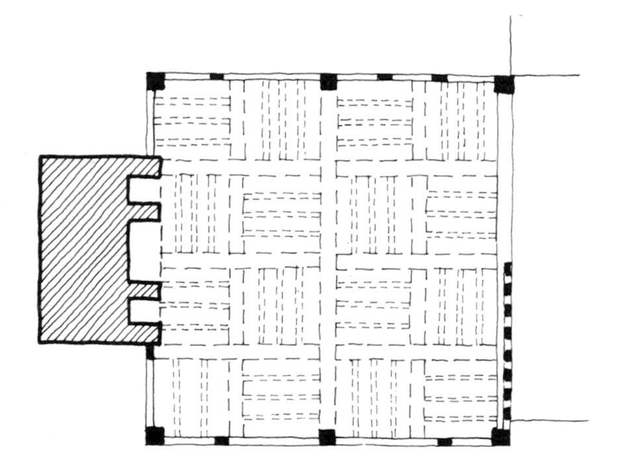

(b) Almley Manor, near Hereford

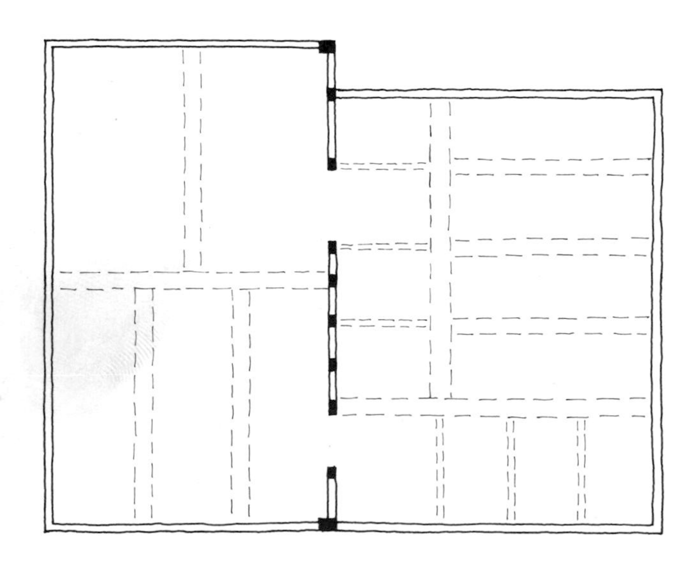

(c) Winwood Farm, Wyre Forest

Figure 100 *Details of floor from demolished house in High Street, Bromsgrove*

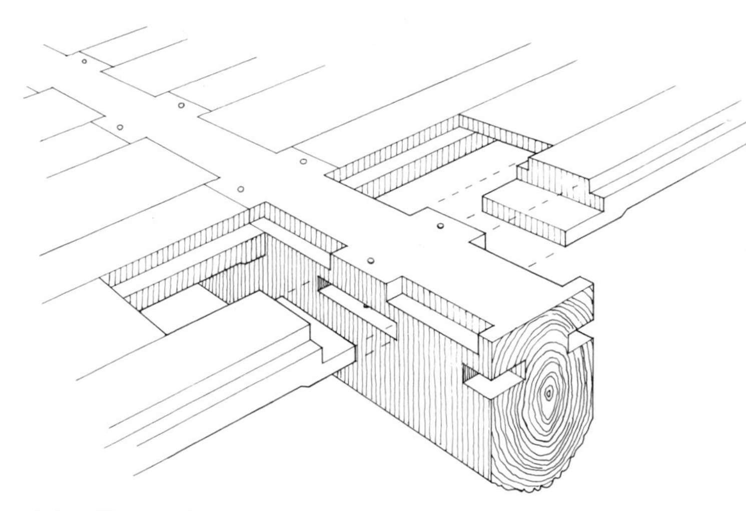

(a) Construction

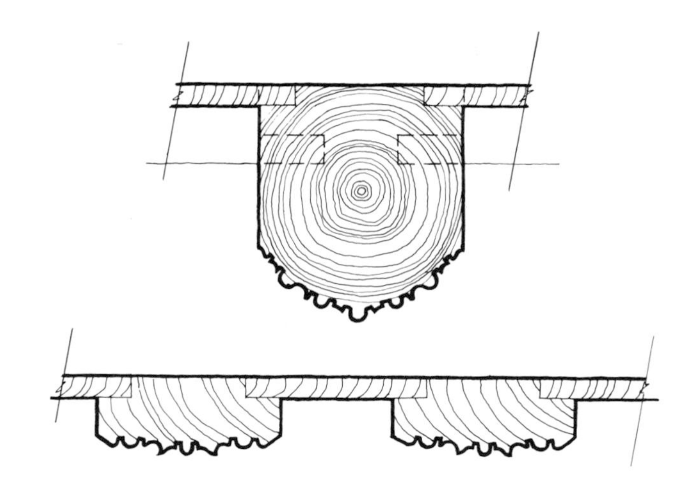

(b) Moulding of beam and joists

100(a) and (b)). An earlier one, *c.*1400, was restored at Chester House (Figure 101).[11] This was the same, only lacking the rich moulding of the former. Harvington Hall near Kidderminster has another such floor in the originally timber-framed, now brick, solar wing. There is yet another in the Bromsgrove House with joists (beams) and planks the full length of the bay. This created acute problems in restoration, since oak planks of such width and length are impossible to obtain in seasoned oak, but it has not seriously shrunk or warped in the twenty years since it was laid. A variant is at Penrhos, Knighton, near Hereford. This has beams about 7 inches square in section laid

Figure 101 *Chester House, Knowle, Warwickshire – restored floor, c.1400*

Figure 102 *Upper Barrow Farm, Cradley, near Malvern – of c.1600 with deep joists*

contiguously the full length of the bay. Since there are no planks or boards it may be typologically earlier than the others, but the building is *c.*1500. All of them are or were jettied, the last example dispensing with the need for a filler above the plate.

Lastly, floors with beams and joists of the same depth were probably designed for plaster ceilings uninterrupted by the beams (Figure 102). This was one reason for the change from joists laid flat to those with the longer cross-sectional dimension set vertically. Moxon[12] also describes such a floor and remarks that the scribing of the joists to the waney edge of the beam is a means of strengthening the joint (Figure 103).

The design of chamfers and chamfer-stops, especially of beams, may also be a guide to dating. Unfortunately they are such an obvious feature for the local carpenter on which to exercise his flair that much more research will have to be done before a reliable chronology can be established. The greatest variety occurs in the sixteenth and seventeenth centuries and many of these have been drawn by Stanley Jones.[13] The most prevalent design is the simple 'run-out', 'ox-tongue', or 'straight-cut' stop (Figure 104), and one or two other extremely simple designs which might belong to any period.

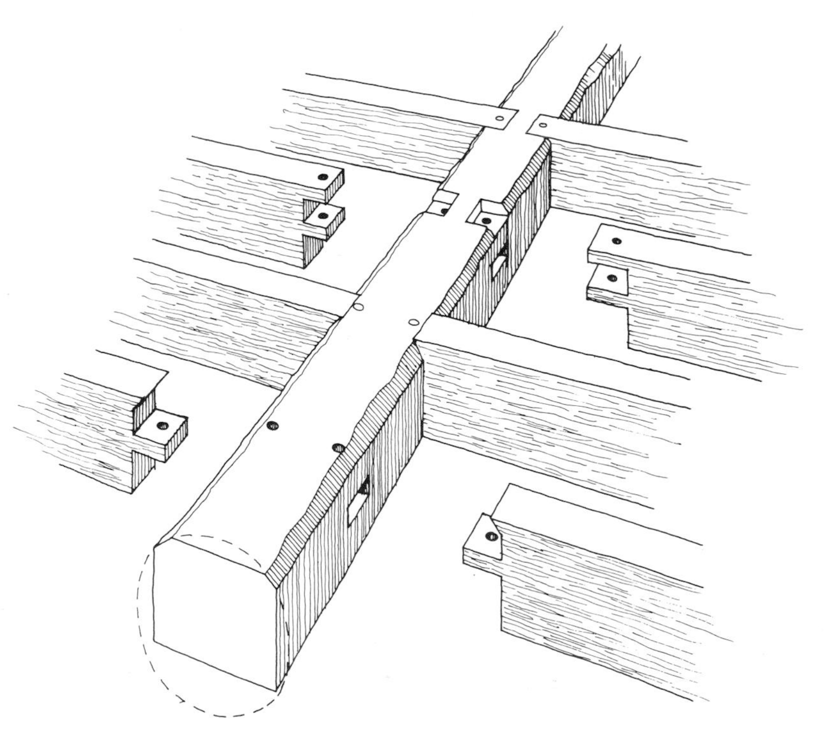

Figure 103 *Floor construction according to Moxon*

Figure 104 *Common chamfer-stops*
(a) Run-out *(b) Ox-tongue* *(c) Straight-cut*

Figure 105 *Ancient High House, Stafford*

The wall-frames have been mentioned earlier in this chapter. They had every kind of decoration, with close-set diagonal, zig-zag, and double-curved braces, and only the slightest structural justification. The small square panels might contain quadrant braces, fleur-de-lis or any other heraldic motif, generally carved in relief on a fielded board that filled the panel, the recessed area being plastered. Following the natural grain now counted for nothing.

If medieval painted timbering had been gaudy, Elizabethan and Jacobean must have been dazzling, the great windows with their hundreds of diamond leaded lights adding to the effect – whether reflecting the sun or letting through the flickering lights of candles. The ranges of stone mullion and transome windows of such houses as Hardwick, Montacute and Hatfield must surely have been modelled on wooden precedent – perhaps Henry VIII's timber-framed Nonsuch Palace.[14] The Ancient High House still has such ranges, under partial restoration (Figure 105).

A detail worth noting is that the moulding, invariably ovolo or an elaboration of it, is now continuous from end to end of the horizontal member; the mullions must therefore be shaped to it, instead of cut off straight as for the former masons' mitre (Figure 106).

The big windows were usually 'planted-on' – that is, pegged to the structural frame – and there was no special way to terminate the head, transome or sill; the moulding simply ran off (Figure 107). The small, invariably high-level windows (sometimes also larger ones) were still integral with the structural frame. Casements were secured to the middle glazing bar with little lead ties; opening lights had a flat wrought iron frame, hung on peg hinges, and highly ornamental fasteners.

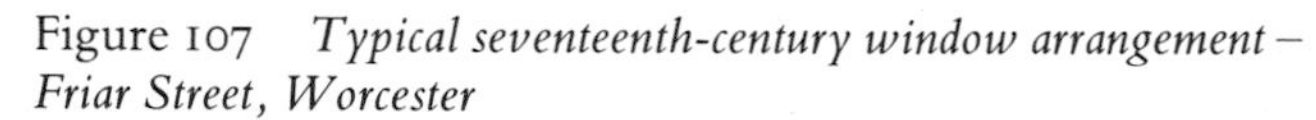

Figure 107 *Typical seventeenth-century window arrangement – Friar Street, Worcester*

Figure 106 *Detail of window showing mullion scribed to sills from Nash House, Worcester*

Internally staircases, fireplaces and panelling were innovations for most house-owners and they made the most of them.

The countryside

The farmhouse has a somewhat different story. New houses at the beginning of the period often followed the old four-bay range with the hall in the middle, or the hall and cross-wing plan or any of its variations, but of course with upper floors and attic throughout (Figure 108).

Figure 108 *Seventeenth-century house at Alvechurch, near Redditch – following medieval hall and cross-wing plan*

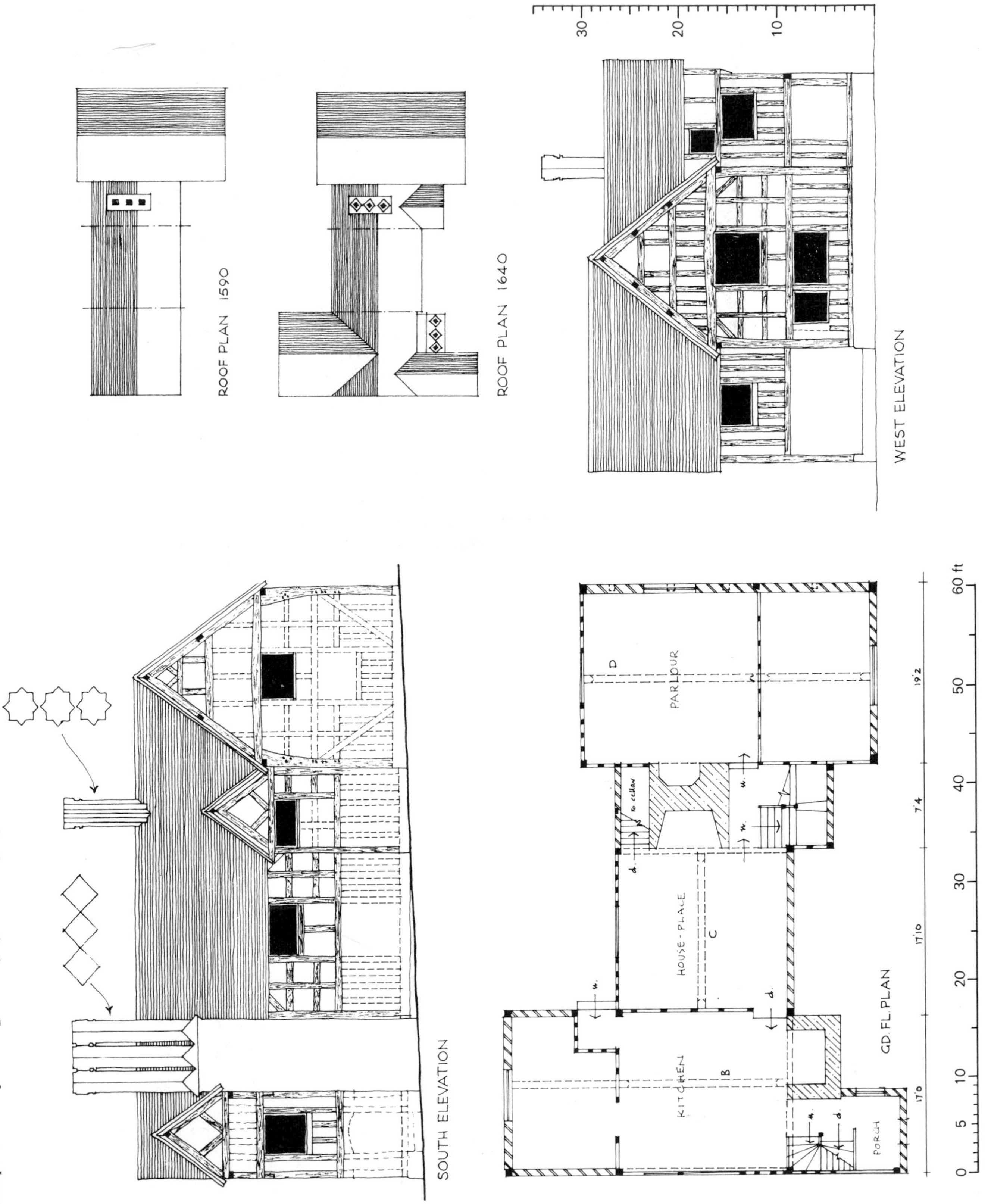

Figure 109 *Tardebigge Farm, near Bromsgrove – with back-to-back fireplace structure and stair turret. Note also eight-pointed chimney-shafts for house-place and parlour and plain diagonal shafts for kitchen fireplace*

Figure 110 *Lane House, Feckenham, Redditch, – with close-studding throughout, original dormer windows and weatherings*

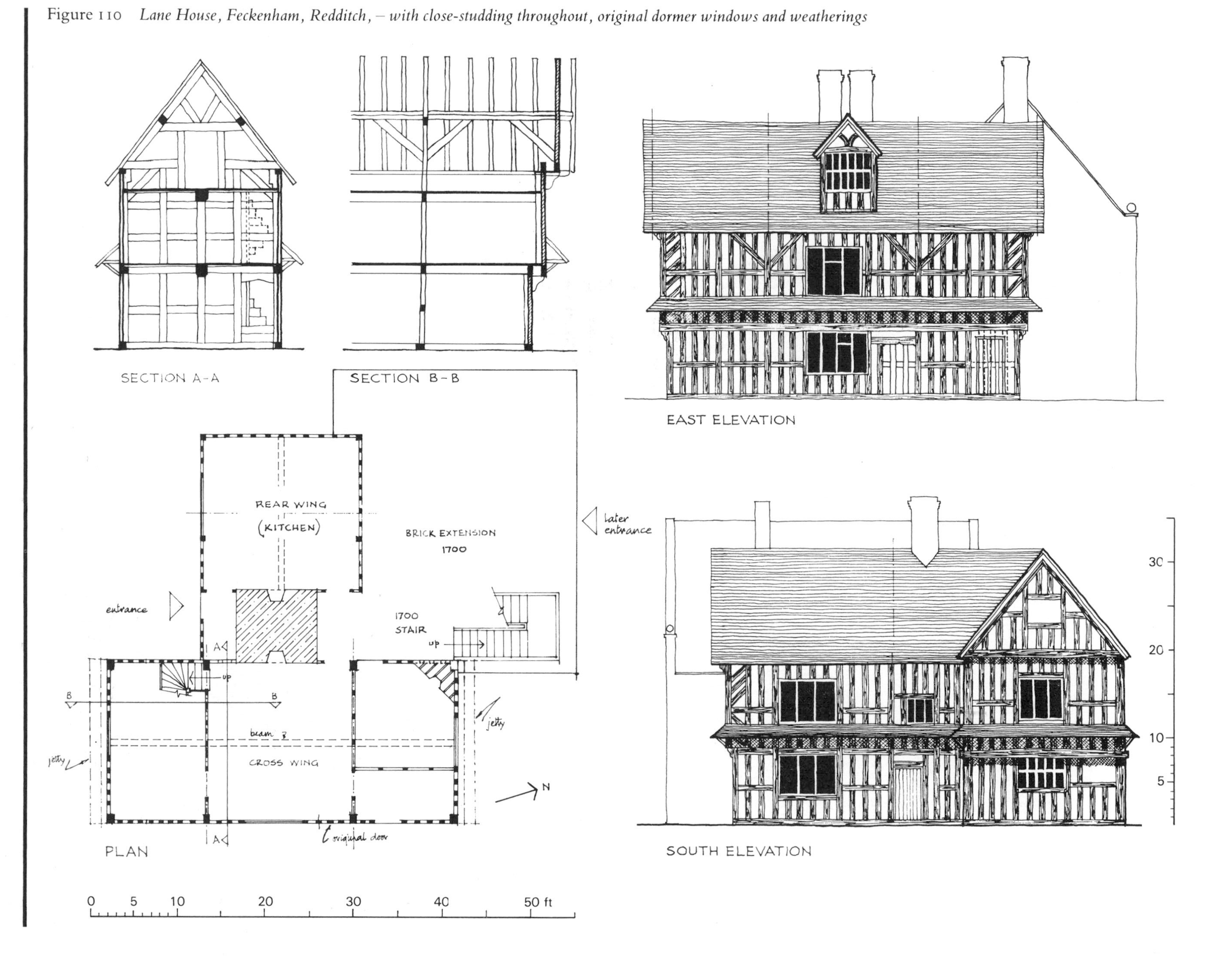

The wall-frames clearly expressed the floors, sometimes by a different pattern of timbering for each storey. The kitchen, having previously been a separate outbuilding near the well, or in an extended service wing with the bake-house, brew-house, wash-house, and other necessities, was brought inside, becoming the house-place or 'inner' kitchen, the 'outer' kitchen, if there was one, still being with the other service rooms.

The fireplace could be placed laterally against the rear wall, or cross-wise between frames forming a fireplace bay. The huge flue structure could then have back-to-back fireplaces at both floors, leaving just enough space in the bay for the stair and a passage between the kitchen and parlour, and for a landing above. The attic was sometimes reached by steps cut into the stonework. A more elaborate stair (and many latterly were little less than monumental even in farmhouses) might be given its own turret, projecting out of this bay, so that the passage became a slightly more spacious stair hall (Figure 109).

More revolutionary plans were, first, the two- or three-storey three-bay range, possibly deriving from the medieval solar, traditionally the most important and well-appointed part of the house and so an appropriate model for the modern one. Such houses often had a rear wing, either centrally placed or to one side, for the kitchen or even the parlour, as at Lane House (see Chapter 11) (Figure 110). Then the kitchen was in the middle, generally longest, bay and the fireplaces and stair were concentrated near the centre of the house. A variation was the T-plan, making the stair and fireplace the core from which projected three wings each with fireplaces. The kitchen was then at the back (Figure 111).

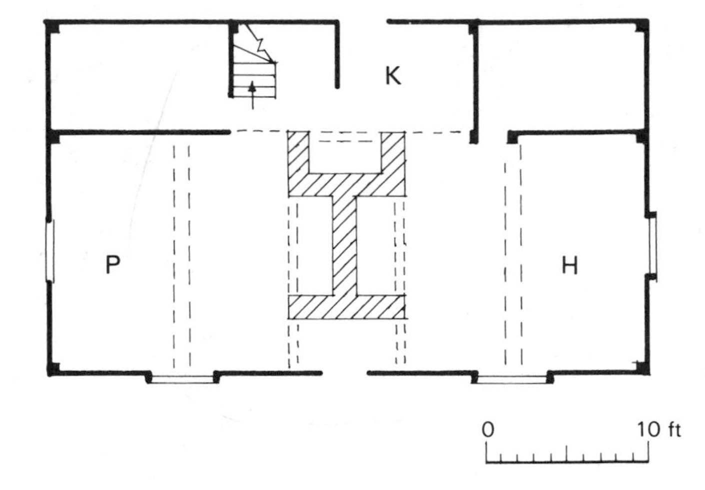

Figure 112 *Norton Green Farm, Knowle, Warwickshire – lobby-entrance plan*

The so-called 'lobby entrance' house, closely related to the T-plan but of three aligned bays only, also broke all connection with medieval plans (Figure 112). This house type was more or less standard in New England,[15] though because of the limited means of the earlier settlers it sometimes had to be built in stages. The initial stage, however, was never without the fireplace bay, for without it life through the winter would have been impossible. Later the typical outshut along the whole of the back, with catslide roof, increased protection against the snow-laden winds, as well as giving more accommodation.

Decoration was as lavish in the countryside as it was in the towns. Though heavily restored at the beginning of this century, the wall-paintings of Dowles Manor, Bewdley,[16] before the house was damaged by fire in 1982, were an exceptionally complete example (Figure 113). Patterns of every description, and the occasional crudely portrayed figure, covered the entire walls of the hall, parlour and two of the bedrooms. They were dated 1622, no doubt also the date of the house. In practically every house of this period, especially in the parlour, where it has been possible to uncover an obscure corner which the Victorian brush with its black paint or thick dark varnish could not reach, some colour has been found. The design

Figure 111 *Moat Farm, Dormston, near Pershore – T-plan with three-bay front range and fireplace incorporated in middle bay*

Figure 113 *Dowles Manor, Bewdley – example of wall-paintings*

Figure 114 *Bromsgrove House as in High Street, Bromsgrove, showing seventeenth-century rendering and cast plaster plaques*

of the patterns may suggest they were done by the son or daughter of the house attending art school, but that hardly diminishes their interest, and the rooms are certainly the duller without them. External painting was no less common. Earth red and ochre were still general; the occasional 'White House' of today may go back to this period, white lime being a rather special finish in the earth-colour regions. Timber was never differentiated from the panels.

East Anglia had been the home of pargetting from the mid-sixteenth century, and by 1600 it had become traditional, with a great variety of pattern and ornamental relief. In the midlands it was plainer, but as early as 1595 Shaw's House, next to the High House, which was built in that year, had been rendered on the side against which the latter was erected. The solar wing of the Bromsgrove House (Chapter 11) had also been rendered by the mid seventeenth century, incorporating plaster casts (Figure 114).

Date plaques also became fashionable; the owner's and perhaps his wife's initials were often carved on either side of the date, and very occasionally the carpenter's below it (Figure 115). Caution in interpreting the figures is necessary as it is easy to mistake 6s and 5s and 5s for 3s, and when a worn figure is renewed by paint or recarving it is invariably the earlier of the possible alternatives that is chosen. Hence the High House was reported in the last century to bear a notice which said, 'Richard Dorrington made this house, 1555'. Boring Mill Cottages (see Chapter 11) had two dates and both of them authentic.

Other features of this final flowering of the timber-frame tradition, such as the huge brick chimney shafts and splendid staircases, are illustrated in the case studies of Chapter 11. Only the false long gallery should be noted now as one of the more quaint signs of the *nouveaux riches* in emulating their betters (Figure 116).

Figure 115 *Tanhouse Farm, Stoke Prior, Droitwich – typical date plaque*

Figure 116 *Middlebean Hall Farm, Feckenham, Redditch – false front of 1637, including long gallery, and porch added to sixteenth-century four-bay house, partially visible at right hand of elevation*

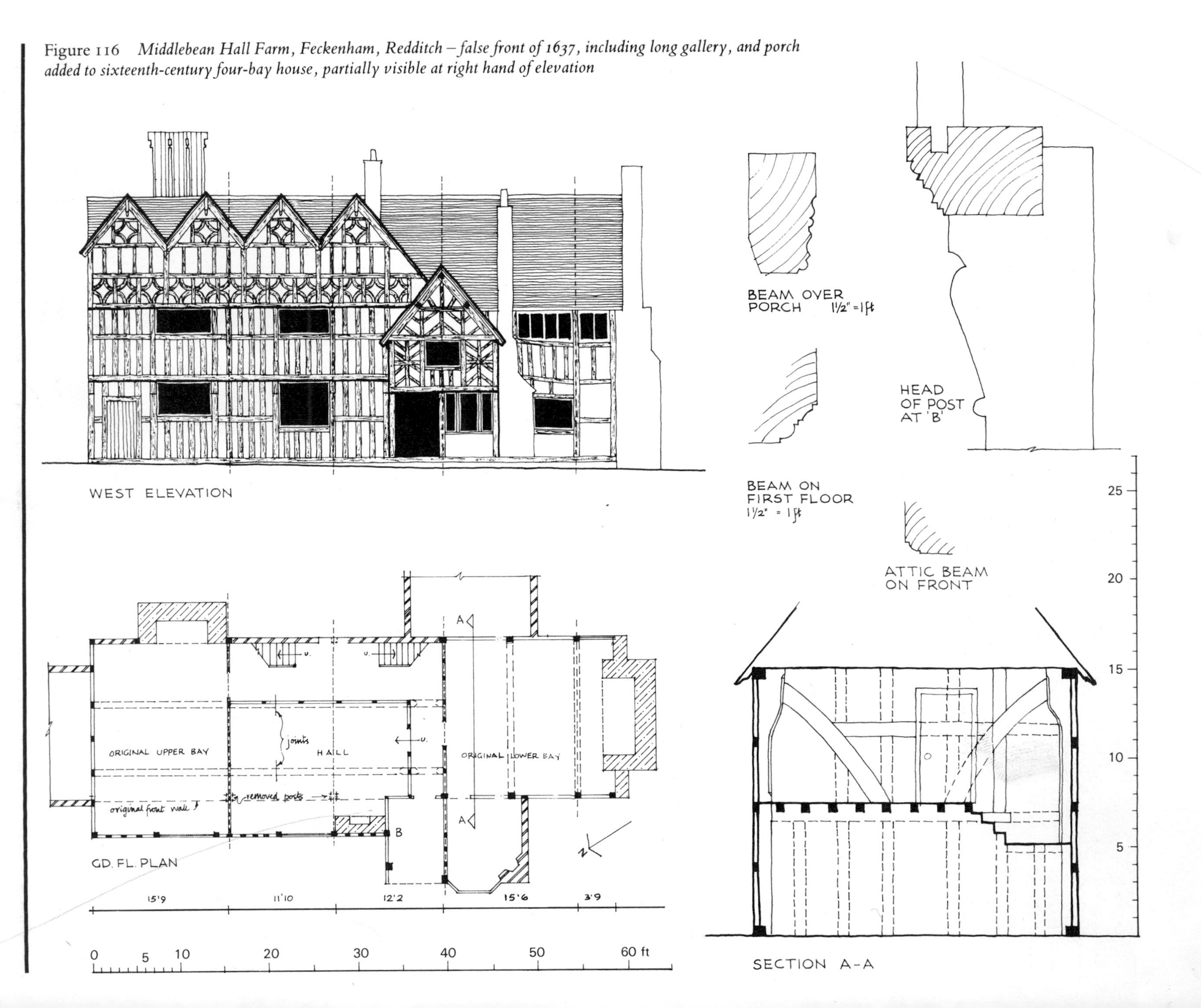

Cottages

This period is also the earliest from which 'cottages' have survived. They were generally the houses of the lesser yeoman farmers, economically independent but probably not much above the level of subsistence farming. The buildings are of both one and two bays. Some are only slightly inferior to the larger houses in their construction, but the diminutive span and light thatch allowed the use of slender poles of oak, alder or poplar, sometimes not even debarked, for rafters. Upper floors too are sometimes of the lightest joists and beams, but the latter are invariably chamfered and stopped and elegantly braced or bracketed to their posts. A favourite method, not without structural reason, was to set the beam of one bay longitudinally and of the other transversely. The kitchen bay was sometimes open to the roof, or half-floored. Its fireplace (there is

Figure 117 *Mud Lane Cottage, Hanbury, near Bromsgrove – typical two-bay cottage, c.1600, kitchen or house-place originally half-floored*

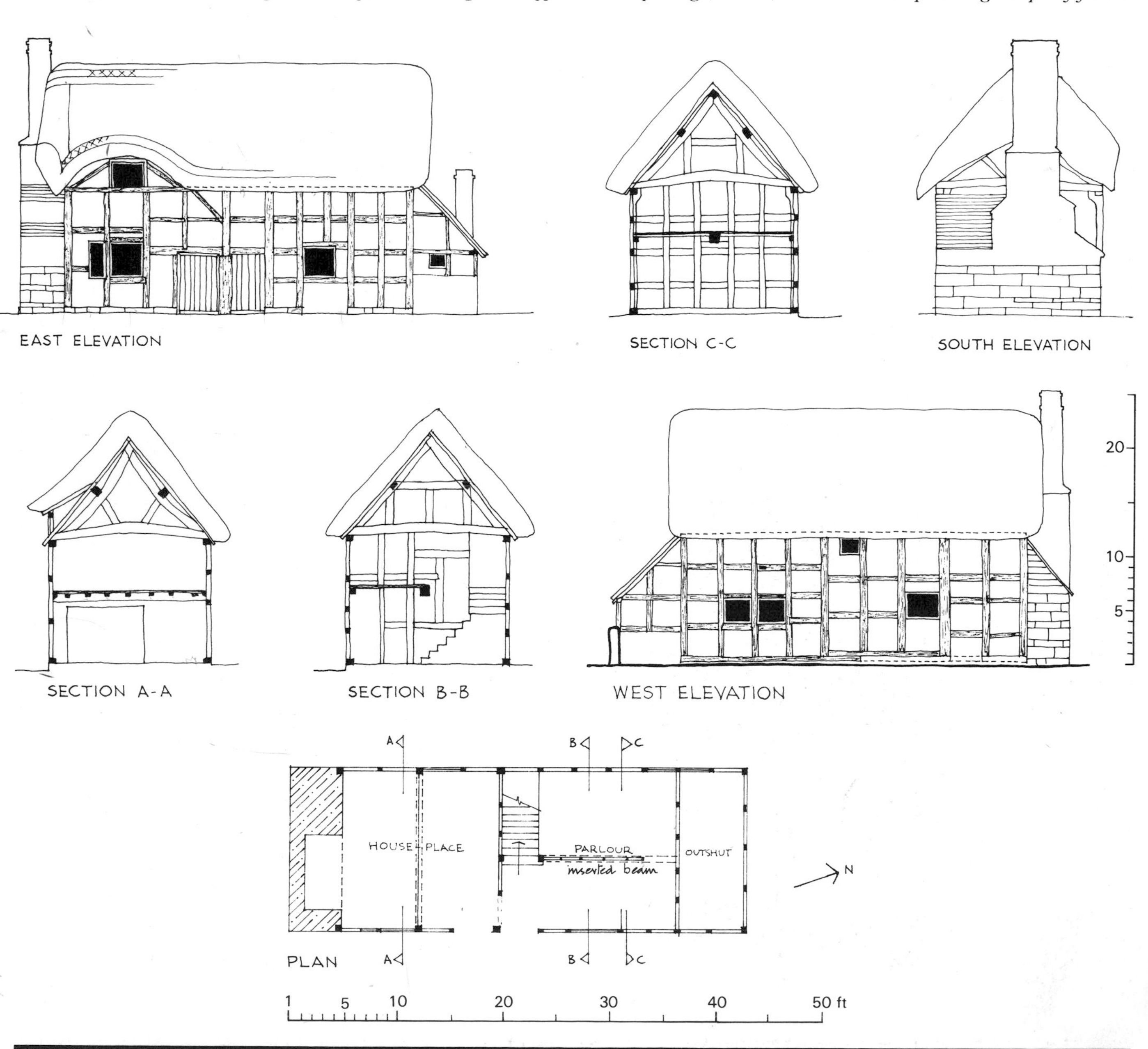

seldom one in the parlour) may be within or outside the gable wall, and the bread oven – baking now of course being freed from the manor – is visible externally, whether the oven door was inside the house or out. Fireplaces often had timber-framed stacks, again inside or out, and it is noticeable that the brick chimneys that have replaced the latter are well clear of the thatch. There were also external stone chimneys, almost the width of the gable, sometimes again with steps cut into them as the only way to the upper floor. More often there was an internal ladder, or even a little spiral stair, the latter invariably alongside the fireplace if the fireplace were within the house (Figures 117, 118 and 119).

The framing even of these buildings was 'architectural', with smaller panels at the parlour end, the superior gable facing the road. Later cottages may be distinguished by

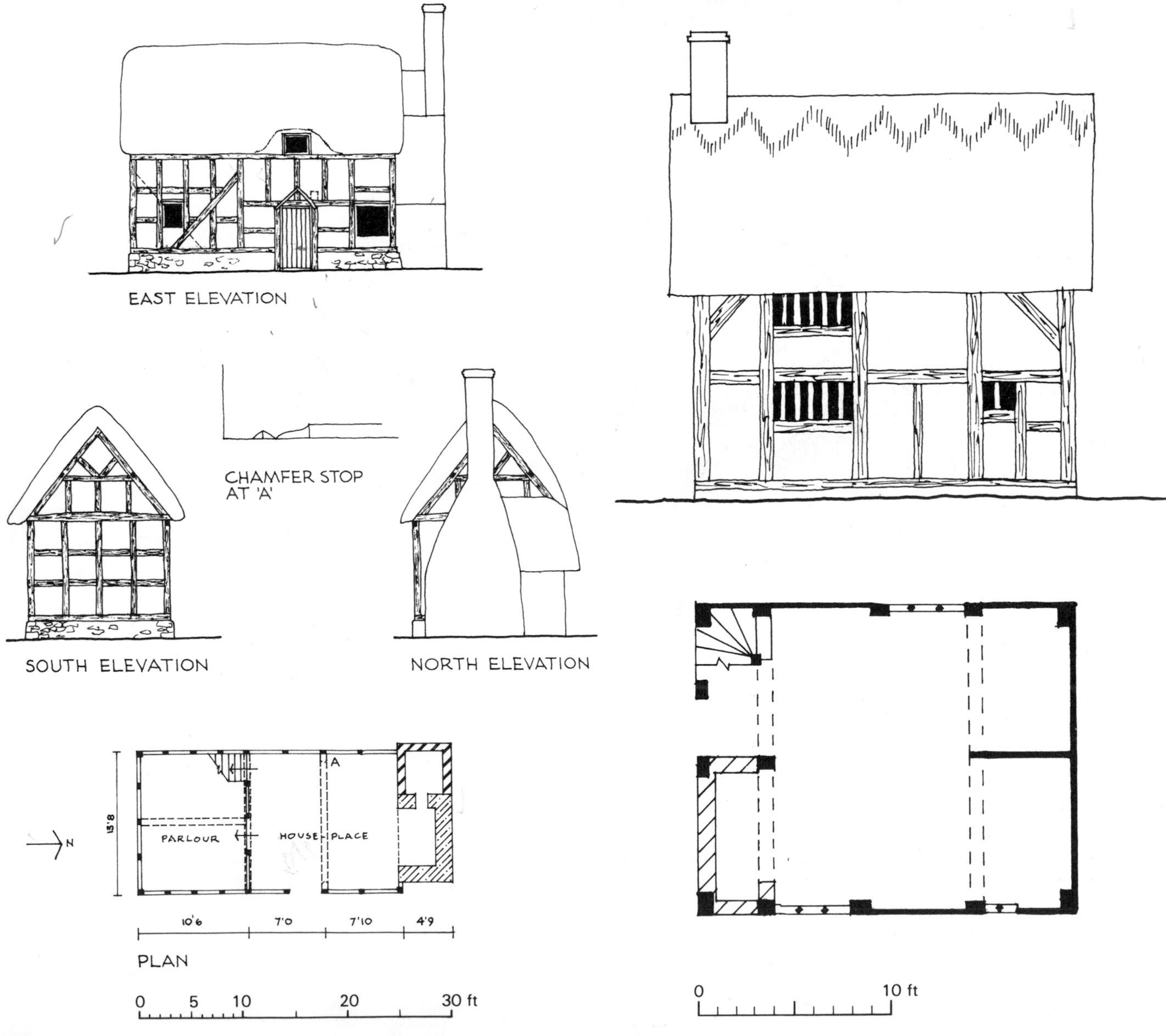

Figure 118 *Albutt's Cottage, Hanbury, near Bromsgrove, c.1650*

Figure 119 *Leaford Cottage, Lea Marston, Warwickshire – now restored as headquarters for nature trail in grounds of power station*

Figure 120 *Libbery, near Pershore – typical late-date cottage, now demolished, with straight slender framing and tall proportions*

their straight full-height braces inclined in the opposite direction in each lateral wall, as in Albutt's Cottage.

The wall-frame pattern is still probably the best guide to dating. Swept braces and relatively large timbers and panels continue down to *c*.1600. After that the straighter and more regular pattern takes over. Many poorly built cottages have also survived from the end of the century and later. Some were built not only with slender timbers but also, because of their minimal sections, with half-lap jointing (Figure 120).

Post-1650

There are still examples of splendid timber-framed houses of the 1660s. Even in towns, framing may have continued to the end of the century. In Gloucester a four-storey-plus-attic building of one-bay frontage was the subject of a recent public inquiry. Its first and second floors were jettied, the upper storeys plain. The timbering was of the sparsest, having only two posts, and two intermediate studs to which the windows were fitted at each storey (Figure 121). In the countryside a few houses may be dated to the eighteenth century. Their framing is of thin, die-straight timber and they were originally lath-and-plastered both internally and externally. Today, with the fashion of exposing the frames (even of the traditionally pargetted buildings of Essex and Suffolk) they have become black-and-white – and hardly weatherproof.

Cottage building continued even beyond the eighteenth century. In the hilly area to the west of Worcester there are land resettlement dwellings of the post-Napoleonic period. The crop, cow and pig are given stone buildings, but the human occupants had timber-framed houses of absolute minimum standard. These are now listed,

Figure 121 *30 Westgate Street, Gloucester – framing of front wall, c.1680. Jettied joists were probably masked by board and whole front was rendered. Evidence of ground-floor framing now lost*

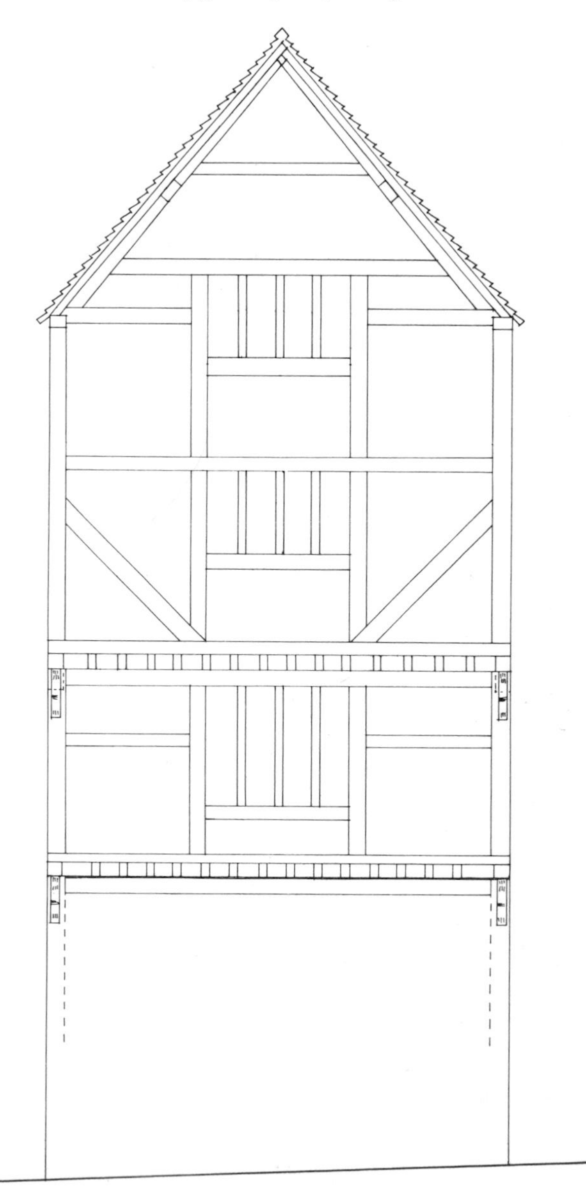

Figure 122 *Abbot's Morton, near Alcester – house with raised roof*

causing impossible problems both for the escapist middle-class owner and the local authority. For nothing can be done short of complete rebuilding in different materials.

Alterations through the eighteenth century consisted mainly of raising roofs, as may be seen in practically every town and village (Figure 122). Those of water-mills, especially, had to be raised, for while two floors were enough for the manor mill, the capitalist miller had to store the corn; hence the invariable 'garner', 'stone' and 'bagging' floors of all surviving old mills. Other alterations were for the sake of fashion, as when a Queen Anne or Georgian brick front was clapped on to the old frame and its jetty underpinned, a cornice and parapet wall replacing the dormers and upright windows the horizontal ones (Figure 123). It is worth noting that half the charm of Georgian symmetry is that, following the uneven bays of timber framing, it is often not symmetrical at all. Weatherboarding and tile-hanging appeared in the south-eastern counties, while in Warwickshire and in most of the midlands brick panels began to replace wattle-and-daub. There are a few timber-framed buildings in which bricks were the original infill material, but fewer than is generally believed, even including those rich Warwickshire houses with herringbone brickwork. It is mostly Victorian.

Figure 123 *Vicarage, Chaddesley Corbett, near Kidderminster*
(a) Front

(b) Rear

Figure 124 *Castle Farm, Bradley, Feckenham, Redditch – with weatherings as also at Lane House (Figure 110) and Moat Farm (Figure 111)*

Figure 125 *Rectory Farm, Tibberton, near Worcester – evidence of late-date timber blacking. The unpainted timber was exposed when a Victorian fireplace and flue were taken down*

Such changes through the eighteenth century were still mainly architectural, and there was little damage to the structure, except for hacking timbers to give a key for plaster. Even conservation was practised, less self-consciously though sometimes more effectively than it is today, for even the complete refashioning of both exterior and interior generally left the structure much as it was before.

But through the nineteenth century deterioration was rapid both through social decline and wrong methods. Timber buildings were treated as no different from any other. Brick was used indiscriminately in the structure, sometimes aided by cast iron columns to hold up the sagging first floor. Partitions, floors, ceilings and fittings were generally in softwood. Everything was plastered and wall-papered. The remaining exposed wattle-and-daub panels were replaced by Victorian bricks, too wide for the timber frame, and instead of the plaster surface being set back a quarter-of-an-inch behind the timber surface, so that the latter would be dried by winds through 180 degrees, the bricks projected beyond the face of the timbers, making water ledges.

Many houses in the midlands formerly also had 'weatherings' (Figure 124) or little pent-roofs, protecting the walls at storey heights or girding-beam and tie-beam levels. These were of tiles, thatch or inclined boards. They were removed probably when the panels were being bricked up, and not replaced. Decaying bargeboards were not renewed. Instead, purlin ends were cut off leaving practically no verge overhang to protect the gable. The eaves were made shallower, either by shortening the rafter feet or through spread of the wall-plates when tie-beams were severed, or most effectively, by substituting tiles or slates for thatch. Thus the walls were eventually exposed from top to bottom. Black-and-white may have been an attempt to make timber and brick weathertight. As a result of coating the timbers with tar, water still got into them but not out, and so decay was accelerated (Figure 125).

At last came filling the timber shakes and loose joints with 'compo' until, with constant patching, filling and painting, the frame became too rotten or depleted to do its job. There are now many buildings precariously held up by the lower brick-filled panels, and very little else. Through all of this, houses were being subdivided, their frames hacked through when they obstructed alterations.

In towns entire ground-floor front walls were taken out to make shop windows, rear plots were swallowed up by workshops, and plumbing and drainage brought water and damp to airless back passages where previously there had been gardens.

Such was the legacy until post-war redevelopment – much of it planned before the war – brought devastation to town centres, especially the centres of market towns and cathedral cities, while on their outskirts, agricultural land with its farmhouses and barns was swallowed up by new housing. Records were confined to those more highly listed buildings whose demolition had to be reported to the Royal Commission on Historical Monuments (RCHM). The mass of smaller houses disappeared without trace. Then, in the 1970s, recession at least stemmed the destruction, and any building with a sign of timber framing was listed.

PART TWO

OFFICE PRACTICE

CHAPTER FIVE

PRELIMINARY SURVEY

There are three parts to every project:

1 The preliminary survey and report;
2 The structural survey and preparation of the restoration contract;
3 Its execution.

There is a further subdivision of 2 and 3, following the RIBA Code of Practice but adapted for the special conditions of restoration. This is set out on the Fee Scale (Figure 126). It is as well to issue it in full, with hourly charges, travelling, subsistence and so on, at the first meeting or communication with a potential client. No doubt commissions have been lost before they have started as a result of this habit, but it is better to lose a client before rather than after a great deal of time has been spent in the dwindling hope that the work done will be paid for sometime. Even acceptance of the fee scale is no guarantee that the contract will go safely through, but it is the best beginning.

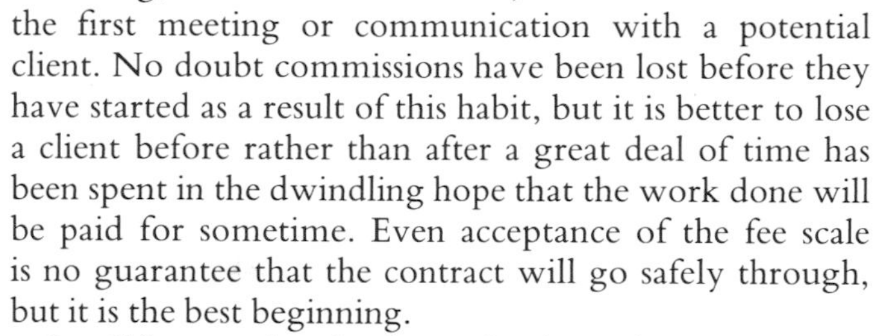

It will be seen that the charge for the preliminary survey and report is not in the percentage scale, but based on time or the floor area of the building concerned. In practice it may be merely a nominal charge, the time spent having to be compensated for in the subsequent stages or written off. But the more thorough the preliminary survey, the easier and quicker will be the subsequent structural survey. In fact, that can sometimes be reduced to hardly more than drawing out information already obtained. The two may indeed be combined if the commission is simply to obtain prices for structural repairs.

The first enquiry should be to the official lists. The Victoria County History may also be worth consulting as well as the Inventories of the Royal Commission on Historical Monuments, should it be in any of the areas the Commission has covered. The two national volumes on vernacular buildings of England and Wales may also be worth a glance; while, for those with access to London, the National Monuments Record (NMR) in Fortress House is also a possible source of information. And lastly Pevsner, though unlikely to mention the building unless it is well above vernacular status, will be informative on the local church and its monuments of old families with whom the house might have had some connection. Further documentary research into the parish registers, seventeenth-century inventories, deeds, nineteenth-century estate maps, even old photographs or written accounts of the building in the county archives may follow, but hardly at this stage, and then the client might also be involved.

The survey itself is an archaeological investigation, not excluding the need for accurate recording. But it is also very different, for there are no 'layers' (except wallpaper and plaster) to be peeled off and each recorded before it is destroyed. Instead the method is preservation, even – or especially – when it comes to stripping. The record is not therefore a series of phase plans, but drawings of the structure in every detail and feature, regardless of date. Anything that must be removed must also be recorded and when of architectural or historic value – from fireplaces down to wallpapers or pegs – preserved, probably as samples in the case of minutiae.

Again, recording is not to metric accuracy. Feet and inches are the basis of the design of any ancient structure and therefore its most appropriate means for interpretation and description. Subdivision of the inch to less than half is seldom necessary in measuring scantlings, and in length and breadth of buildings, feet and maybe six inches are not only accurate enough but as accurate as possible, for such dimensions largely depend on the level at which they are taken. And when it comes to restoration nobody can take the right measurement except the carpenter – a fact that must be firmly stated in the contract. His preference is also invariably for imperial, despite new training and the law, and it is likely to remain so for most carpentry and joinery, as it has for decades in Danish furniture design. In our field, one-eighth and half-inch scales are the most appropriate, the former for plans, sections and elevations of the building as a whole, the latter for trusses and cross-sections.

Drawing elevations during the preliminary survey – indeed, drawing any part of the building – is preferable to taking photographs, as there is always something to learn, but there may not be time. The only measurements needed for an elevation are from end to end and from the bottom of the sill-beam (if it survives) to the top of the wall-plate. The bays are plotted within those total dimensions by visual proportion and the panels, windows, doors and timber sizes – and of course, joints – are drawn by eye. Indeed, dimensions, so long as the totals have been put down to scale, can actually be taken off the drawing! Similarly, the visual proportions of the external roof trus-

SCALE OF CHARGES FOR SERVICES ON HISTORIC BUILDINGS

CONDITIONS OF ENGAGEMENT - Subject to revision every 6 months

A. PRELIMINARY SURVEY	B STRUCTURAL SURVEY	C APPLICATIONS	D WORKING DRAWINGS	E SIGNING OF CONTRACT - COMPLETION
Preparation of draft scheme in sketch form only; preliminary survey of structural condition and necessary repairs, assessment of possibilities of bringing building or buildings up to standard for modern use, advice on how work will probably have to be done, on form of contract and likelihood of grants, all set out in form of illustrated report, to include, if required, provisional and approximate estimates for structural repairs only.	Survey of structure in greater depth, following stripping of internal finishes, drawing each structural frame to ½" or 1:20 scale, and providing complete set of survey ("S") drawings; confirmation or otherwise of findings of above preliminary survey; working out, with or without aid of Quantity Surveyor, provisional estimate of total cost based on draft scheme on which percentage charges will be based until tenders are received or negotiated contract sum accepted. Any contractor's charges for stripping, scaffolding or propping that may be necessary, are in addition.	Preparation of final scheme to ⅛ or 1:100 scale showing whole of proposed work in addition to structural repairs, including application for Listed Building and Planning Consents, and grant applications but not necessarily obtaining them.	Preparation of complete working drawings including structural repair drawings ("R"), repair schedules, and specification (or outline specification for QS if Bills are to be prepared), preparing contract documents and obtaining contract sum by competitive tender or negotiation; obtaining Building Regulation Approval.	Supplying information to the contractor, arranging for him to take possession of the site and examining his programme. Making periodic visits to the site, issuing certificates and other administrative duties under the contract. Accepting the building on behalf of the client, providing scale drawings of other services as executed, giving initial guidance on maintenance and maintaining normal responsibilities until end of Defects Liability Period.
PROPORTION OF AGREED FEE AT COMPLETION OF EACH STAGE				
Charge per sq ft area including each floor, or as agreed	One-sixth	One-third	Two-thirds	Remainder

NOTES:

1. The sequence of work stages depends on circumstances and may not be exactly as shown.
2. If work should be stopped by the Client or for any reason outside the Architect's control, or by agreement between Client and Architect, the total fee shall be based on the total time spent by the Principals and Assistants from inception to the point of termination.
3. In the event of the cost having to be cut following receipt of tenders the fee up to Stage D will be assessed on the tender, but the fee for Stage E will be assessed on the revised contract sum plus the cost of variations.
4. In all respects, except as given above, the RIBA Conditions of Engagement as prior to January 1981 should apply. Mileage will be charged according to current RAC rates.
5. All charges are irrespective of Local Authority fees for Planning and Building Regulations and any other official charges.

Figure 126 *Fee scale*

ses will give the correct roof pitch – if observed far enough away from it to avoid perspective distortion.

So, to revert to the preliminary survey, the building may still be occupied and furnished so that mere signs of its construction may have to take the place of a view of the structure itself. After a brief walk round the exterior, the roof-space is one's first destination. There are always surprises on climbing through a trap door in the ceiling; the worst is the discovery that the original roof structure has gone – but that is fortunately rare – the best that it is medieval.

The assembly numbers and directional joints will give the key to its erection. The trusses must be drawn and main dimensions taken – all of it generally in the vilest conditions, involving climbing through trusses and crawling on weak nailed-on ceiling joists, probably under flickering torchlight. But accurate and clear survey drawings at this stage are essential, and must be comprehensible to people other than the surveyor.

On the way down, drawing the upper floor over the roof plan and the ground floor over that, many other features will be picked up, 'shadows' of braces under the plaster, the depth and possibly shape of tie-beams and main posts, evidence of original windows or doors, whether floors are inserted, size and position of fireplaces, chamfers and chamfer-stops and everything of interest whether original or not that can be seen or divined.

If the house is post-1600, it can be put down with practical certainty that its plan is original, apart from obvious additions. With a medieval house, on the other hand, the plan and its features may be thoroughly contradictory and confusing. Everything must be drawn as it is, remembering that no part of a structure is ever accidental. Because something does not fit the conventional medieval plan or the kind of building it appears to be, it must not be written off as 'later insertion'. Everything may ultimately make sense, but if it does so at this stage the exploration has probably been too cursory.

Generally, if the building is a house, its internal arrangement and services must also be noted. The following might be used as a check-list of things to be indicated on the survey plans:

1. Room partitions, door-swings and windows. The partitions which follow bay division should be obvious. It is not necessary to draw timbers in elevation, whether exposed or not, unless they have some special features such as door-heads or exceptional shapes or mouldings. Then photographs may be the quicker.
2. Ceiling heights and changes of level, shown on the plan by a suitable convention giving the measurement.
3. Fireplaces and flue structures.
4. Staircases.
5. Ceiling beams – which should already be obvious from the roof plan.
6. Floorboards, noting whether they are oak or elm, and their age, direction and nailing centres.
7. Bathroom and sanitary fittings.
8. Construction of external walls and approximate thickness, some of which may have been bricked.
9. Signs of damp or decay and firmness, or otherwise, of floors, noting sagging, spring and loose plaster. Methods of testing, especially jumping, should be done with caution!

Plotting the ground floor may be more complicated than the upper floor or floors. Alterations will have been more numerous. There will also have been extensions of different periods, not always easy to date but often invaluable in preserving some earlier feature of the timber frame, such as an original window or wattle panels.

Again the services must be noted, especially in the kitchen, the hot-water system, plumbing and drainage, and the state of fittings.

The condition of the structure will generally be visible externally and points to check are:

10. Whether or not the sill-beam has completely rotted or disappeared below ground.
11. Where gutters and downpipes have been leaking and caused decay.
12. The state of purlin ends, verges and eaves.
13. The pointing of chimneys.
14. Slipped and broken roof tiles.
15. Whether any wattle-and-daub still exists.
16. The state of brick panels, fractured joints of the frame.
17. The amount of 'compo' or pitch and tar patching and filling, and the state of timber thus treated.
18. The style, date and condition of windows and doors.

Lastly, a walk should be taken round the site and outbuildings to locate, if possible, the original well, the present drainage layout and septic tank, the water and electricity mains, telephone, fences and gates, and perhaps to discover a timber-framed earth closet. Even the layout of the garden, the kinds of trees, the orientation and views over the countryside or threat of future development – not always ascertainable from the local authority – should be sketched or noted on a 'rough' site plan.

Time is always what decides the amount of information that can be gathered, and there is never enough. But it is better to spend it at the building than in the office, trying to remember what should have been put down on the survey.

The report, whatever its uncertainties and limitations, should turn out to be an education both in the writing, for its author, and in its being read by others. It will give a reasoned surmise of the building's history, and set forth its potential for modern use or living, often very different from the client's first ideas. It may be necessary to point out that he or she is more adaptable than an old building, and that it is better to change one's ideas than to try to force it into a false mould, especially that of 'period'. Its true character, if only it is allowed to express it, will demand no false values.

History must neither be destroyed nor introduced. If the architect really knows his building he need have no scruple in saying 'no' to his client – for the client's own good, as well as the building's. But he must also say 'no' to his own architectural predilections.

Sample survey report

A typical report is that on the Old Crown House in Birmingham done for Ansells' Brewery. Beginning with the documents, this building's more recent history was discovered in the fulsome writings of its nineteenth-century owner, Toulmin Smith,[1] who 'restored' it. However, the house itself, or what he had left of it, proved the more reliable source of information. The rest of this chapter is taken directly from our own report on the house.

Report on the Old Crown House, Birmingham

The original buildings comprising the Old Crown House were grouped around a courtyard (Figure 127), with the hall at the back of the yard, solar wing on the left, service wing on the right, and front range along the street. Its date, to judge by the architectural style of the surviving front range and design of roof trusses, is *c.*1450, rather than 1368 as advertised on the building today. This spurious date doubtless derives from the fact that Toulmin Smith was writing in the 1860s and constantly referred to the house as being 500 years old.

It had been subdivided, according to this author's paper, into three properties in 1693, having already become two a little earlier; but Smith himself was responsible for most of the alterations which account for the building's present form and appearance. Finally, its conversion into a public house did not take place until recently, probably the 1930s.

The main questions are:

1. Is the surviving original structure of sufficient architectural merit and in good enough condition to justify expenditure in its preservation?
2. If so, how much restoration will be necessary to achieve structural soundness and what will be both the external and internal appearance of the building as a result?
3. How far would the restored building be likely to overcome the present disadvantages of the site in attracting clientele?
4. How much would such a scheme cost to carry out and what financial aid might be expected?

Some knowledge of the stages by which the building has reached its present state should help to answer the first question. There is no doubt that the original house was outstanding and that it has always attracted admiration

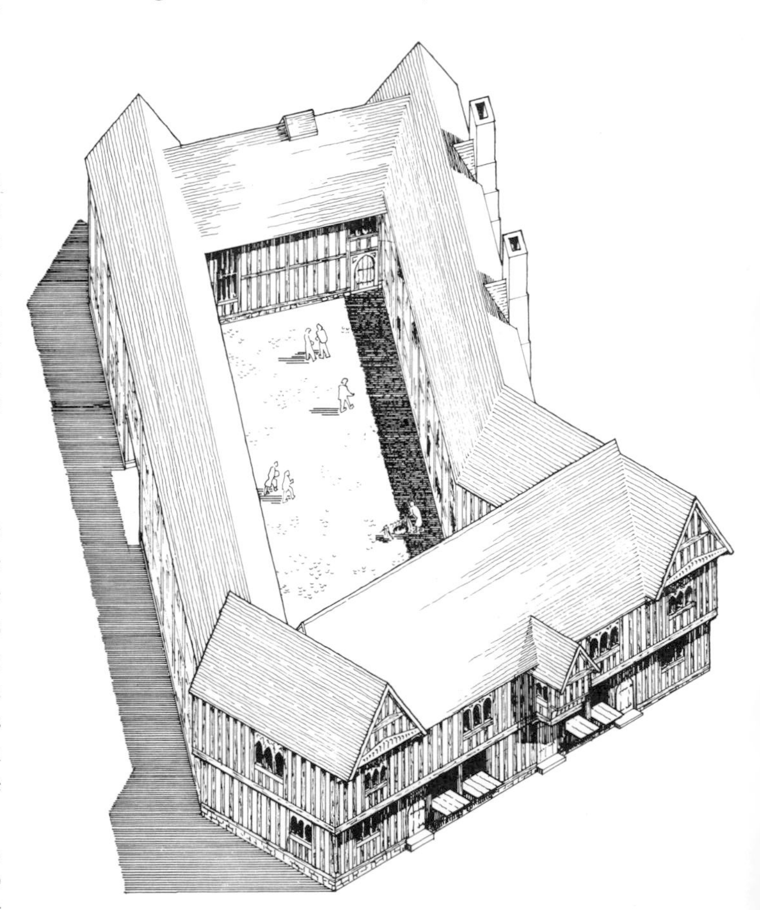

Figure 127 *Old Crown, Deritend, Birmingham – reconstruction of original house*

and comment. Even though it does not appear to have been a manor house, its owners, according to Smith's reading of the deeds, were closely connected with the Lords of Birmingham through various common interests, especially land-owning and guild affairs, and would rank as their near equals. It is indeed possible that the Old Crown House was superior to the latter's moated manor house that formerly stood just to the south-west of St Martin's Church.

In 1538 Leland remarked on it as the 'Mansion House of tymber'. Two hundred years later, according to Bradford's plan of Birmingham of 1751, it still stood very much as first built with an orchard beyond the hall twice as long as the courtyard (Figure 128).

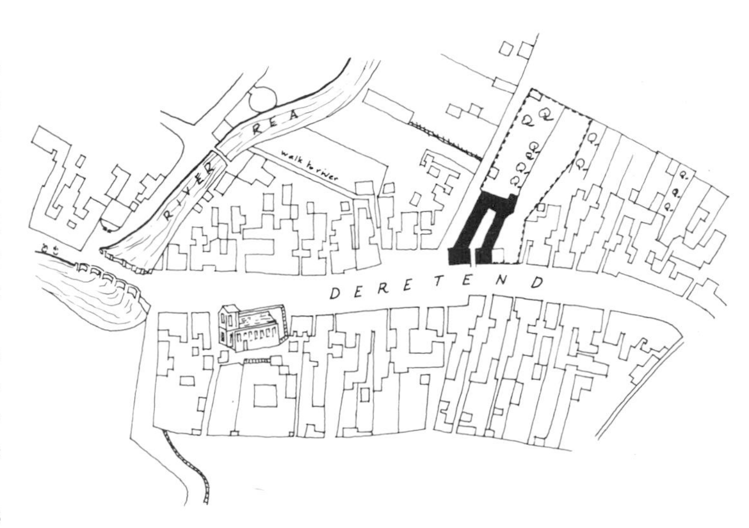

Figure 128 *Bradford's map of Birmingham showing Crown House*

In 1750, according to Smith, Heath Mill Lane was widened by the town authorities and the Old Crown House thus lost 7 feet of its garden along the west side. Then in 1847 the estate was severed by the Oxford and Birmingham Junction Railway, and in 1851, 'The Corporation of Birmingham included the Old Crown House in the schedule of their Bill, then before Parliament; designing to destroy the House in order to "improve" a street'. Smith managed to resist not only that threat, but two further attempts by the Corporation to demolish it in 1856 and 1862. Thus the house survived, though Deritend, described by Leland as 'a pretty street as ever I entered', became the industrialized squalor that it still remains.

Figure 129 *Toulmin Smith's 'Tudor' block*

Unfortunately the house did not escape unscathed, and Smith was responsible for destroying and altering probably considerably more than he preserved, believing none the less that he had restored it 'to more of the likeness that it bore in the days of its youth, than it has had for some centuries'. The buildings at the farthest end of the courtyard and on its right-hand side were pulled down. He suggests that they were rebuilds of 1830, yet adds in a footnote, 'It is a cause of great regret to me, that the works lately done were obliged to be done while, being absent, I was unable to make observations on what was disclosed as they proceeded.' This would surely not have mattered had they been only thirty years old. The range of the left-hand side was also demolished, 'having become

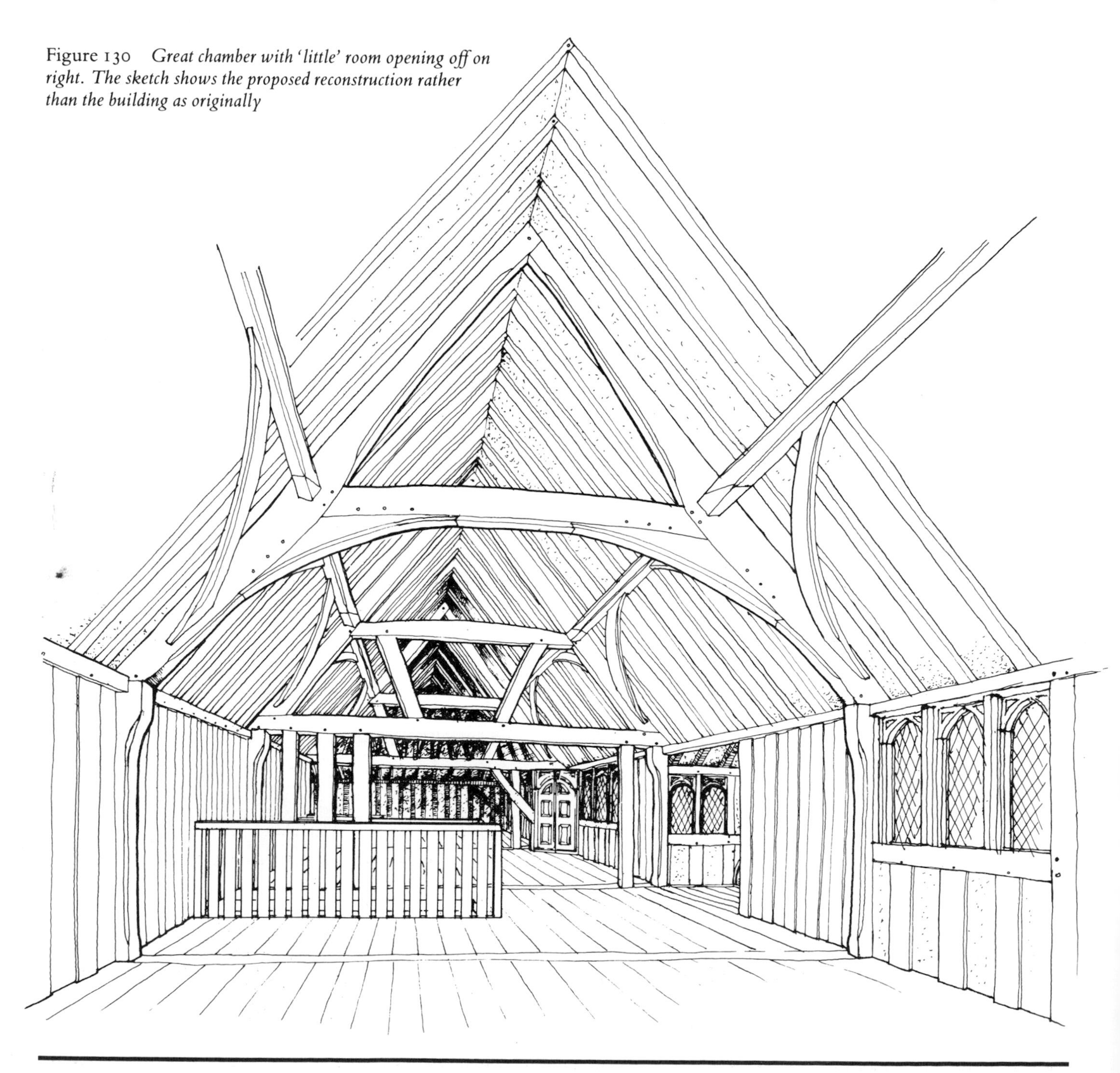

Figure 130 *Great chamber with 'little' room opening off on right. The sketch shows the proposed reconstruction rather than the building as originally*

very ruinous'. This he replaced with the present 'Tudor' block facing Heath Mill Lane which he maintained (rightly) was 'much more substantial' and (highly questionably) 'more in keeping with the character of the Old House' (Figure 129).

Thus the original hall, the service wing, the rear wing and the solar were lost without record, except that there may still be some vestiges of roof trusses in the present rear wing. This will be further investigated, but even if original they have been too severely mutilated to be of much more than academic interest.

That all these buildings had survived, albeit altered and probably in poor condition, until Smith's time, is borne out, not only by a reading between the lines of his paper, but also by the number of other medieval town houses in which great roof trusses and timber frames are only now being rediscovered.

Victorian owners, it seems, never bothered with the rear buildings, probably servants' quarters, and seldom penetrated into roof spaces. Thus, these elements, if they survive at all, have generally escaped 'improvement'. The Old Crown House was at least fortunate in that respect, for the original roof structure of the front range, though containing some botched repairs, is still practically complete.

Smith believed that the front range consisted of the hall with parlour at one end and kitchen at the other. In fact it was the commercial part of the house. This is clearly indicated by the original bay arrangement and the projecting porch-room over the central doorway, leading through a passage into the courtyard. The porch-room was the owner's office and look-out for visitors, of some importance to a merchant or man of wealth.

This little room was part of his great chamber (Figure 130) where he could best display his wares or impress potential creditors. The plaster reliefs (Figure 131) on the porch-room ceiling probably advertise his trade or trades. Smith connects the double triangle with a deed of 1517 and suggests Freemasonry as another possible source. The only certainty is that they were cast long after the house was built, probably after 1600.

The ground-floor room beneath the great chamber is likely to have been the owner's shop. The west wing must also have been for commercial use, but whether it was rented as a separate unit or used by the owner is not yet known. The interior of the ground floor was imposing on account of the dragon-beam. The upper floor, open to the roof, was an even more splendid room (Figure 132).

To summarize, the owner occupied all the rear buildings and probably bays 1 to 4 of the front range (Figure

Figure 131 *The porch room as in Toulmin Smith's time. The plaster reliefs on the ceiling still exist*

133), certainly bays 3 and 4; a shopkeeper rented the ground floor of bay 5 – this latterly became a blacksmith's shop; and another tenant occupied a combined shop and domestic unit in the rest of the front range. Except for the demise of the blacksmith and the expansion of the latter tenant into the rear wings, this arrangement has not radically altered. Thus there may already have been three tenancies of the Old Crown long before the seventeenth-century deeds quoted by Smith, in fact from the very be-ginning, the only difference being in the internal arrangement that must have been altered from time to time.

Condition and restoration

While the front range now stands out in Deritend as an architectural oasis, the former courtyard has completely disappeared beneath the unplanned chaos of its surroundings. Not only has everything of historic and aesthetic value been destroyed, but ill-designed buildings of the post-Smith era have now been built up against and on to the whole of the timber-framed rear wall of the front range, even including that part of which Smith proudly

Figure 132 *The upper room, as originally and as it would be reconstructed. Note diagonal jointing of boards following dragon-beam*

says, 'its ancient timbers are now properly picked out, and it is thus restored to the colour and sight to which it has long been a stranger'.

How much of this timbering has survived is not yet possible to say. However, it is likely to be restorable if the offending buildings are removed. Smith also made much of his front elevation, taking out earlier, if not original, windows and replacing them by the existing windows within entirely new brick walls. The new front wall was built immediately beneath the plate, to preserve the jetty, encasing the plate or perhaps replacing it complete. If it still exists the full evidence of this wall-frame could be discovered, including the type of dragon-post (Figure 134).

The upper walls are still wholly timber framed with a certain amount of restoration. Some repairs may be

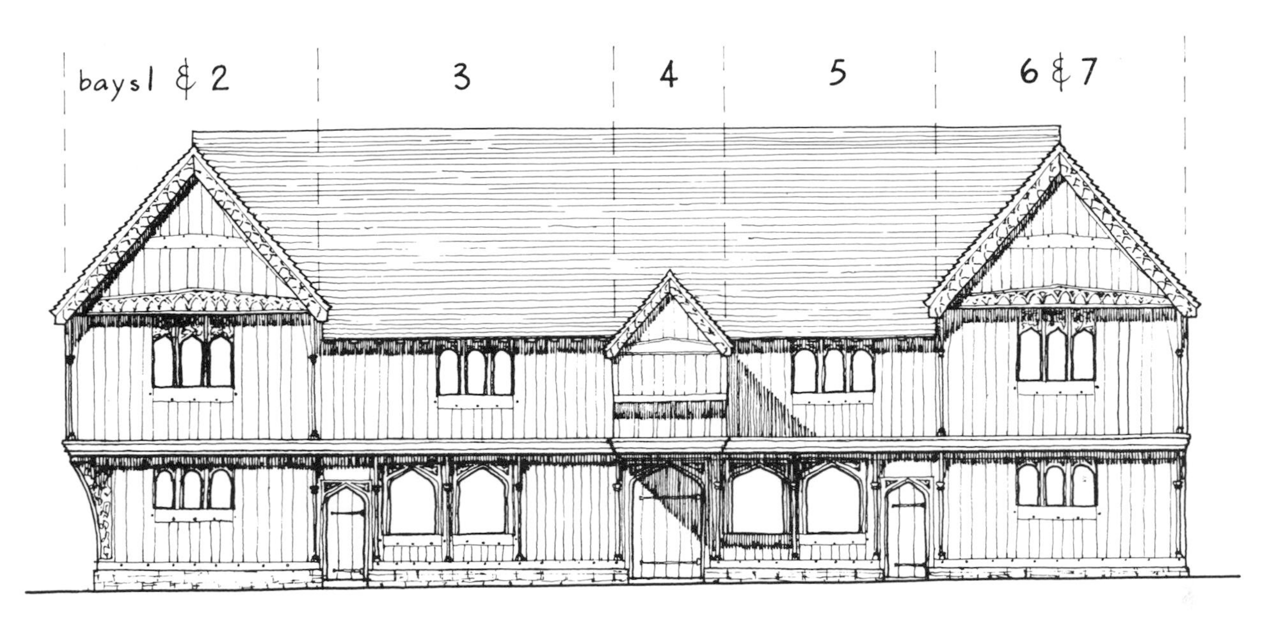

Figure 133 *Front range, as originally, showing bay numbers*

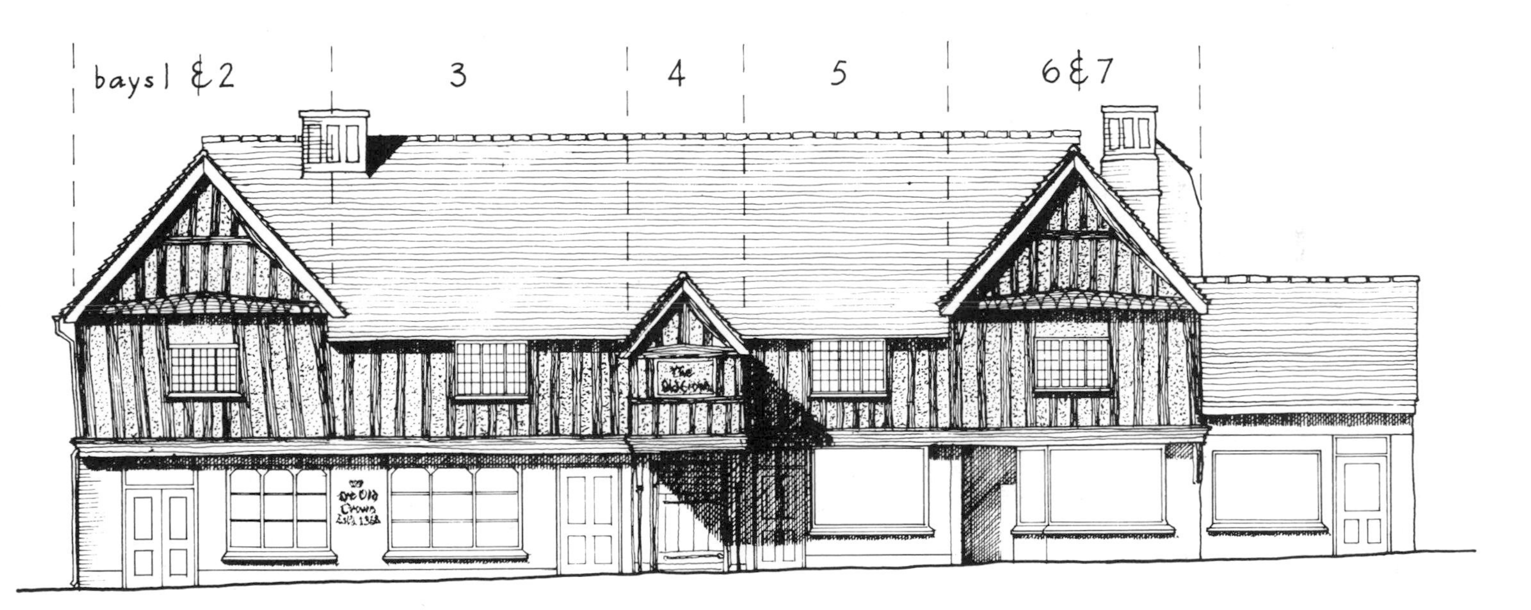

Figure 134 *Front range, as existing, showing Smith's reconstruction in brick of the entire ground floor*

necessary, especially along the bressummer which was entirely replaced by Smith, 'precisely in the original shape and size'! He adds, 'the massive boldness of this moulding, as now restored, cannot fail to strike the eye'. There is no doubt about the latter comment, but considerable doubt about the former. The jointing of the studs and corner posts and its relationship to the jettied joists will prove whether or not the restoration was authentic. But the whole bressummer may have to be removed, not because of its appearance, however bizarre, but rather because of water penetration behind it, affecting both joists and studs, a defect that would not have happened with the original member.

The construction of the upper floor itself should be generally sound, judging by the exposed beams and joists. Only the floor of bay 5 is suspect. Here the joists are ceiled except where their ends appear beneath the jetty, and these are new, suggesting that the whole floor has been reconstructed. This may have some connection with the fact that bay 5 at the beginning of this century was a blacksmith's shop, for which the first floor may have been taken out in part if not altogether. If the date of its restoration was the same as that of Ansells' acquisition of the property the question of what was done will be answered by the records; otherwise it must await stripping. Incidentally, most of the wall-frame of this bay has also been renewed relatively recently.

The roof is fully visible above the inserted ceilings. As already mentioned the structure has suffered nothing worse than age and slipshod repairs. The latter include not only strapping up and bolting the fractured principals of truss V, architecturally the most important truss in the building, but also inserting new rafters above the old, thus altering the roof pitch and falsifying the wall-plate levels. The extent or structural consequences of this are not yet known, but when the ceilings are removed to expose the roofs, repairs and some structural restoration, not confined to truss V, will be necessary.

It is certain that the roof structure must be attended to before very long, and doing it properly now is the only alternative to replacing it completely in a few years' time. Further stop-gap repairs would be worse than useless. So also would chemical treatment of the timbers.

The other inserted elements are the two fireplace structures in bays 3 and 7. No doubt these were put in at the same time as the ceilings (*c.*1600) though the external stacks are Smith's. The fireplaces may still exist behind the wallpaper and plaster; they are no doubt damaged, but if the damage is not too severe they should be well worth reopening and repairing.

Evidence of the original windows will be found, but whether or not they should be reproduced or the existing ones preserved depends on a number of factors not yet considered. It is most likely that decay will have necessitated replacement.

The ground-floor structure of the front range, as rebuilt in brick by Smith, is still serviceable, but the passage entrance should be given back its proper use and the passage incorporated as additional space in bay 4.

Of the existing rear buildings only Smith's Tudor block facing Heath Mill Lane is in any way worthy of preservation. It represents a fashionable style of the nineteenth century, as unlike timber-framed architecture, regardless of Smith's views, as any building in present-day Deritend. But it should still have a useful span of life. The rest are of no value except as stores or sheds which have eliminated open space and fresh air, essential to the preservation of the front range.

Restoration of the Old Crown House will consist of three operations:

1. *Roof.* This must be attended to as a matter of urgency right through the front range. It will involve taking down all the inserted ceilings, repairing main structural members, removing inserted rafters and replacing damaged or missing original rafters. The roof covering will include insulation, and all existing undamaged tiles will be reused. The ceilings will not be replaced.
2. *General repairs.* These will also be necessary throughout the structure. They will include wall-timbers, a number of floor joists though (it may be hoped) not floor beams, some windows and all the wall panels, using the method devised in this office to ensure adequate insulation. It is not necessary to demolish or alter Smith's ground-floor structure of brick, nor the cellars. The plaster reliefs and any other historic details still to be discovered would be preserved.
3. *Courtyard.* Except for the Tudor block, this must be cleared by demolition of the existing buildings. Then a manager's house and stores must be built, leaving adequate open space for a garden, large enough at least for some grass to grow, and a servicing yard (Figure 135).

These proposals are the minimum that will assure preservation of the front range. It is invariable that Victorian restoration, because of its extraordinary ignorance of building history (as we have seen), its total disregard of traditional timber-frame technique, its use of cheap materials (brick and softwood), and its almost exclusive concern with superficial effects and covering up the structure, has become the main cause of trouble in ancient buildings today.

The Old Crown House is typical. The attempt to cure a leak, replace a rotting timber, even put in some new service pipes or wiring, leads to discovery of defects previously concealed, generally worse than the known one – until the whole job has to be undertaken piecemeal and without a plan or means of controlling costs.

To avoid this the next stage must be internal stripping, followed by a complete structural survey, then detailing and specifying every repair, preparing contract documents and finally going out to tender for a fixed-price contract.

The question will be asked, 'Does the Old Crown House warrant such attention and expenditure?' First, as

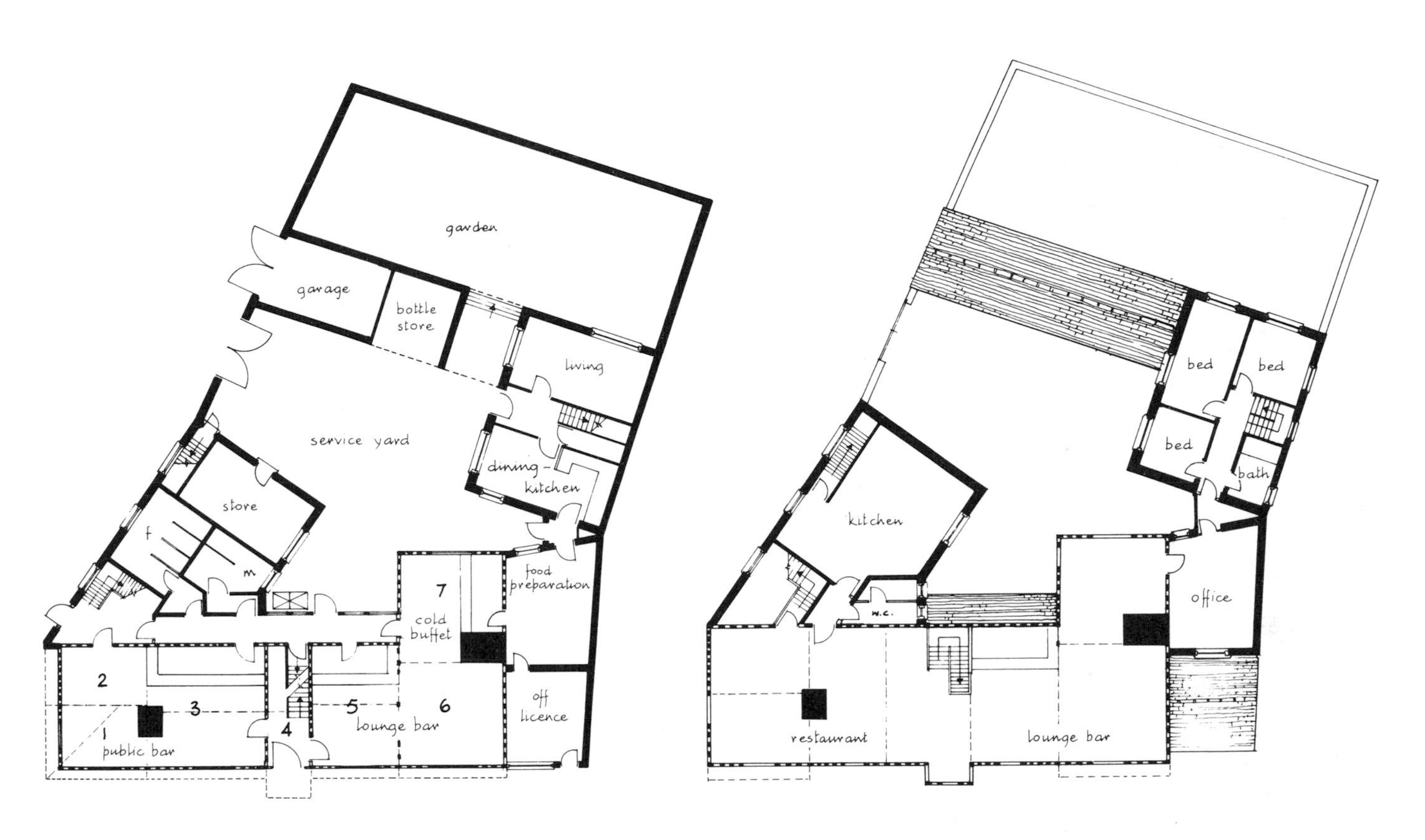

Figure 135 *Replanning of ground and first floor to preserve Old Crown in its present use*

a listed Grade II* building, the owner has the responsibility of maintaining it in good condition. Short of selling it, he therefore has little choice. The real question is thus one of how to do what is necessary in a way that will realize the building's potential. There are two comparable examples, West Bromwich Manor House[2] and the Old Wellington Inn owned by Bass Charrington at Manchester. Both were restored according to the principles of absolute respect for the original structure, in sharpest contrast both to Victorianism and today's, perhaps even worse, 'olde worlde' sentimentalism. Both buildings have become widely known and enormously popular.

The Old Crown House may seem to be at a disadvantage on account of its surroundings and lack of an adjacent car-park. As regards the former, an outstanding building inevitably asserts its influence, even to the point of changing the environment. Neither West Bromwich Manor House nor the Wellington would lose clientele if placed in Deritend. On the contrary, by making use of the exceptionally broad pavement in front of the building to create a sympathetic setting, and by floodlighting in an imaginative way the front elevation, the present site would be no less favourable than that of either of the others. All that is necessary is that the Old Crown House should be seen for what it really is, its proper exploitation in terms of use being left to do the rest.

CHAPTER SIX

STRUCTURAL SURVEY AND REPAIRS

The frame

This stage includes the preparation of all drawings and documents leading to the contract. Since much of this is common to all architectural practice it need only be briefly mentioned. There is, however, one important difference. The one-eighth scale drawings must be duplicated to show 'before' and 'after'. Though the local authority has no power to insist on this, but may only reject an application for lack of information – unfortunately seldom done – it is essential to record both exterior and interior before any stripping is done. The drawing, 'as existing', is then accompanied in the application by the 'as proposed' drawing, showing the same plans, elevations and section. The scheme is thus worked out before the state of the structure is known, except for what has been learnt about its general condition in the preliminary survey.

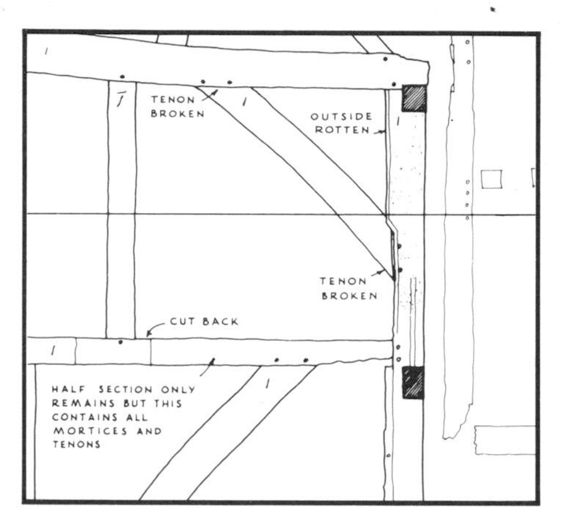

Stripping the internal finishes requires awareness, anticipation and care. To come upon a building already stripped by the owner or his builder – to make one's task easier – is to suffer a sense of deep loss, even though the loss may be of no more than lath-and-plaster and some layers of wallpaper, of which he may even have kept samples. It is generally a contractor's job, though to do for oneself the initial peeling of wallpapers and breaking through to the timber frame is a wise and often exciting preparation. The follow-up must be under the architect's constant supervision, armed with a camera as well as pencil and paper. External rendering may also have to come off, not only for repairing the frame, but because the internal face often belies the external condition, and members that look sound inside may be rotten for half their depth on the outside.

The building may also be so lop-sided that measurements of the extent of its lean and subsidence may have to be taken. The former is easy enough by dropping a plumb-bob from the wall-plate to the floor (or to the ground outside) or from the tie-beam to the foot of the post. For measuring the subsidence, a horizontal string stapled or cup-screwed to the timbers is the easiest to set up, and this gives a datum for measurements to both ceiling and floor.

As a check-list before setting out on a structural survey the tools and equipment that will most likely be needed, in addition to the obvious requirements of drawing and measuring equipment and torch and camera, are:

1 Claw-bar;
2 A strong screwdriver, for holding open the claw-bar's initial levering-up of a board;
3 Spirit-level, of 3 feet minimum length;
4 Hacksaw blade, for penetrating joints and discovering whether there are tenons;
5 Plumb-bob;
6 Profile gauge, for mouldings;
7 String, staples, nails, hammer;
8 Demountable aluminium ladder.

Even with the help of all this and the contractor's men, stripping will be slow and laborious. Fourteen layers of wallpaper, and two separate ceilings over the same room, have been encountered. Rare wallpapers can probably be dated at the Victoria and Albert Museum, London. The layers of paint in descending order are generally brown (latest), green, blue, yellow and red (earliest), often with white somewhere. Wall and ceiling paintings can seldom be preserved except in the form of colour photographs, willingly accepted by the National Monuments Record, but they may also be worth trying to preserve for a little longer by covering them with polythene. Plain timbers and wattle-and-daub panels, when at last they emerge, probably mean that the wall was oak-panelled. Chalked messages behind lath-and-plaster may give important dates.

But our objective is the timbers, their condition, and the whole of the evidence of the original frame. Mortices often still contain the tenon of the sawn-off member and will establish whether the latter was a stud or a brace. Absence of mortices where they should be expected is as important as their unexpected presence. Wattle grooves, mullion mortices, mouldings and peg-holes must all be recorded.

The S (survey) drawings to half-inch scale show each truss and cross-frame, the side-wall frames with relevant

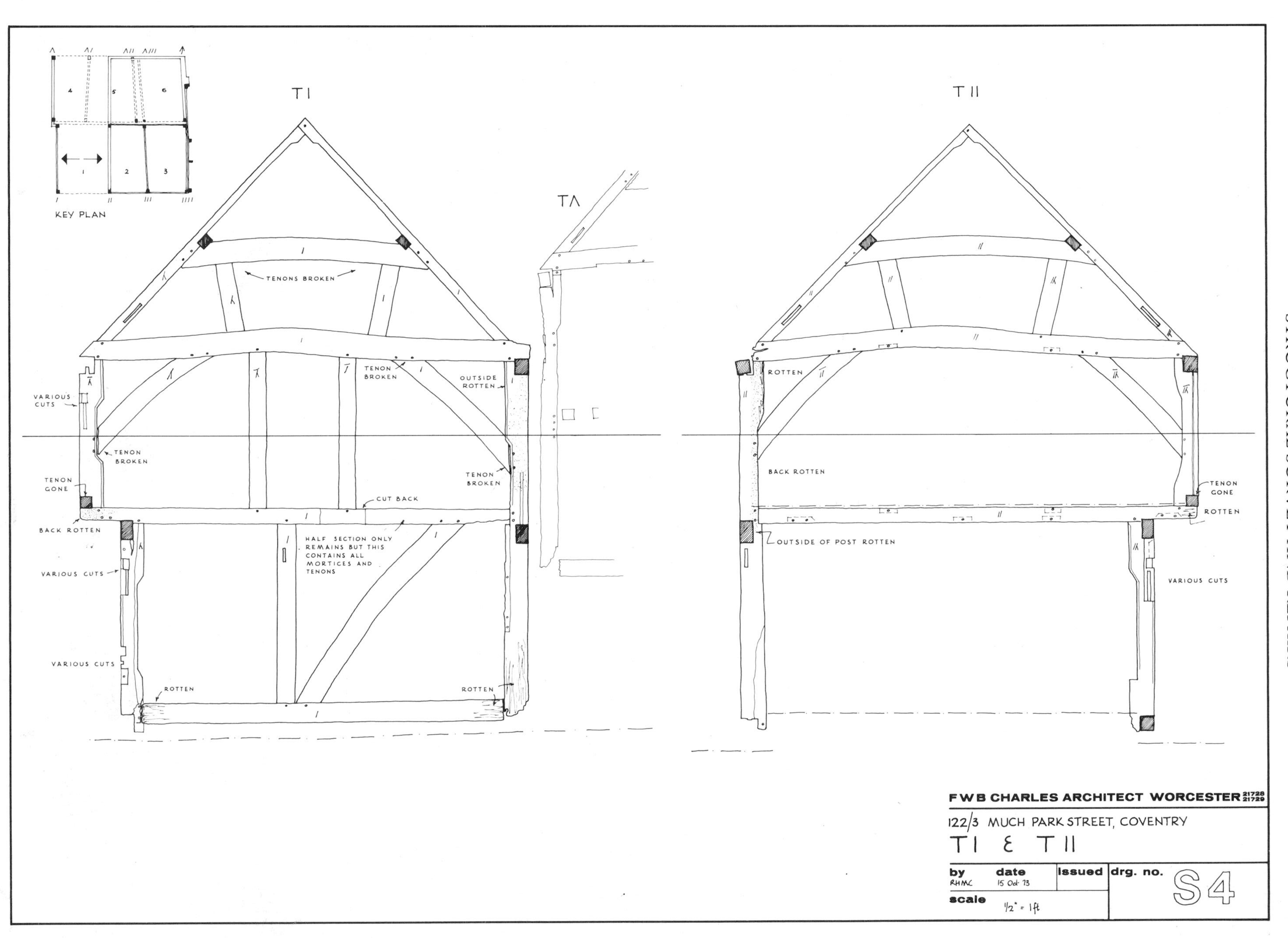

Figure 136 *122–3 Much Park Street, Coventry – survey drawing of typical cross-frames (S4)*

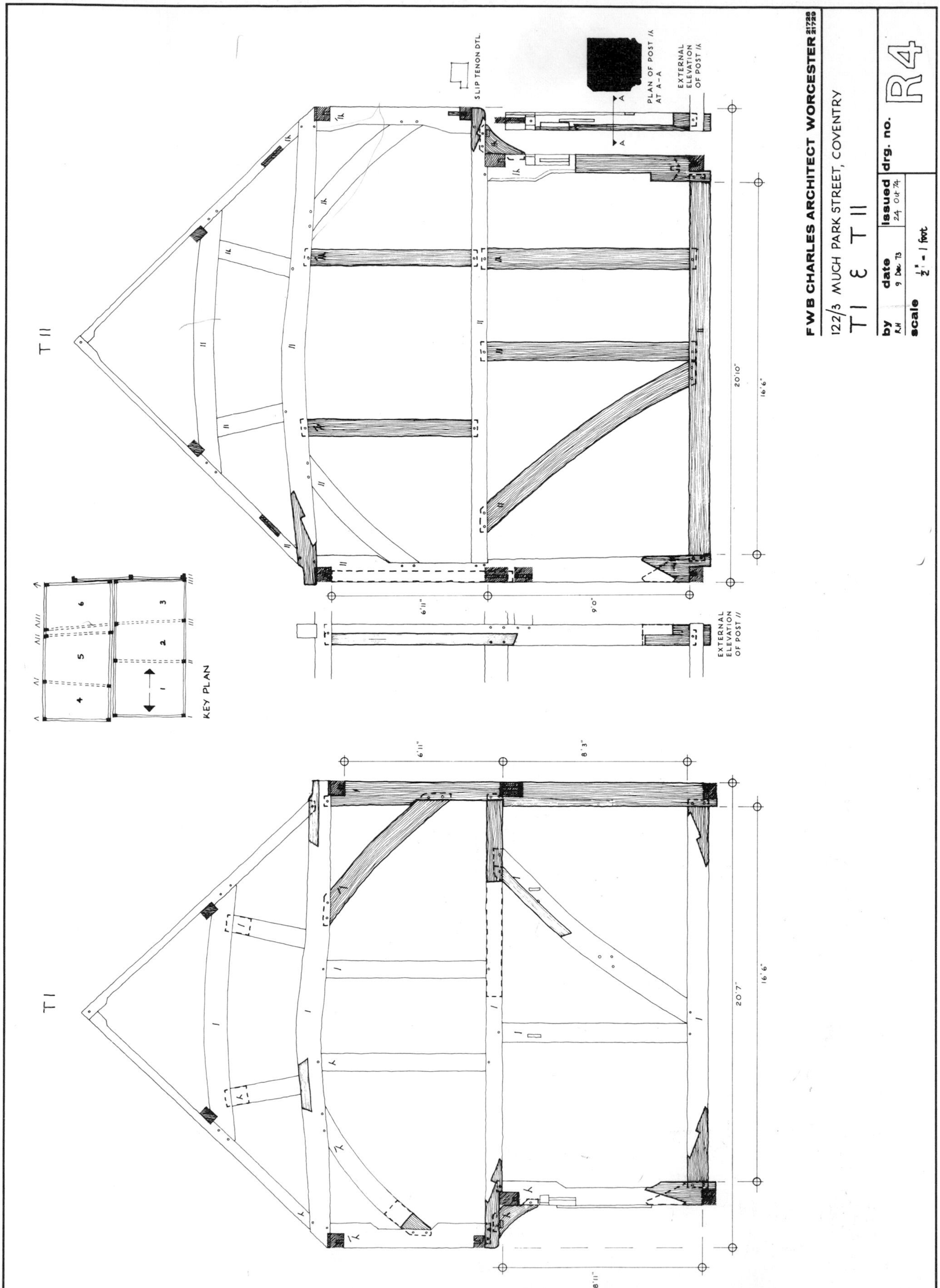

Figure 137 *Repair drawing of the same frames (R4)*

Figure 138 *Repair schedules for the same frames*

R E P A I R   S C H E D U L E

122/3 MUCH PARK STREET, COVENTRY — 9th December 1973

DRAWINGS S4 and R4 - T I

Item	Mark	Description
Rear-post	/	12" x 10" x 17' 6" reduced to 10" x 10" between girding-beam and brace, tenon for tie-beam and wall-plate at head, mortice for brace, two girdings and cross-beam, tenon for lateral and mortice for transverse sill-beams.
Front-post	λ	12" x 5" x 3' 0" scissor-scarf at foot, tenon for lateral and mortice for transverse sill-beams.
Jetty bracket	λ	10" x 3" x 2' 6" swept, tenon both ends.
Sill-beam	/	11" x 6½" x 3' 6" table -scarf front end, squint butts, tenon.
		11" x 6½" x 3' 0" table-scarf rear end, squint butts, tenon.
Girding - front end	/	9" x 10½" x 4' 0" table-scarf, rebate over plate, mortice for bracket and two posts, quadrant-rounded end.
Girding - rear end	/	9" x 10½" x 9' 0" vertical half-lap scarf, tenon to post, mortice for brace.
Tie-beam	/	6" x 4" x 3' 0" patch to mortice, squint-butts.
Tie-beam - rear end	/	4" x 7" x 3' 0" patch, squint-butt mortice for principal rafter.
Brace - lower wall-frame	/	5" x 4" x 4' 6" patch, squint-butt, tenon.
Brace - upper wall-frame, front	λ̄	10½" x 3" x 2' 3" vertical half-lap scarf, tenon.
Brace - upper wall-frame, rear	/̄	10½" x 3" x 8' 0", bare-face tenon both ends.
Struts	λ, /	2 - 9" x 3" x 1' 0" slip-tenon.

DRAWINGS S4 and R4 - T II

Item	Mark	Description
Rear-post	//	14" x 11" x 3' 6" foot, scissor-scarf to existing, tenon for lateral and mortice for transverse sill-beams.
Rear-post	//	6" x 6" x 8' 6" patch, squint-butt, tenon to wall-plate, mortice for girding.
Front-post	/λ	14" x 9½" x 6' 0" combined foot and 3" deep side patch, tenon with squint-butts to existing, tenon to lateral and mortice to transverse sill-beams. Mould angle as detail.
Jetty bracket	/λ	10" x 3" x 2' 6" swept bare-face tenon both ends.
Top post	/̄λ	1' 1" x 9½" reduced to 9" x 9½" slip-tenon for bressummer.
Sill-beam	//	11" x 7" x 17' 0" tenon both ends, mortice three times.
Girding	//	6½" x 10" x 2' 0" table-scarf end, mortice for post, quadrant-round end.
Tie-beam	//	12" x 7" x 4' 3" table-scarf end, mortice for principal rafter and post, dovetail over wall-plate.
Brace	//	12" x 4" x 12' 0" swept, tenon both ends.
Stud	//	9½" x 4" x 9' 6" tenon both ends.
Stud	/λ	11" x 4" x 9' 6" tenon both ends.
Stud	/̄/	9" x 4" x 7' 9" tenon both ends.
Stud	/̄λ	10" x 3" x 2' 6" swept, bare-face tenon both ends.

roof-slope, and the floor structure (Figure 136). Corner and intermediate posts are shown with their cross-frames, and omitted from the side wall-frames. The drawings also differ from normal half-inches in showing no horizontal members in cross-section (unless by shading); but the mortice in the supporting member is shown. Fractures, areas of decay and other defects are hatched or in some other way indicated. Cross-frames are drawn generally from their upper face, though frame I and the final frame, being surveyed from inside the building, will show their lower face. This also applies to the side wall-frames. Floor frames are preferably shown from below.

The S drawings are accompanied by corresponding R (repair) drawings (Figure 137). These show each frame as restored, the new timbers that must replace missing or unusable old ones, and all repairs with their appropriate repair joints. Each timber is numbered. The system of numbering follows the traditional one but it must be completed – the numbering in most standing buildings being anything but complete. Adjacent timbers in any one frame may however have the same number if they are interchangeable.

A further set of documents necessary at this stage is the repair schedules, cross-referenced to the R drawings (Figure 138). Their purpose is to enable the quantity surveyor to calculate the labour and cubic content of new

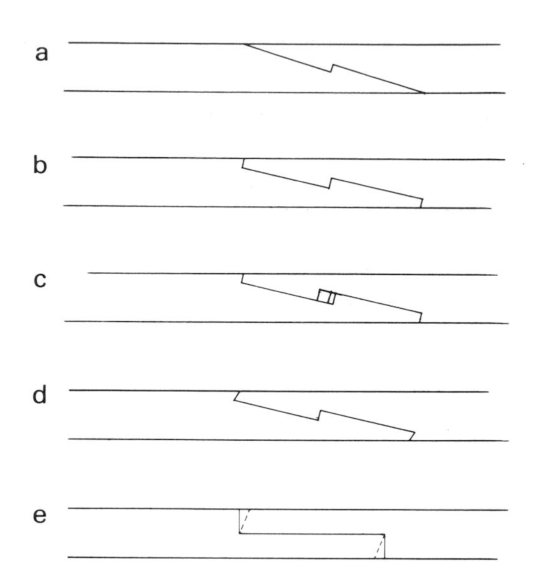

Figure 139 *Simple scarf joints*
(*a*) *Splayed tabled-scarf*
(*b*) *Tabled-scarf with stepped butts*
(*c*) *Tabled-scarf with folding wedges or 'key'*
(*d*) *Tabled-scarf with undercut- or 'squint'-butts which may have the same variations as above. This joint has to be engaged by sliding the two members together on their sides*
(*e*) *Half-lapped scarf with straight- or squint-butts*
All the above joints must be multiple skew-pegged

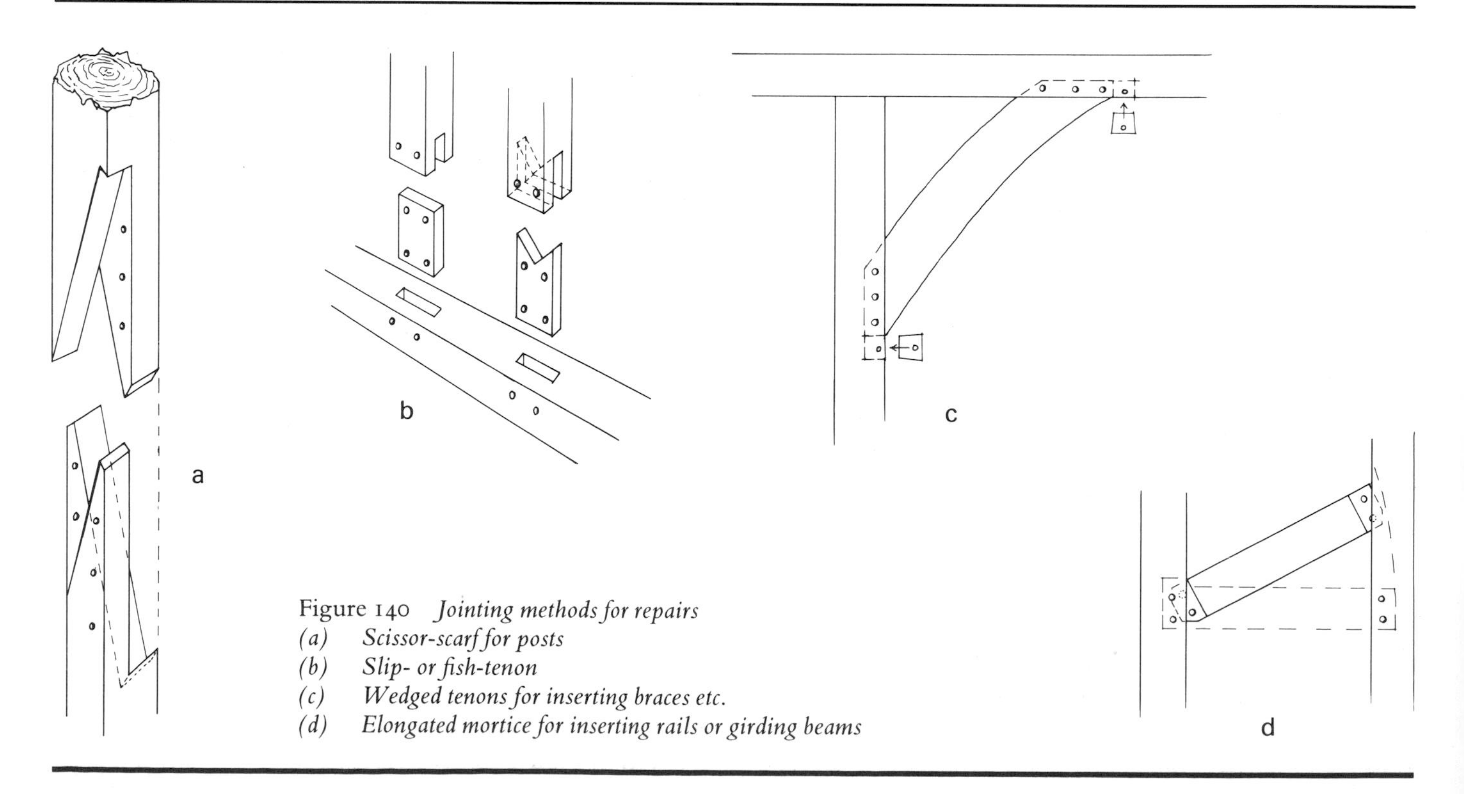

Figure 140 *Jointing methods for repairs*
(*a*) *Scissor-scarf for posts*
(*b*) *Slip- or fish-tenon*
(*c*) *Wedged tenons for inserting braces etc.*
(*d*) *Elongated mortice for inserting rails or girding beams*

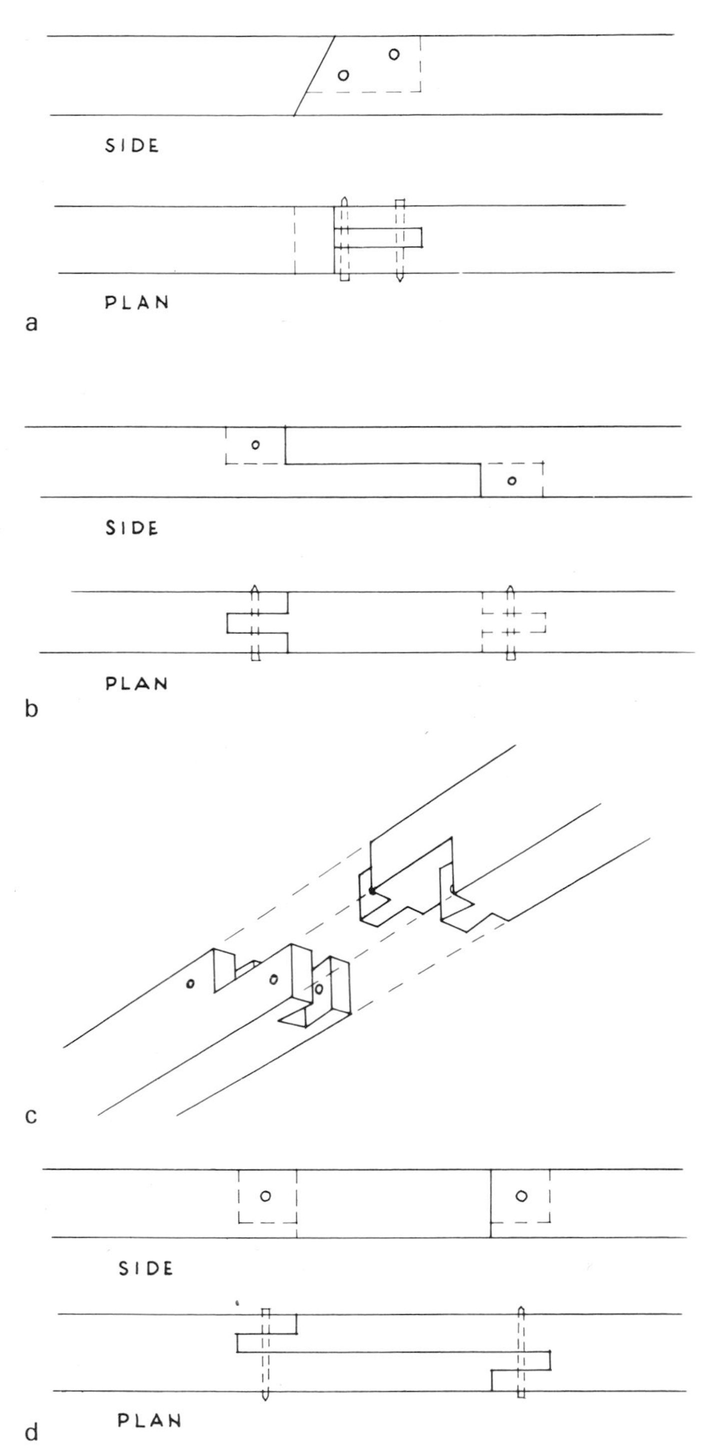

Figure 141 *Further scarf joints*
(a) *Bridled-scarf*
(b) *Double-bridled half-lap scarf. This scarf may be splayed*
(c) *The same with lower bridle dovetailed to resist tension*
(d) *Bladed vertical half-lap scarf*

oak in various scantlings for each type of joint. Both S and R drawings and schedules must accompany the bills when going out to tender. And it is as well to warn each contractor to time his carpenter in making trial joints before pricing them. The schedules give the scantling and length of each *new* timber, the latter being safely generous and 'to be checked by the contractor'. Also noted are the number and types of joints it must have for receiving or being connected to adjacent timbers and, if a repair, the type of scarf for jointing it to the parent member. A detail of repair joints is also issued (Figures 139, 140 and 141).

The scarfs are of various designs mostly found in standing buildings (Figure 139). Even the more elaborate scissor-scarf (Figure 140(a)), the perfect compression joint, was first found at Middle Littleton Barn (Chapter 11), dateable therefore to the thirteenth century. A variation of the half-lap scarf is to provide bridles, preventing sideways slip (Figure 141(a)), and a further variation, especially valuable for members in tension, is the half-lap scarf with a dovetail bridle in the lower half (Figure 141(b)). It must, however, be borne in mind that the dovetail or any joint narrower at the neck than at the end is liable to split the walls of the receiving member if tension stress is too strong. Simpler joints for beams with compression at the top and tension at the bottom are strapped half-lap scarfs. Undercut or 'squint'-butts probably look better but they do nothing structurally if the stress is taken by the strap. Any joint likely or actually *beginning* to split the receiving member must also be strapped, but this will happen only with long unsupported spans. Cross-beams and tie-beams of external frames are always supported throughout their length, and so their joints will need no strapping. The straps are U-form with lugs to tighten each half against the other (Figure 142).

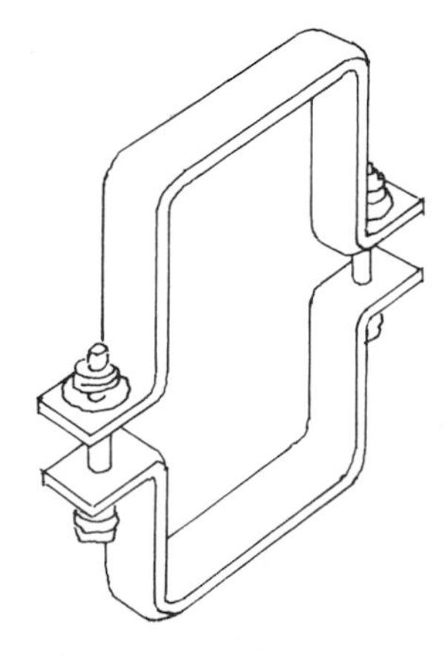

Figure 142 *Typical steel strap*

The repair of jettied beams, or joists, is assisted by the reverse bending moment, and a splayed-scarf with squint-butts hardly even needs to be pegged, but it is as well to do so (Figure 143(a)). For directional joints where the stress is mainly compressive, such as in arch- or arcade-braces, the bare-face tenon with squint-butts is often useful (Figure 143(b)). Where the members are too slender for this, the half-lapped 'sallied' bridle, pegged or bolted, or even plated, may be the most suitable (Figure 143(c)). For rafters the vertical scarf is safer than the horizontal, even though the shear must be taken by the pegs (Figure 143(d)).

A rule of thumb for the length of scarfs is three times the depth of the scarfed member, but this may be reduced to twice the depth if the joint is taking only a minimal bending moment, as in external frames or framed partitions. All scarf-joints must be skew-pegged, with a different angle of skewing for each, and they should be randomly aligned. The pegs may also be driven in from the sides as well as top and bottom.

In principle, repair joints should not be concealed or disguised. The behaviour of visible repairs can be watched. Straps are thus generally preferable to bolts, and there may be some risk of corrosion with the latter. But steel plates on either side of a fractured member bolted through to each other are safe for centuries. The only time stainless steel should be resorted to is where a plate, as in flitches, must be entirely bedded in the oak – but such a necessity is extremely rare.

Glues are to be avoided except under controlled conditions impossible to obtain on a site, and must of course never be used in construction joints, of which the essence is slight 'give', both in the construction and stressing of the frame. Having regard to the punishment timber frames have had to withstand through history, there is hardly a building that would have survived without such flexibility. Epoxy and polyester resins are to be condemned both for this and other reasons. With plastic patching of partially rotted timbers, moisture collects at the back of the impervious filling, and the timber at the interface then rots.[1] For restoring beam-, joist- or rafter-ends or for letting in a new piece between the ends of an existing timber, new oak is preferable, both economically and as a matter of principle – that of preserving the nature of the original structure. There are however special circumstances that may justify plastic repairs. For instance, where beam-ends or post- or cruck-feet *must* be bedded in damp masonry or brickwork, or where a plaster ceiling must not be disturbed, though the beams and joists above it are rotting or decayed, the strengthening of their affec-

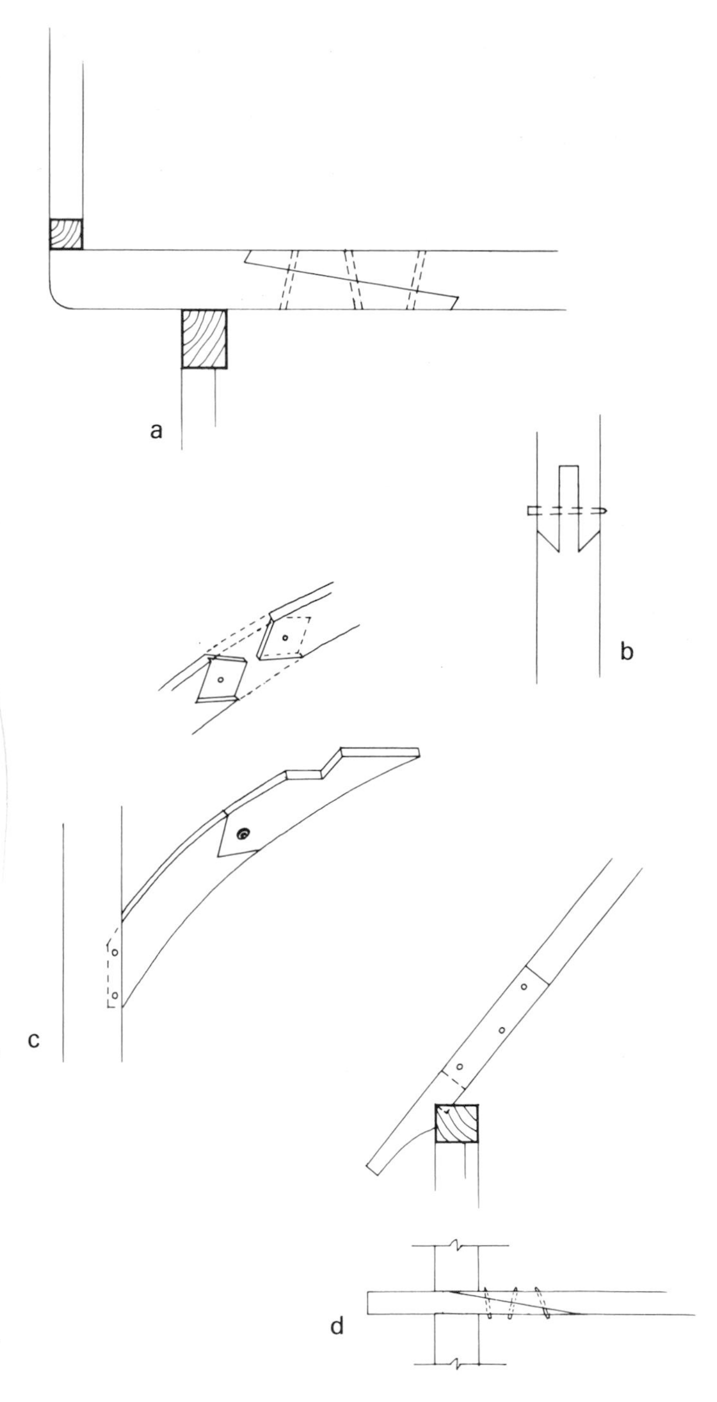

Figure 143 *Repairs*
(a) Splayed-scarf (not tabled) with squint-butts for jettied beams or joists
(b) Bare-face tenon with squint-butts for compression joints
(c) Half-lapped 'sallied' scarf for more slender members such as arcade or wind-braces, either bolted or pegged
(d) Vertical scarf for rafter feet – shear taken by pegs

ted sections by impregnation or insertion of glass fibre reinforcement with epoxy resin moulded round it *in situ* may be the only answer. But these are problems seldom, if ever, met in a timber-framed structure. As for the appearance of the material, it no more resembles oak in colour or texture than it does in physical properties – particularly when 'grained to match'.

Finally, the difference in principle between repair and structural joints cannot be over-stressed. The former are rigid; the latter allow slight movement in the total frame without exerting stress in the members that could lead to their fracture, a risk far more prevalent in timbers that have had centuries to harden and shake, particularly for instance around shrunken knots. If the structural joint at either end is made rigid the condition at any such weak point must worsen.

Not only must glues and resins be avoided (except as already noted) but also concealed plating and bolting. This was a Victorian speciality, of which the great roof of the Guesten Hall of Worcester Cathedral was an example. It was completely destroyed as a work of carpentry in order to change its pitch. In its reassembled form it was a steel-jointed frame, with tie-rods passing right through the collar-beams and arch-braces to join them to the principals. The same method was used in Westminster Hall roof in the 1920s, before it was subjected fifty years later to epoxy resin repairs as well. The late A. R. Powys's apt simile for the former was, 'to liken the present state of this, the finest piece of carpentry ever done by man, to a living lion caged for show as opposed to the beast roaming freely the foothills of Kenya Mountain'.[2] Now the lion has also been stuffed. The most recent example is the White Hart at Newark where every post-head joint has been diagonally drilled and bolted with bolts over 4 feet long. When such 'strengthening' is demanded, as generally by the structural engineer, the architect must insist on physical stressing of the frame or floor concerned as the *only* means of testing its strength. For no calculations could ever take into account the transference of stresses through organic components and pegs and arrive at anything better than a guess.

The fabric

So much for the frame. The rest of the contract drawings are the standard details, half-inches and full-sizes that go with any properly designed project. Standard details must usually be adapted for each job. To begin with the roof, fibreboard is laid *on top* of the rafters, and then 2 inch by 2 inch battens over each rafter throughout its whole length. The battens both secure the fibreboard and can greatly strengthen weakened or bowed rafters if the two are screwed together. Glass-fibre and felt complete the insulation before fixing the tiling battens and laying the tiles (Figure 144).

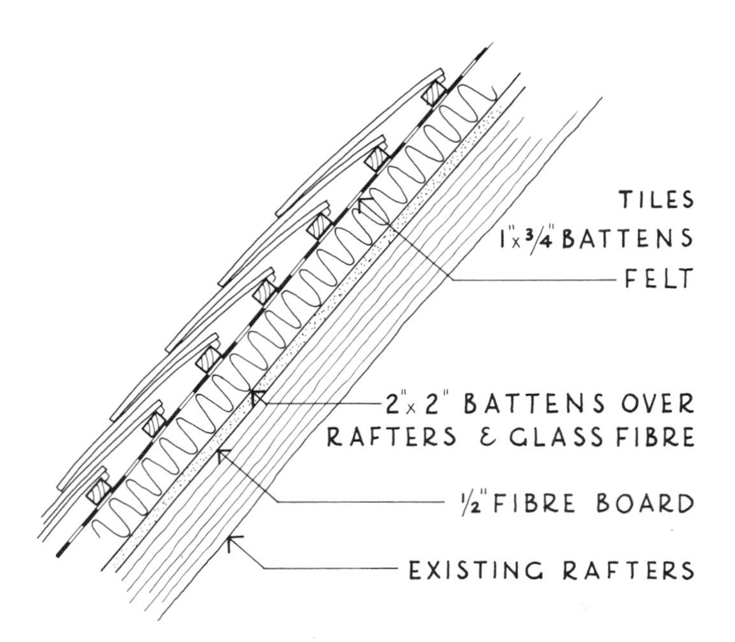

Figure 144 *Insulated roof covering*

But this thicker roof covering, from the upper face of the rafters to the underside of the tiles, must be protected at the gables. Bargeboards may be adjusted or reintroduced or possibly introduced for the first time. Also the purlins and wall-plates may have to be lengthened to extend the verge overhang, which should never be less than 9 inches. Each building presents its own problems, possible solutions to some of which are illustrated in Figure 145(a), (b) and (c).

With single or rafter roofs it may be necessary to extend the collar-plate as well as the wall-plates, and set another rafter couple. Bargeboards will again be necessary to close the extra roof thickness. Methods of extending the horizontal members depend on the conditions of each roof, which are seldom the same. But when the roof covering is off, there is little difficulty in scarfing on new ends, even though the tie-beam (or collar for collar-plates) will have to be temporarily lifted off and the wall-plate dovetails reproduced (Figure 146).

The eaves, a most effective source of permanent ventilation, generally found stuffed with loose bricks, rags or newspaper, must also be filled. A simple triangular arrangement of boards between each pair of rafters, fixed

Figure 145 *Methods of extending purlins at verges*

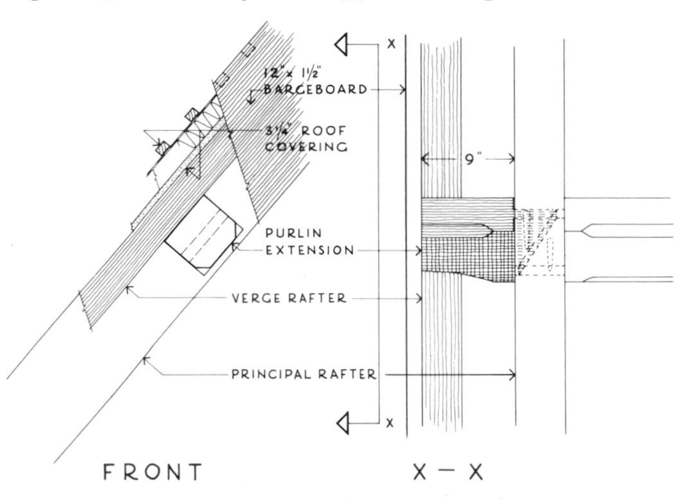

(a) *Tenoned-purlin – including bargeboard*

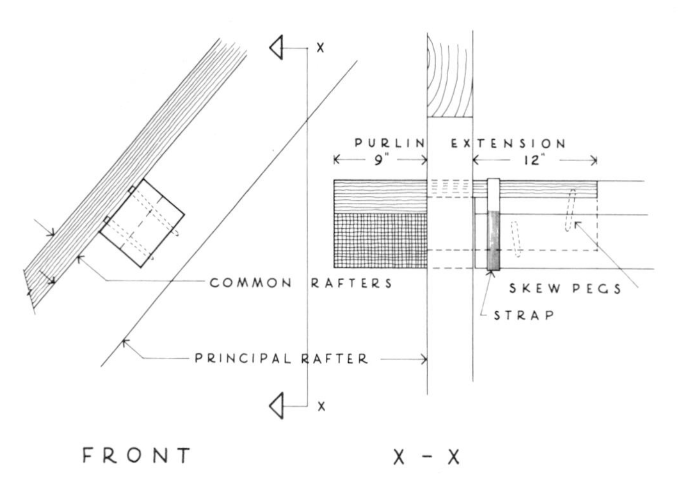

(b) *Trenched-purlin*

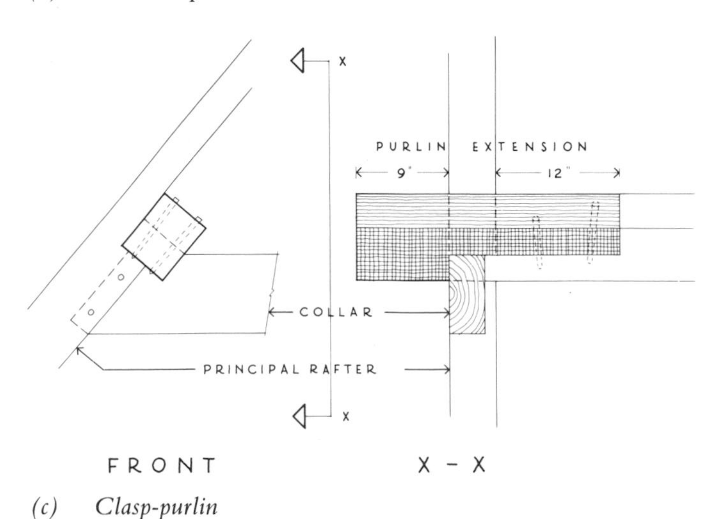

(c) *Clasp-purlin*

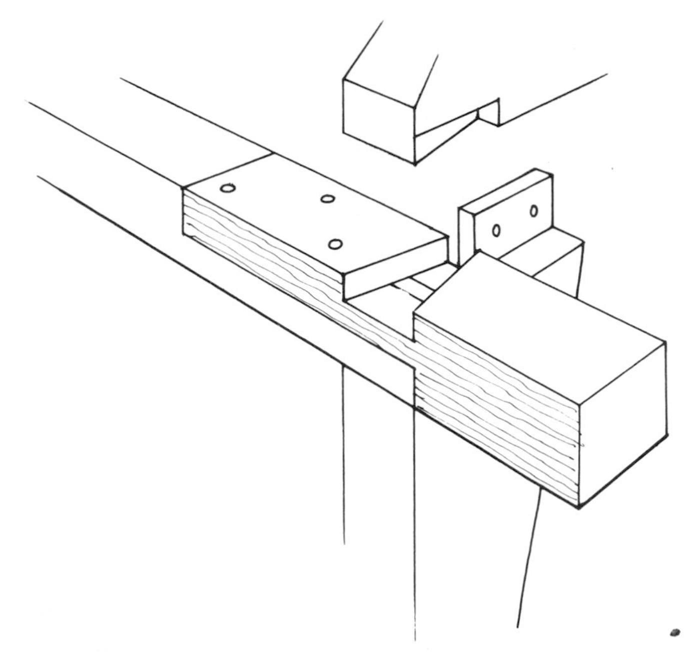

Figure 146 *Wall-plate extension piece – collar-plate may be similarly extended*

to the wall-plate and packed with glass-fibre, is our normal detail. But sometimes the vertical board must be a continuous one notched beneath each rafter (Figure 147(a), (b) and (c)). Both for verges and eaves the soffit board must of course be an external quality board such as plywood or blockboard (Figure 148).

Storey-height pentroofs or weatherings (see Lane House, Chapter 11) are often worth replacing or adding to a house. Whether they formerly existed on a particular building is irrelevant. For most of them they were probably a later improvement, as they still would be. Tile-covered weatherings, projecting 1 foot 6 inches to 2 feet with oak brackets, wall-plate and 3 inch by 2 inch rafters laid flat are shown on Figure 149(a) and (b).

It is necessary to use lead flashing, not only for weatherings and pentroofs in general, but for sloping roof to wall abutments and sometimes for the cheeks of dormer windows. The cover flashing should be turned into a continuous groove not less than two inches by half-an-inch deep, cut into timbers and panels, the rendering of the latter being mastered by a standard metal plaster-bead. A hardwood weathered fillet is let into the groove and masters the turned-in edge of the lead. Fear of acid corrosion is unfounded as the area of contact is too small, and there is no condensation as usually compounds the problem when lead is laid directly on to boards.[3]

Figure 147 *Methods of filling eaves*

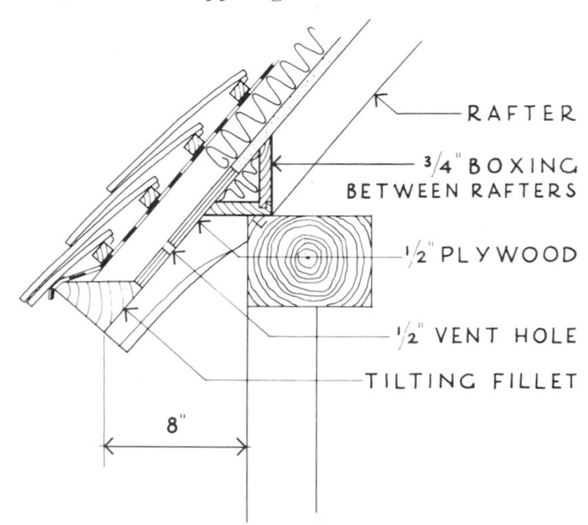

(a) Usual condition with a rafters bird's mouthed to plate

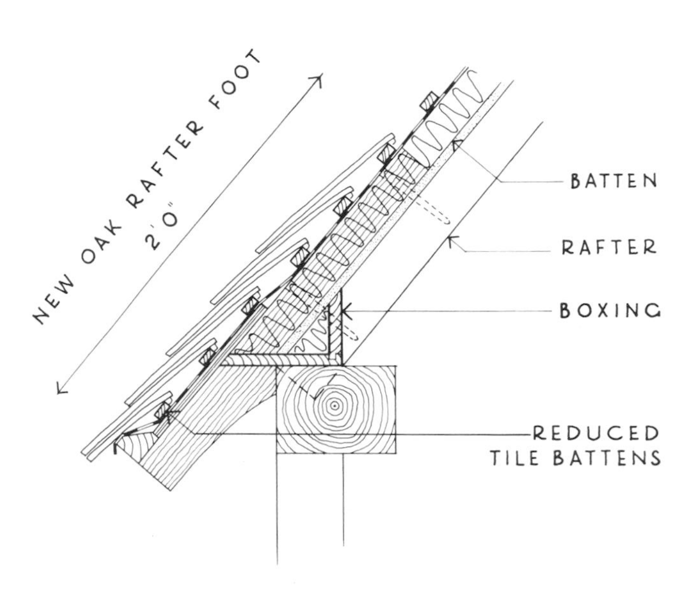

(b) Where rafters terminate at plate as for originally thatched roof or sprocket rafters

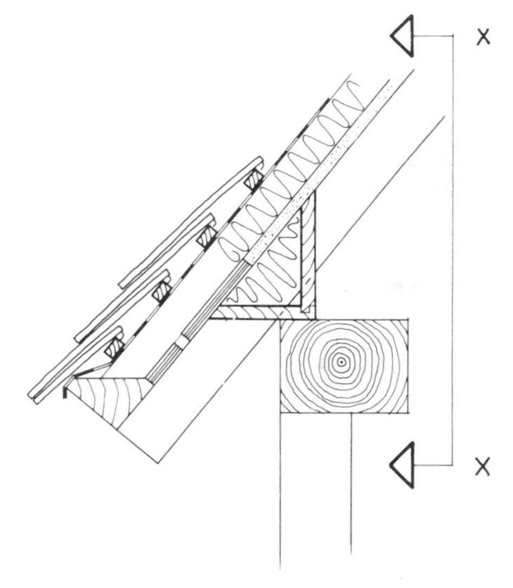

(c) Where rafters ride over plate – eaves filler upstand scribed to rafters, baseboards in short lengths between rafters

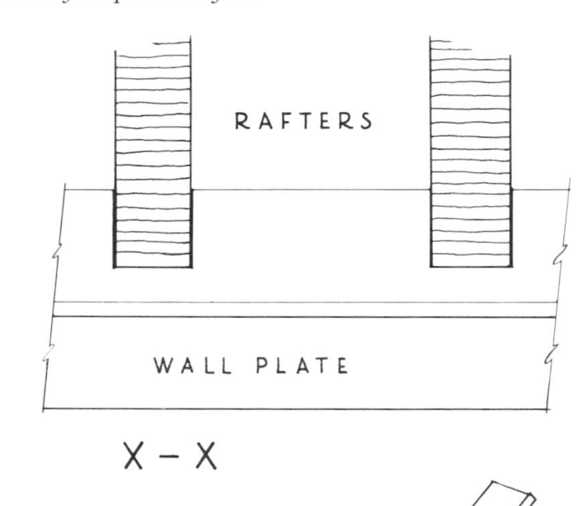

Figure 148 *Adjustable gutter bracket for screwing to rafter battens*

Foundations for timber-framed buildings, unlike all others, may be laid last, and the plinth built under the sill-beam so that the fit is exact. It certainly cannot be built until the sill-beam, almost invariably decayed if not completely buried in the risen ground and rotted away, has been repaired or replaced by offering it up to the tenons of the posts and wall-frames. These may also have rotted at their feet, and the decision must be made whether to cut them and form new tenons higher up or to repair the timbers. Generally the former is preferable, certainly economically.

Figure 149 *Baston Hall, near Worcester*
(a) The house after adding weatherings

(b) Detail of weatherings

In taking out the plinth and footings, and preserving their stone for the reconstructed plinth (below ground this may as well be brick), it is worth looking for signs of posts or post-holes of earlier buildings. Similarly, if the ground floor must be taken up, earlier occupation levels may be found if the work is done carefully.

Generally the subsoil is dry, and a new damp-proof membrane, often demanded by the local authority, can result in a damp floor, probably for the first time in the building's history! Condensation is the cause, for rising damp from a site that has had the protection, weight and warmth of a building on it for centuries is a practical impossibility.

Infill panels are one of the trickiest problems of restoration, perhaps only finally to be solved by reverting to traditional wattle-and-daub. Even with this, however, the shrinkage of green oak, of up to half-an-inch in a 7 inch stud, must also have concerned the first builders. Yet there is never a sign of the habitual crack between plaster and timber when an original undamaged panel is re-exposed. This can only be because in less hurried days the plasterer came back with his bucket of lime to 'make good', perhaps not only at the end of the 'maintenance period' but each year; or more likely the owner did the job, just as may still be seen in the whitewash countries of the Mediterranean, or even in any coastal village where traditional cottages are still lived in by traditional owners.

But both wattle-and-daub and annual maintenance are now impractical. There are many substitutes. Brick has already been condemned (page 105). Expanded metal or metal lath rendered both sides is perhaps the nearest approach to wattle-and-daub, but with modern hard renders of sand and cement, instead of the straw-bonded earth and lime, cracking is inevitable and there is no insulation or means of dealing with timber shrinkage. Polystyrene or any of the increasing number of insulation materials, some of which may be sprayed on to a board nailed on to the framework, may improve the U-value, but these must also be rendered and they too are inert. Moreover, the quilt-like bulges, the ugliest of all the errors of modern restoration, are as endemic in the use of such materials as in rendered brick.

The sandwich panel with double subframe alone overcomes the problem of shrinkage and provides fair insulation (Figure 150). Each frame consists of 1 inch by $1\frac{1}{2}$ inch softwood pieces half-lapped at the corners but not nailed. If the structural oak is new it can be grooved on the bench and the pieces of the subframe driven into the groove to a depth of half an inch, after erection. Where they must be fitted to the existing wall timbers, grooving is impracticable, so that the tightest possible fit must be obtained by widening and squaring the wattle groove with an inch chisel and fitting the subframe as tightly as possible to the oak, so that no hair crack will open between them. This

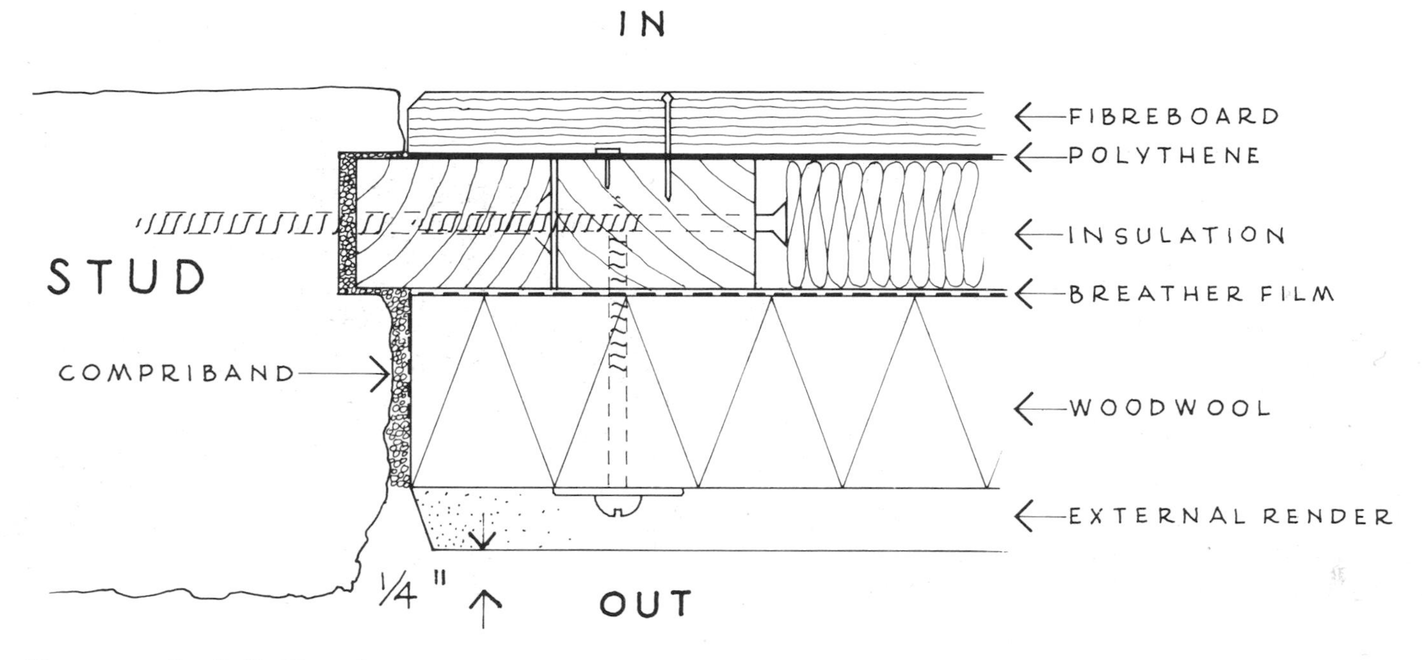

Figure 150 *Detail of infill panel*

subframe must be fixed with galvanized screws at every 6 inches. The next subframe, again in its four separate pieces, is then screwed to this frame, but the screws are left to project a quarter inch all round, so that when the inevitable gap opens it will be between the two subframes, and so will not be seen.

The external skin is $1\frac{1}{2}$ inch or 1 inch woodwool depending on the thickness of the studs and rails. The edges are dipped in cement slurry to seal them and each panel is wrapped at the back and edges in 500 gauge polythene breather film. The sides of the structural frame are lined with an elastic sealant (Compriband)[4] and the woodwool panel slipped in. Its face must be set back behind the oak surface to allow for the traditional recess of the finished surface behind the face of the timber. And the woodwool is then screwed, with washers, to the second subframe. Finally, it is rendered in two coats, never proud of the frame, nor covering or even touching any of the oak. Its edge should be a neat bevel, and the joint between the oak and the woodwool is taken up by the Compriband left as a thin black line of varying thickness according to the irregularities of the old timbers, but perfectly even and about one-eighth of an inch thick where the timbers are new. In practice this has been unexpectedly difficult to achieve, the plasterer invariably plastering over the Compriband, if not parts of the timber as well. Only in the latter case must he be told to put it right, though correction of the former might also be necessary at the end of the maintenance period. The internal skin may be fibreboard or plasterboard against a sheet of 500 gauge polythene and the cavity filled with a building slab insulation.

As a final point about the external structure, weatherboarding must sooner or later become acceptable for casing timber-framed buildings. This is no revolutionary suggestion. The seventeenth-century houses of New England are all clapboarded and their framework, despite a far harder climate than ours, is today in much better condition. It is a paradox that in this country the buildings of our drier eastern and southern counties are often weatherboarded or tile-hung, while in the wetter midlands and west the frames are generally exposed. Or if they are covered it is much more often with plaster or brick, both moisture holding, than weatherboarding. And the frame rots faster than it does when completely exposed. However, weatherboarding (especially oak or elm) is traditional for barns, and it could do nothing but good on houses. Protected outside and exposed inside, the timbers would be dry and ventilated and insulation could be as efficient as in any structure. A great deal of money would also be saved in reducing, though not eliminating, structural repairs. But these would be minimal as external battening greatly strengthens the framework. The main saving, however, would be in the panels. Having had competitive tenders for alternative schemes for the same building, it seems that the saving would be about one-third. In the event, the full repair and panel scheme was chosen, partly because of resistance by the local authority. Their conservation officer said that he would advise his council to refuse an application for listed building consent for weatherboarding as it was 'not indigenous to the region', while for full repairs no application would even be necessary. He overlooked the many barns in the region, but they are disappearing so fast that his comment may very soon be true.

Secondary elements

There remain the secondary elements. There is no purpose, except again in museums, in trying to regain the original design down to a building's last details. There are too many unknowns, and the original house could not in any case today be lived in. But it is no easy task to design doors, windows, staircases and so on that settle in as if they had always been there. Principle is far more important than superficial appearance. 'Plant-on' windows, for instance, secured to the outside of the frame, as were the great Elizabethan ranges, follow the same principle and have the advantage that they can be rectangular even though the frame of the building is not (Figure 151). In appearance they will generally be much plainer than their forebears, without complicated mouldings or leaded lights and certainly without blown glass. And they will be fixed with coach-screws or bolts rather than pegs. Even full reproduction may sometimes be justified, as at the Ancient High House in Stafford (Chapter 11).

In other respects, the houses of the sixteenth and seventeenth centuries had so many straightforward and honest pieces of design that to disregard them would be a double loss, both for history and for modern use. Such details are window shutters, door latches and bolts, stair handrails and the staircases themselves. But it must be remembered, especially in the design of staircases, that it is their material and construction, as much as decoration and form, that give them their superb quality. They cannot be reproduced except for the simplest examples, such as the solid-tread 'vice' or spiral stair (see Lane House, Chapter 11). But the stair that makes no attempt at copying, using different materials such as imported woods or softwood, may, like reinforced concrete for a fireplace lintel, be beautiful in an ancient context.

Central heating is often said to cause such shrinkage of the timber and joints that structural stability is threatened. If so the house must surely have been so damp beforehand that it was under even greater threat from that source. The main problem with central heating is of course the pipes. The sandwich floor may be of help in this. It consists of fibreboard laid over the beams and joists as the ceiling for the floor below, then 2 in by 2 in battens, firred if necessary to regain a level floor, and finally the old floorboards relaid, or new ones if necessary, but preferably wide. This construction gives space both for small-bore pipes and electric wiring. Never should timbers be drilled for pipes. A box-skirting may also be a duct. Incidentally, a board planted flat is better as an edging to a floor than a vertical skirting, as it can be scribed to the wall's irregularities.

All of this, of course, is new design, as indeed is everything that goes on to or into the frame and the framed structure. Only that is sacrosanct. But if the design principles of the Bauhaus[5] or the aphorism of Mies van de Rohe, 'the less, the more' ever applied, it is in timber-framed buildings. Perhaps the new vogue for wood-burning stoves might, if they are put in the place of the central hearth, help to restore life as lived again in the open hall, as indeed William Morris foretold.[6]

Execution

The contractor must be interviewed before he is accepted, preferably before being invited to tender. He must be prepared to appoint a carpenter as foreman and to keep him on the job from beginning to end. And the carpenter must also be interviewed. If, as in one instance already noted, the carpenter is also the owner of the firm, the standard of work will be the highest and the cost lowest. Unfortunately the capacity of this firm is precisely one building at a time. But the contractor typical of county or slightly smaller towns is generally perfectly able to undertake restoration, and there is no need to look for the specialist.

There will be variations, however thorough the survey and design. Nevertheless, with bills of quantities drawn up by a quantity surveyor fully informed by the architect, a fixed-price contract is both practicable and necessary, and can be obtained either by tender or negotiation.[7] The only latitude is a 10 per cent contigency sum in the contract but it should be possible to deduct most of this at the end of the day.

With the contractor's taking possession his first job is to screw on to every timber a 4 inch square plywood plaque with the timber number scribed, not painted, on to it. Next, if all the timber can be ordered immediately so much the better, as the longer it stays outside under all weathers the more stable it will be when the framing

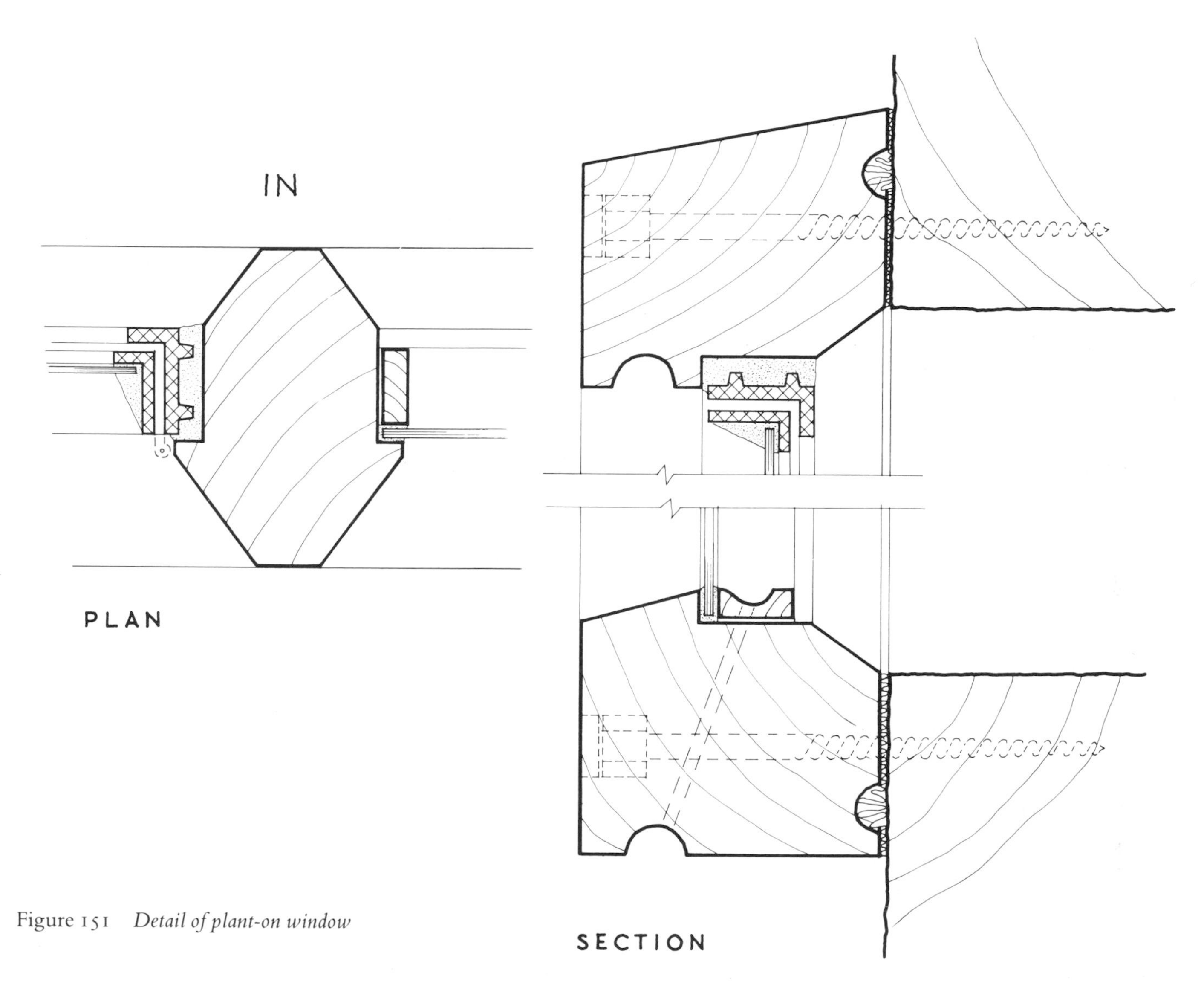

Figure 151 *Detail of plant-on window*

is ready for it. A visit to the merchant's yard with the contractor and foreman will save time, money and trouble (even a forest expedition may be necessary). For selection at this stage must be little short of ruthless. Though the architect must have all the answers by dint of knowing the building and what is to be done, it is the contractor, the foreman and men on the job who will decide how.

Dismantling begins by hammering out pegs, never drilling them out, and as the timbers are lifted down their size and weight is appreciated probably for the first time. Carpenters who have never had the experience of a restoration may start with a degree of scepticism. However, with the first sight of a post-head joint, when the tie-beam is lifted off and the perfection of the cutting of the tenons by his predecessors several hundred years ago can be seen, when the hardness and strength of the oak is first discovered and the first of the new timbers is worked, then scepticism changes into respect and soon into self-respect when he finds that his skill is the equal of theirs. And this is not only in making accurate joints, but in handling heavy members in restricted spaces.

It is essential to take time rather than to get the job done as quickly as possible. The boss is the one most likely to object to apparent lack of progress and, through no willing fault of the carpenter, a poor piece of workmanship may result. The architect's instruction to do it again, properly, is likely to be considerably more expensive for the contractor than would have been giving his men more time in the first place.

Absolute firmness is the architect's first need, but he must also defer to the contractor's knowledge on his own subject, and learn from his carpenters. The puzzles and vicissitudes that are met with are never the same twice, but a fair range of them are given in the case studies of restoration contracts which follow.

PART THREE

CASE STUDIES

CHAPTER SEVEN

SHELL MANOR

Himbleton, Droitwich

Shell Manor was first surveyed in 1958 on account of its interest as a historic building. The commission to restore it came in 1960. Its plan is that of the medieval hall and double cross-wing, the H-plan (Figure 152). The only surviving medieval part, however, is the solar. The original hall and service wing were entirely rebuilt in the sixteenth or seventeenth century, but the plan was exactly followed in the reconstruction, even to the passage entrance with its porch and little chamber over, originally no doubt gabled.

The site layout (Figure 153(a)) is typical of farms of old Worcestershire, with the house facing the cattle court, or fold, surrounded by barns, cowhouse and cart-shed. The lane in front of the fold leads eastward down to one of the numerous fords which cross Bow Brook on its way to the Avon and thence Severn. Across the lane there is a hipped-roof open shed or 'hovel' of the eighteenth century. Shell had been a considerable farm until it was bought by the Bearcrofts at the beginning of the century; they put in a series of tenant farmers, having removed from the house all its internal panelling to their family home, Mere Hall, a few miles away.

Figure 152 *Shell Manor House before restoration*

The seventeenth-century house-plan had also been extensively altered (Figure 153(b) and (c)). First the timber-framed extension of the service wing was built at about 1700, the brick stables were added in the eighteenth century and the kitchen in the nineteenth century. There had also been more recent rearrangements internally. The front door had been blocked. The exact date of this was recorded in a message chalked on the inside of the door and discovered when we came to unblock it. It read, 'I, John Gittus, carpenter, bricked up this door on 25th June 1874 to prevent the wind blowing up the housemaid's petticoats.' The new doorway was made on the east side of the house, entering a stair hall with typical Victorian staircase. In the 'house-place', occupying the site of the original hall, the stair was no doubt in the corner between the rear wall with its fireplace and the side wall of the solar. Indeed, the original hall must have had a stair in this position to give access to the upper floor of the solar, the doorway still being marked by a four-centre arched lintel. In the seventeenth-century house the bedrooms no doubt intercommunicated via the same doorway. The attics of this house would have been reached by ladders, but after the Victorian alterations there was no access to them except from the stables. One had to scramble through no less than six roof trusses at varying levels to reach the roof of the solar.

The ground-floor rooms of the service wing extension had retained their seventeenth-century materials – lias stone slabs for the floors and exposed timbers in the walls, partitions and ceilings. The larder in particular, with its slate settlas and walls, whitewashed over timber and panels alike, was well worth preserving in the same state.

The unique part of the house, however, was the solar wing of three bays with the main room, at each floor, occupying the two front bays (Figure 154(a)–(d)). The ground floor was the parlour entered from an antechamber through a wide four-centre arched doorway (Figure 155). The upper floor was the great chamber with its antechamber communicating in the same way, but the door from the hall entered the great chamber directly. There was another door in the opposite, external wall, formerly approached by an outside stair, and this led into the antechamber.

The structural and decorative design of the entire wing is architectural down to the last detail. The wall timbering is broad and closely spaced; the trusses dividing the wing into its three bays are massive, the open truss having a deep cambered tie-beam and strongly curved knee-braces. In contrast, the arch-braced intermediate trusses, purlins and pairs of wind-braces are all delicate, a unique feature being the series of little Gothic points where the wind-braces meet above the purlines.

Figure 153

(a) Site layout

The common rafters equally spaced at three to each half-bay are 6 inches by 4 inches in section, laid flat and half-lapped to each other at the apex, without a ridge-piece. All the rafters are original, with scalloped feet at the eaves. The purlins, of only half-bay lengths, are tenoned into the principals with splayed abutments of the tenons allowing the joint to be entirely concealed within the thickness of the principal (Figure 156). Such exacting method means that the roof structure was erected bay by bay in strict sequence from upper to lower end.

The quality of oak is outstanding. The owner of the house, when he saw the solar roof structure for the first time, could not believe that its timbers were by far the oldest. They had never been painted and fortunately were protected from Victorian blacking by the inserted ceiling.

This ceiling was carried by longitudinal beams tenoned into the tie-beam of the open truss (II) and dropped into notched-out recesses in tie-beams I and III. In bay 3, inserted joists spanned crosswise, bearing on the wall-plates. At the same time truss II was closed by inserting wattle panels, plastered only on one side. The cramped roof-space became three attic bedrooms, leaving only one panel open in truss II for a truly child-size door (Figure 157).

(b) Ground-floor plan before restoration

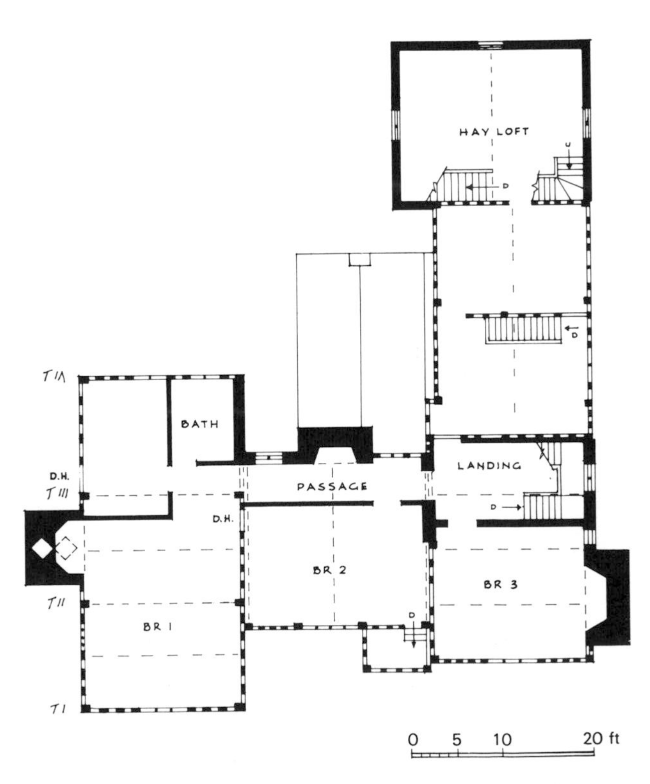

(c) First-floor plan before restoration

Figure 154

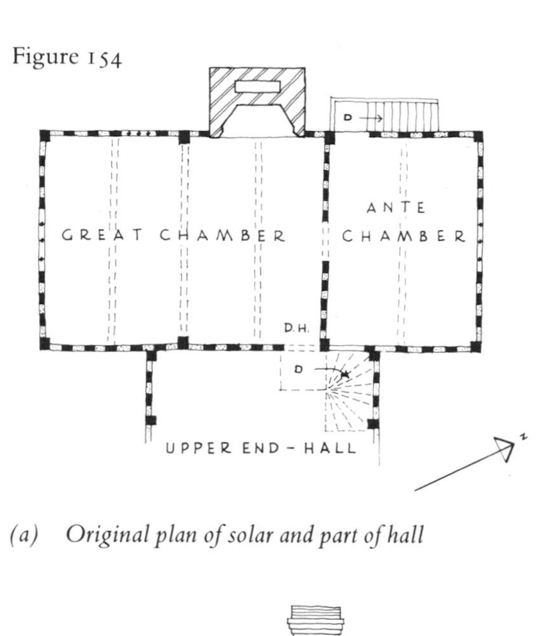

(a) *Original plan of solar and part of hall*

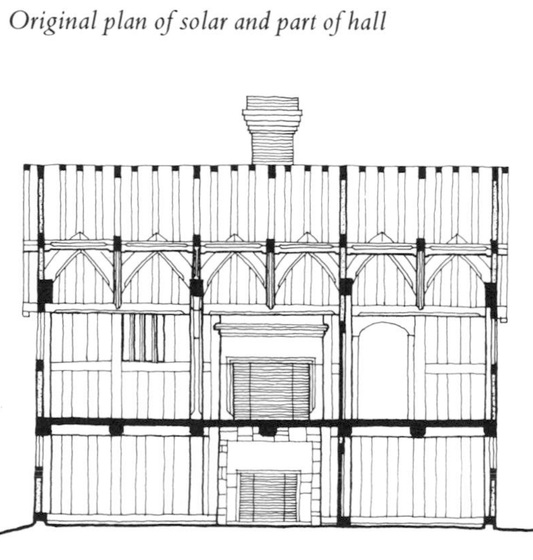

(b) *Long section*

(c) *Cross-sections*

(d) *Interior of great chamber*

Figure 155 *Partition between parlour and antechamber, with medieval door-head and brace and simulated timbering*

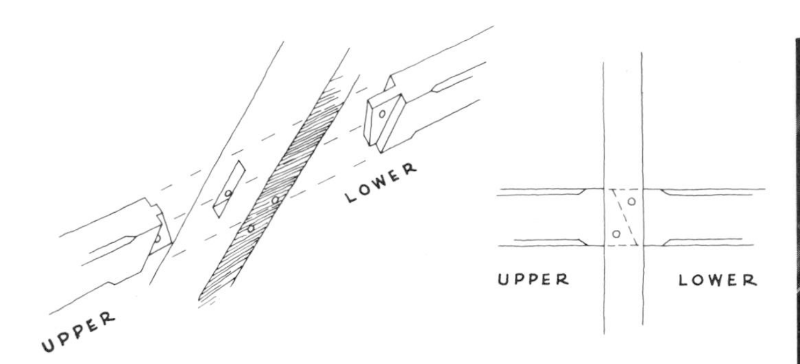

Figure 156 *Purlin-to-principal detail*

Figure 157 *Solar roof showing medieval features and seventeenth century filling of open truss and mortice in tie-beam for inserted ceiling beam*

The original first floor is an example of floor construction in which the joists provide part of the floor surface (Figure 158). It consists of 10 inches by 4 inches joists laid flat and tenoned into the cross-beams including the front and rear external beams. The joists are grooved on their sides, exactly as studs for wattle-and-daub panels and then plastered flush both for the floor and ceiling. All of this however had gone, leaving only the joists. The removal of the modern boards and suspended ceiling showed that all the original wooden surfaces had been painted in typical Elizabethan deep red-earth ochre (Figure 159). This was also found on the posts and studs, including the simulation of close-studding on the cross-partition already mentioned (Figure 155). There is another example of make-believe timbering at Harvington Hall, Kidderminster, and at the White Hart at Newark which has already been mentioned (page 86).

Lastly, the solar chimney stack of Shell is perhaps the most impressive feature of all (Figure 160). Its lower portion is of sandstone, 8 feet square, with the two flues arranged at right angles to the longitudinal axis of the building, the flue of the lower fireplace rising outside the upper fireplace. Such an arrangement saves the builder the need to gather over each flue sideways, and so may be easier to construct. That the stone base is medieval and contemporary with the solar is proved by the fact that the bay which it fills at each floor level never contained wall-framing. It was also designed for smoke outlets quite different from the present great chimney shafts. The form of the original outlets is reflected in the lower portion of the latter. The stack probably terminated just above the eaves with a pyramid top, with little gabled lancets projecting from each angle. The brick shafts are 16 feet high and each face is of three bricks in width instead of the usual two-and-a-half. The cresting, beginning with a dog-tooth course and interlaced with tiles at each of the four oversailing brick courses above, is also much richer than

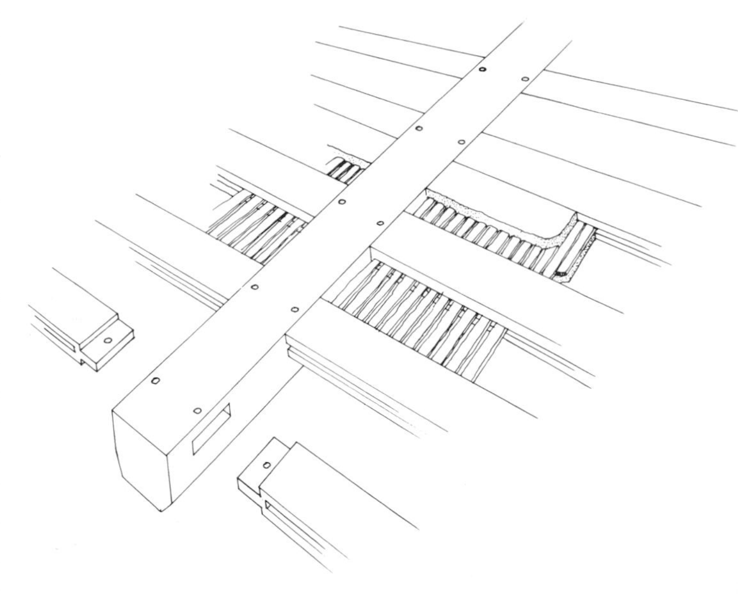

Figure 158 *Detail of first floor of solar*

the more common shaft. Their date is probably nearer to 1500 than 1600, and so nearly contemporary with the only example in the county of a highly decorated chimney, that of Huddington Court about a mile away (Figure 161).

Figure 159 *The floor as exposed from beneath when the modern ceiling, covering both joists and beams, had been removed*

Survey and preliminary repairs

Service wing

The house was surveyed elevation by elevation and room by room by John Giles whose sketches are more vivid than words or photographs.

The illustration (Figure 162) shows his portrayal of the front of the service wing. The lias stone plinth, about 4 feet 6 inches high, had practically collapsed. The brick buttresses had merely compressed the soft ground that runs the length of the whole front elevation, so helping to pull the plinth still further over. The original sill-beam had been severed in the last century to insert a larger window, and the entire frame below the tie-beam had bellied outwards, its loosened joints being filled with bitumen – accelerating decay. A tie-rod, threaded at ceiling level through the post on the right of the window back to the inner floor beam, did neither harm nor good.

The most gratifying operation was allowing the frame to realign itself. It had to be relieved of all dead weight

Figure 160 *The great chimney*

Figure 161 *Chimney at Huddington Court – the plain brickwork between the shafts is later*

Figure 162 *Service wing – sketch showing partial collapse of front framing. From Giles's survey*

Figure 163 *Preparation for easing the front frame back to its former alignment*

by taking out the brick panels. The first floor did not bear on it, its support being a modern stud partition without any connection with the wall-frame. The corner posts could be relied upon to support the tie-beam and truss. The mid-rail or girding-beam of three sections spanning between the intermediate posts and corner posts, was supported on a simple scaffold with three wooden needles greased on their upper surface and sloping slightly downwards towards the inside (Figure 163). The middle needle at the centre of the window was set slightly lower than the others and a large wedge was inserted to make contact with the rail. It was intended to knock this wedge in when the wall-frame had been released from the sill-beam and so ease it back. That, however, was unnecessary as the frame was already straining to resume its proper alignment and did so the moment it was freed – leaving the tie-rod projecting about 9 inches.

Figure 164 *The completed frame of the service wing*

The plinth was rebuilt on a deep enough foundation. The new sill-beam, a prefabricated laminated beam with preformed mortices, was offered up to the waiting tenons and halved and pegged under the lateral sill-beams. Then the plinth was completed by laying its top few courses beneath the new member. The reason for the choice of a laminated timber instead of oak was simply its greater stability. There was little difference in cost (Figure 164).

This was our first use of plant-on windows, the design being subsequently modified only to match local peculiarities in the design of mullions and mouldings. Occasionally, in deference to local advice (though the Elizabethans never bothered and there seems to have been no trouble in the ensuing centuries) the projecting ends of the sills were bevelled off to help any water trapped behind the moulding to escape.

The weathering at girding-beam level was also a first use, but this, in the case of Shell, represents restoration, not innovation, as an old photograph shows weatherings almost exactly as they now are. Lastly, the choice of black-and-white was the owner's, but it requires no apology as the black is merely a dark preservative.

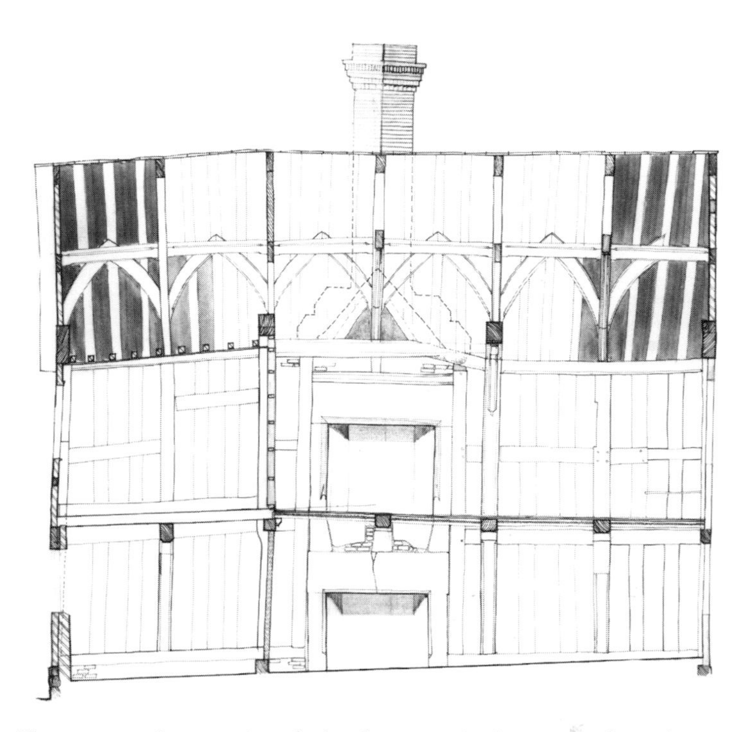

Figure 165 *Long section of solar showing subsidence towards south end*

The solar

The problem of the entire solar wing was that it had subsided no less than 1 foot 6 inches from back to front, taking the chimney with it (Figure 165). A list of contingent defects had also been noted in Giles's survey (Figure 166): 'Considerable sagging between trusses. Tiles have slipped in many places. Thin rails have been inserted in timber frame at points A, apparently during attempts to straighten up the wall. Studs have been cut right through and some short studs above the rails are of a later period. The rail at B has had a new piece spliced into it. The brick nogging is cracking against timber . . .', and so on.

There was no question that preservation, as opposed to restoration, would be the only practicable course, for while the frame could have been repaired and to some extent realigned, the leaning chimney could only have been taken down and rebuilt upright on new foundations. But since the stack penetrated the frame, its sides being held by full-height studs on either side of the chimney breast, the one operation was dependent on the other. The absence of the wall-frame in the chimney bay had also not only left a weakness in the lateral structure of the solar as a whole, so permitting differential subsidence as between the end bays, but also made certain that if the frame subsided the chimney would go with it, and vice versa. Both therefore had to be restrained and secured as they stood.

Figure 166 *Sketch of solar from Giles's survey. Note depth of verge as one of few instances where original purlins and wall-plates had not been cut back*

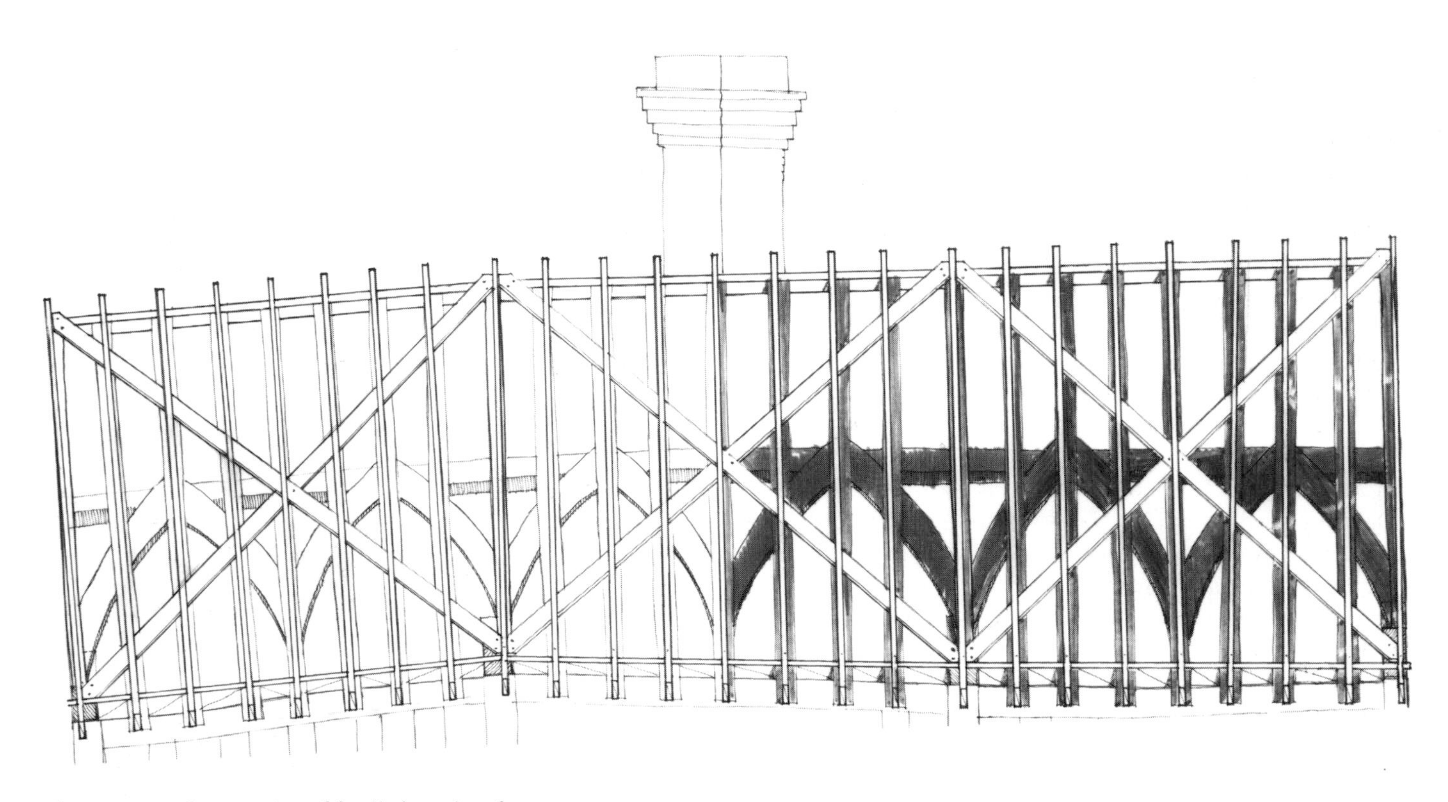

Figure 167 *Long section of finally braced roof*

The chimney was shored and several tons of concrete placed under its foundations along its south wall, underpinning it at intervals of 2 feet to a depth of nearly 6 feet. This, as well as the work on the front wall of the service wing, was done by John Giles with his own hands, helped by the owner's gardener and the rest of the office when needed.

The timber structure, on the other hand, had to wait nearly twelve months while a complete scheme was prepared, permissions obtained and a contract let. Then the framework and roof structure could be tackled. First, the front gable, sinking not only because of the softness of the ground but also because the sill-beam was entirely encased in cement render and had practically rotted away, had to be stabilized. New foundations from the chimney stack round to the inner end of the east wall of bay 1 were laid, the whole bay being propped by needling-through beneath the girding-beam. This also permitted new studs to be inserted, and with the sill-beam set 9 inches above their original level, the rotten feet of the posts and of the studs to be preserved were cut off and new tenons formed to engage them.

The plinth was rebuilt beneath the repaired frame, a method adopted in other cases by preference, as it assures exact conformity of the plinth with sill-beams. This time the replacement sill-beams were oak, together with all other repairs and renewals.

Next, the fractured wall-plate of bay 2 above the fireplace lintel had to be reinforced by means of a new 9 inch by $2\frac{1}{2}$ inch oak plank coach-screwed to the wall-plate wherever it happened to coincide, and picking up the two studs, one on each side of the chimney breast. In addition a $\frac{1}{2}$ inch diameter steel rod was threaded through the foot of each of the principal rafters above the wall-plate to stiffen, with the plank, the wall-head and prevent further differential movement of the two trusses.

The whole roof structure was also braced. First the joints between the principals and purlins were tightened by gently tensioning wire ropes, drawing the trusses together. Then the whole structure was anchored back to the ground by further wire ropes from truss IV to a point about 25 yards from the north gable. These too were tensioned so that while the original structure was thus stressed, without however forcing it to move more than fractionally, diagonal braces were coach-screwed to the existing rafters and halved across each other at their intersection in each bay (Figure 167). Over these, new 4 inch by 2 inch rafters were set and notched over the braces.

Lastly, the wall-frames were strengthened by means of 2 inch by 1 inch horizontal battens nailed 2 feet apart to the inner face of the studs. For the loss of the half-timber appearance internally, many more original timbers were saved than would otherwise have been possible.

During these operations the evidence of an original diagonal-mullioned window was found, two panels to the left of the fireplace in the upper chamber. This window was restored. For the gable wall of both floors ovolo-mould mullion and transome windows were fitted into the structural frame.

That the owner insisted on a new ceiling at purlin level and the covering-up in plasterboard of the rafters is the most unfortunate part of Shell's restoration (Figure 168). Dust, cobwebs and heat loss were the reasons for his insistence. With modern insulating materials the last is not arguable, but he would not be convinced. The other two have to be set against the unfailing delight of being able to see a medieval roof.

Figure 168 *Interior of upper chamber as completed*

Two further points concerning the solar: first, the first floor had to be substantially firred up over the existing joists to counteract the steep slope from back to front. Second, the ceiling of the ground floor parlour was plastered flush with the joists as originally and their red paint retained. Needless to say, the false studs in the parlour cross-frame were not only preserved but also protected by applying a sheet of glass.

House-place and service wing

Except for the gable frame of the service wing, already noted, no such drastic reconstruction as that of the solar was required in the rest of the house. The two other great chimneys were merely repointed (Figure 169). That these have three flues for only two fireplaces is not uncommon. Generally the larger fire was given the benefit of two flues, on the perfectly correct principle that the larger the flue the more smoke it would take. But perhaps architectural pride also played its part. The eight-pointed plan of the shafts is sometimes looked on as earlier than the diagonally placed square shafts. Not only Shell disproves this. The two designs are contemporary and when they appear on the same house, it is the former that marks the more important rooms.

The rear service wing was not altered on the garden (east) side; even the false timbering of the brick-built stable block was retained (Figure 170). However, on the other side the Victorian kitchen was demolished revealing some timbers which had decayed as a result of the valley gutter and had to be replaced, and a glazed lean-to was constructed, returning along the back of the chimney of

Figure 169 *The kitchen chimney, with three flues for only two fireplaces*

Figure 170 *The rear service wing with false 'black-and-white' on the brick-built stable block to right of the diagonal brace*

the house-place to provide the boiler room whose flue could go straight into the chimney (Figure 171).

The house-place was also not altered except for unblocking its front door. Its huge fireplace was also reopened and its cracked lintel repointed (Figure 172). It is worth mentioning that stone fireplace lintels almost invariably have a vertical fracture near the middle, but very seldom is there any sign of movement, either in the brickwork or stone above the lintel or in the jambs. There is sometimes a relieving arch or, in this instance, a beam or bonding timber a few courses over the lintel. This and the sheer weight and volume of material used in the construction of the fireplace and stack are a pretty reliable assurance that nothing will collapse. Nevertheless, a simple precaution is to drill two holes diagonally from the soffit of the lintel and insert $\frac{1}{2}$ inch diameter stainless steel rods 12 inches long, having filled the holes with an epoxy resin adhesive (Figures 173 and 174).

The whole contract was carried out by C. & L. Walker of Kidderminster between May 1961 and October 1962. Their estimated price on plans and specification only was £4397 14s. 9d. The final cost was £11,500, the extras resulting from client's changes of instructions as well as additional structural repairs. These were anticipated by a

Figure 171 *The backyard after demolition of the Victorian kitchen and construction of glazed lean-to and boiler room alongside chimney*

Figure 172 *The house-place as completed*

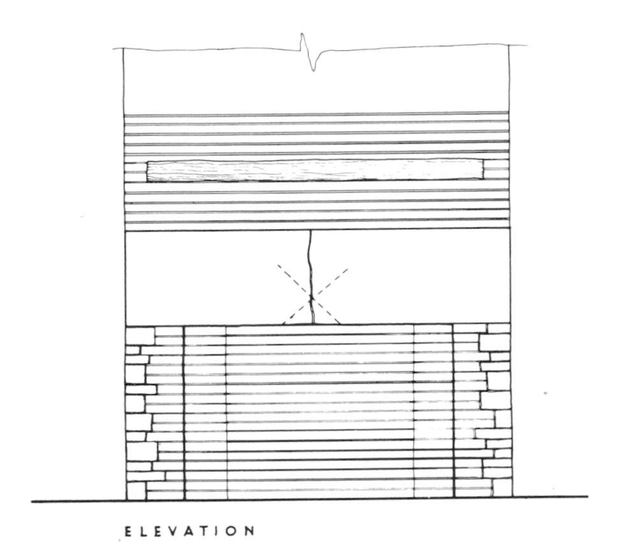

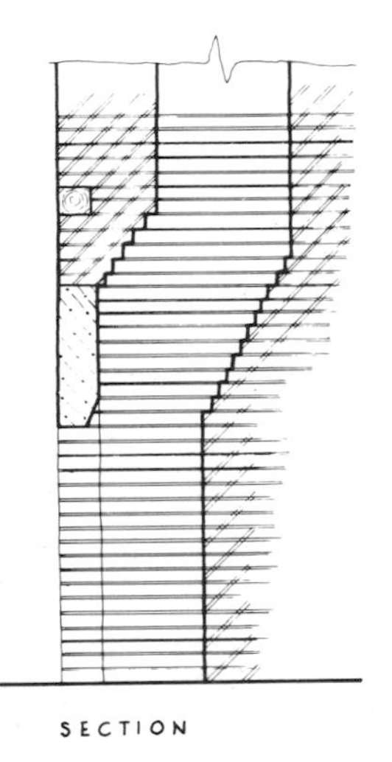

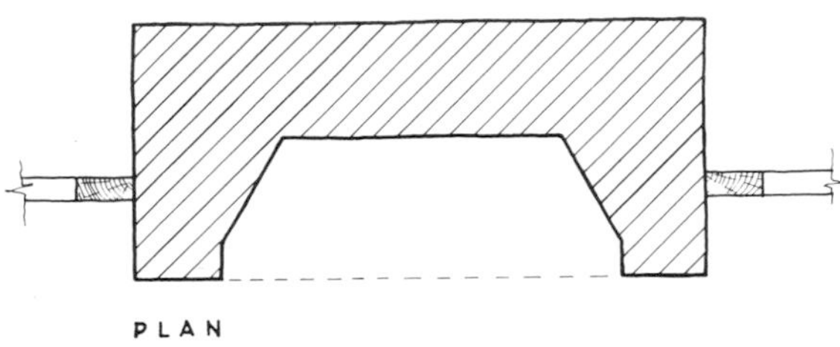

Figure 173 *Typical fireplace with stone lintel and method of repair*

note in the specification: 'Where demolition is required which may uncover structural defects these will be dealt with as extras and will be paid for under daywork rates or as otherwise agreed.' The fact nevertheless remains that there had been no adequate preparation; this was a lesson for subsequent commissions, though still not always possible to apply.

A grant was sought without success from the Historic Buildings Council. Since the Local Authorities' (Ancient Buildings) Act 1964 was still in the future, there was no contribution for structural repairs from the County or District Council either. However, under the Housing Act the owner did receive from the latter, as an Improvement Grant, the princely sum of £647 0s 0d.

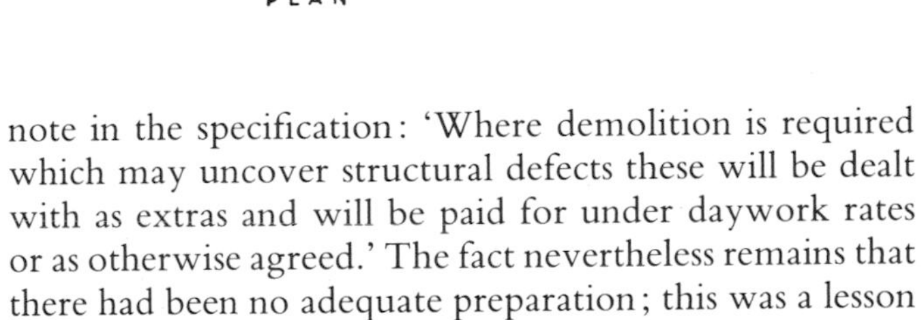

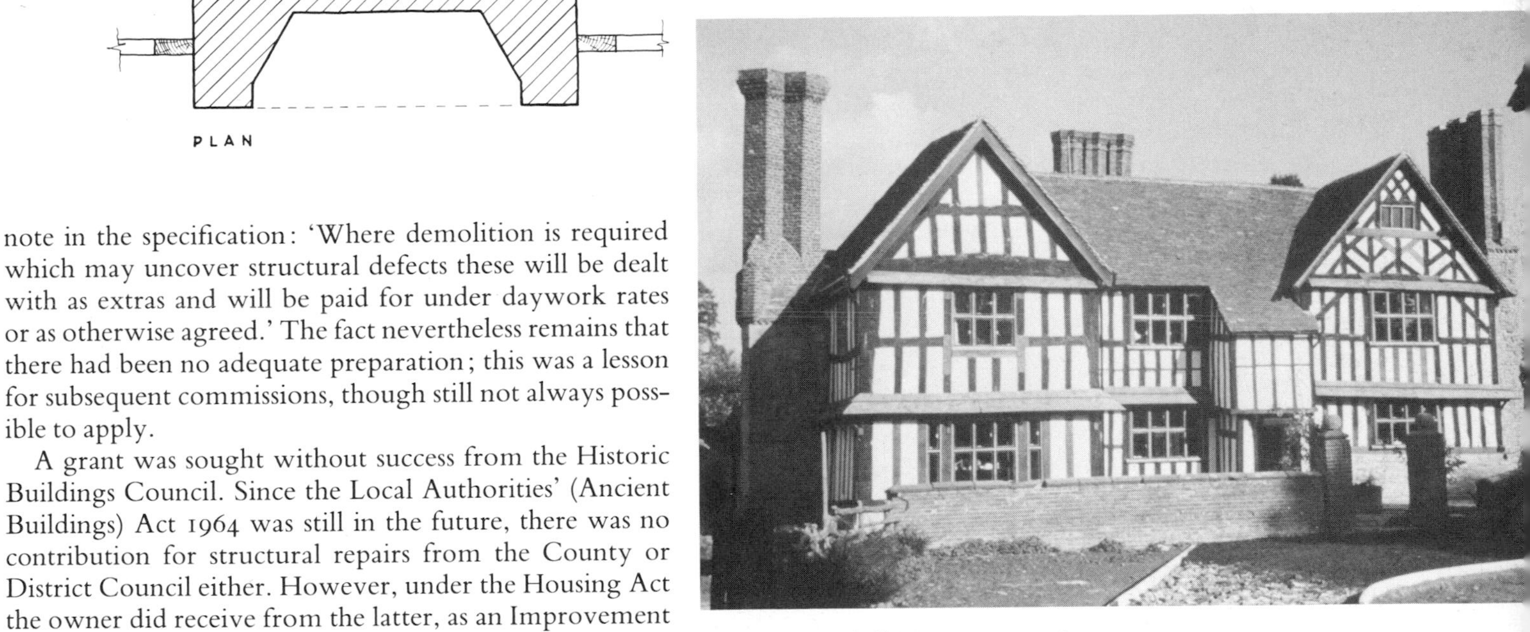

Figure 174 *Shell Manor as restored*

CHEYLESMORE MANOR HOUSE

Coventry

Introduction

The gatehouse and two cross-wings, all that had survived of Cheylesmore Manor House, were restored and converted into the City Registrar's marriage rooms by Coventry Corporation between September 1966 and December 1968 under a contract with the Nuneaton firm of Parker & Morewood. It was the first of several contracts with that firm for restoring Coventry's ancient buildings. It was also the largest, calling for a bill of quantities and competitive tenders. The quantity surveyors were Silk and Frazier of Birmingham. The winning tender was £31,757.00. The final account was £37,043.00. The total internal floor area, excluding the carriageway and open porch in the west wing, is 3845 square feet.

The contract should have run in conjunction with another, separate contract under the City Architect for the Registrar's new offices. This, however, was postponed on account of government cuts in local authority spending, and construction did not start until the marriage rooms were complete and in use. The only consequence was that in the meantime the restored building had to provide some office accommodation and a temporary strongroom, the latter inappropriately built as an independent structure within one of its rooms. It required its own foundations, 18 inch thick brick walls and reinforced concrete roof, all constructed *after* the timber-framed walls and first floor had been completed. When eventually the new offices were also built, this was taken down and the room, none the worse for its experience, made into one of several waiting rooms.

The project for the two buildings, ancient and modern, presented the challenge of designing the glazed link that was to be their common entrance hall. Its design was at first a joint undertaking, with somewhat more attempts being made by us than by the City Architect (Figure 175). The delay in his project, however, righted the balance and finally reversed it, as he could still work on it when we had finished. By this time the concept had also somewhat changed, the entrance hall becoming a much larger space

Figure 175 *Proposal drawing showing elements of old building and link with new offices, the main front and sketch of the interior of the link*

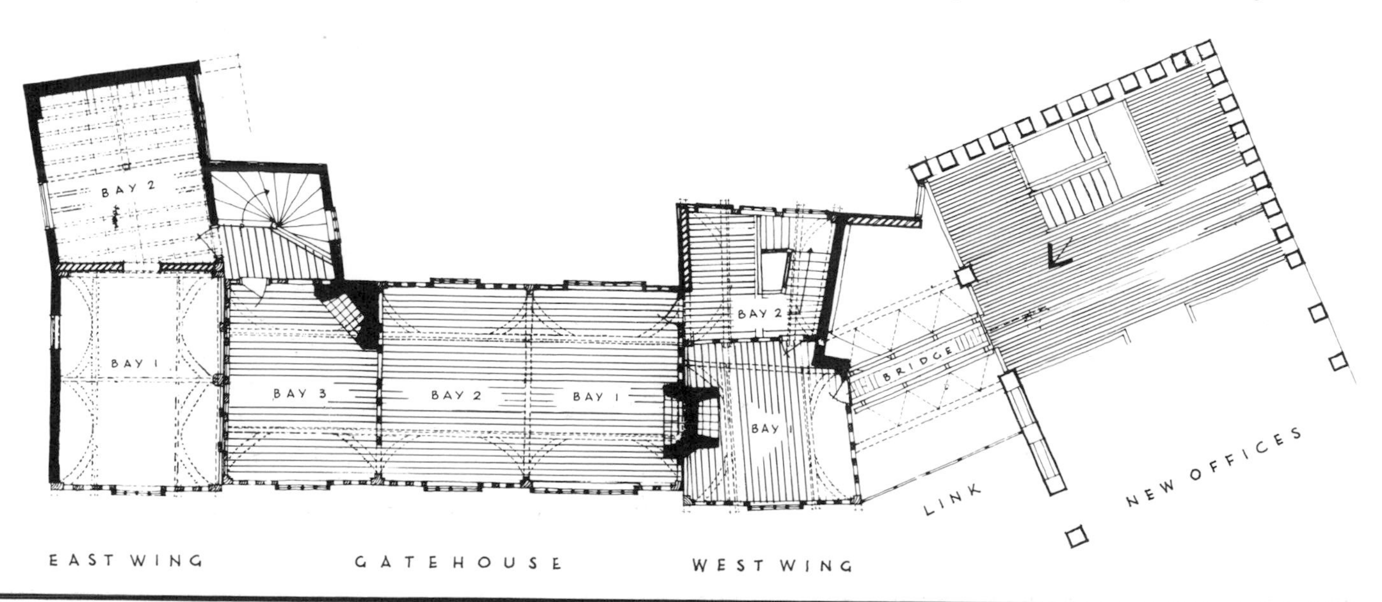

than was at first envisaged; it therefore had to be planned as part of the new building, with only secondary access to the old, the marriage parties making use of one of several other entrances.

Our preliminary survey was done in December 1964 and January 1965 and presented to the local authority in the following month. It was in the usual form of an illustrated report and included a skeleton specification. It ended:

> The above clauses describe the method of dealing with the structure. Its fundamental purpose is to leave exposed to air and ventilation the whole of the original structure and honestly to distinguish between this and new work and materials. The policy is to retain everything that can be usefully retained and never to introduce facsimiles. Finishes will be the natural throughout. This also helps to distinguish new from old but avoids too stark contrast as happens in the 'black and white' treatment – to be avoided at all costs.

History

Cheylesmore as a royal manor is well documented. Moreover, the Victoria County History for Warwickshire was being revised before and during the two years of the restoration. The opening up of its structure meant that the historian, Margaret Tomlinson,[1] was able to identify at least one important feature previously known only by documentary evidence, while we were able to use her considerable knowledge in interpreting several others which would otherwise have remained a mystery.

According to her, the etymology of the name Cheylesmore is doubtful and can be traced with certainty only to 1250. The park in which the house stood, however, was in existence a hundred years earlier. It lay south of the town and may have extended as far north as the ditch of the castle (Figure 176). Much Park Street was developed in the twelfth century just east of the castle and led to the park; then, as the castle fortifications disappeared, Little Park Street became another route between the town and the park. It was probably a portion of the park that was granted in about 1230 to the newly arrived Grey Friars by Earl Ranulf, recorded among their founders as 'Lord of Cheylesmore'. A document of 1275 mentions the capital messuage of Cheylesmore, the park and the mill. Under a settlement made in 1327–8 by royal licence, Cheylesmore and the service of the prior passed to Queen Isabel, wife of Edward II, for life. Under Edward III, however, the manor on reversion was settled on his eldest

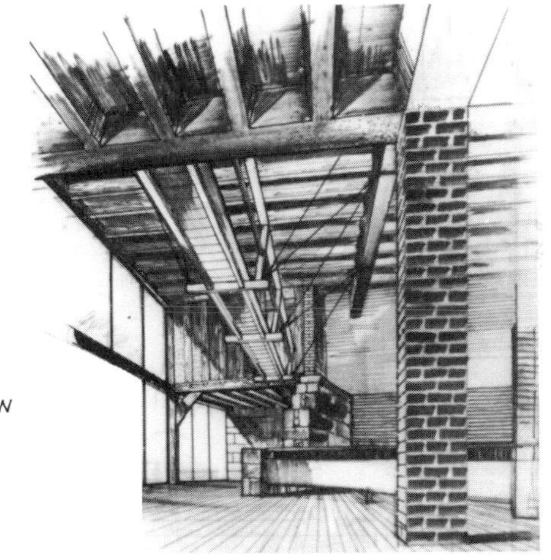

VIEW FROM ARROW – GROUND FLOOR

NORTH ELEVATION

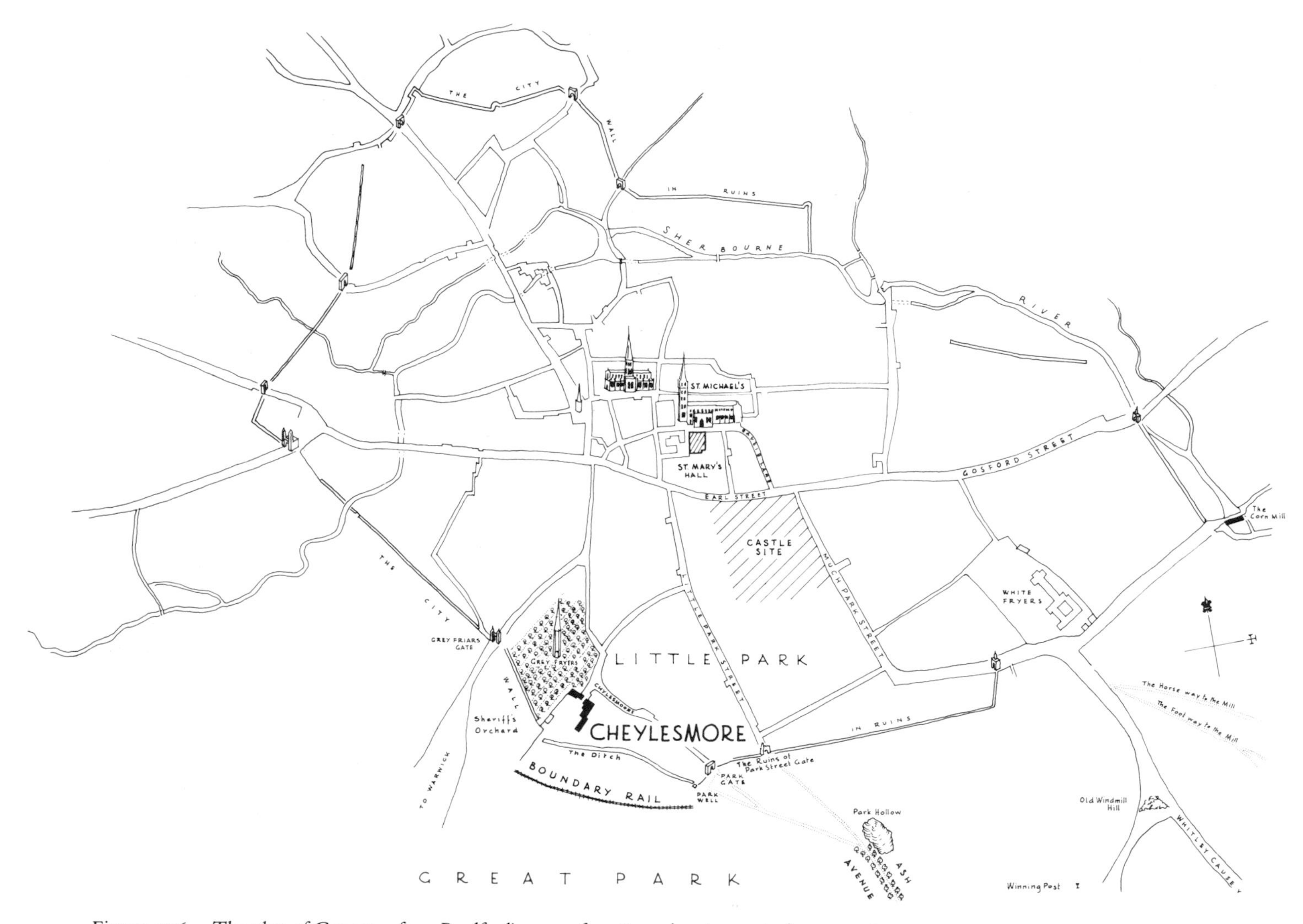

Figure 176 *The plan of Coventry from Bradford's map of 1748–9, showing main features only*

son, the Black Prince, and the queen, his grandmother, and the prince spent a considerable time together at the manor house. It is said the City's motto *Camera principis* owes its origin to this close association. On the death of Isabel in 1358 the prince held his council in the hall of Cheylesmore.

When the city walls were built during the second half of the fourteenth century it seems that the northern boundary of the park was intended to be used for the line of the south wall of the city, but the petition of the mayor and bailiffs in 1385 for a licence to complete the walls was granted only on condition that they enclosed the king's manor of Cheylesmore within the city – with the help of stone from the quarry in Cheylesmore Park. Thus the manor house and its immediate precincts lay within the south-west angle of the wall, and the park skirted its southern boundary.

By this time there were the Great Park and Little Park. According to a mid-seventeenth-century survey the Great Park then contained 480 acres, extending southwards from the city wall for about a mile and was roughly half a mile wide from the Warwick road on the west to the road leading to Witley Common and the common itself on the east. The Little Park (25 acres) lay by the city wall on the north of the Great Park, from which it was separated only by a rail, and was bound on the east by a lane from Coventry to Park Mill. The park was an asset to its owners for it provided timber, stone, water, fish, venison and other game.

The manor remained with the Crown until the nineteenth century. From 1549, however, when Edward VI granted the manor and park to John Dudley, Earl of Warwick, who leased it to the mayor and bailiffs for ninety-nine years at an annual rent of £9 on condition that they allowed pasturage in the park to the poor of the city, there followed a tug-of-war for possession between the city and the Warwicks. In 1609 Henry, later Prince of Wales, claimed the manor on grounds that it was parcel of the unalienable Duchy of Cornwall, but in 1620, on the petition of the mayor and bailiffs, Prince Charles granted them a lease for twenty-one years. A second lease was granted in 1628 to run for eighteen years from 1641. Coventry was on the side of the Commonwealth during the Civil War and it is not clear what happened to the property between 1659 and 1661. But after the Restoration a petition to Charles II, claiming that the park was used as pasture for cattle by nearly a thousand poor families of the town, was refused. The city, however, still had physical possession, for in 1661 Charles leased the park and manor to an influential Royalist, Sir Robert Townsend, on condition that he would recover them for the Crown. He was successful both in ejecting the mayor and bailiffs and as a consequence in denying the citizens the right of pasturage. However, popular agitation apparently forced even the Townsends to allow pasturage from time to time, and Anthony, Robert's son, finally assigned the lease of the Great Park to the mayor and bailiffs in 1705. In the very next year the city bought the manor house and its precincts, though not the park.

From the end of the seventeenth century, when the park had become a place of fashionable resort, decline seriously set in followed by dismemberment of the estate. Most of its trees had already gone but there was a fine avenue of ash extending from the Little Park Gate nearly to Quinton. Freemen of Coventry were still allowed to pasture their cattle, but horse-racing resulted in booths being set up, and in 1787 the avenue of trees was felled. Then began the process of dividing the park into gardens and enclosed fields, until the Prince Regent in 1819 sold it to the then tenant, the Marquess of Hertford.

By 1837 a row of weavers' houses had been put up beside the manor house and by the 1860s the Quadrant had been built on what was then the Sheriff's Orchard, previously a nursery garden. In 1871 H. W. Eaton, created Lord Cheylesmore in 1877, purchased the Coventry Park Estate from the fifth Marquess of Hertford. He gave land for the site of St Michael's Vicarage and for a road from Greyfriars Green to Coventry Station. The railway had been cut through the northern part of the park in 1838. It was not until the twentieth century, however, that the whole area was built over. The extensive factories of Armstrong Siddeley, Maudsley Motor Works and several others occupied the north-east corner of the park, close to the railway, and in 1934 a private development company, London and Home Counties Property Investments Limited, having bought out the fourth Lord Cheylesmore, developed the rest of the park for houses.

Most of the features mentioned are to be seen on Bradford's map. First the kink in the city wall from Friars Gate to Park Gate encloses the manor house within the city and severs it from the park. Only the tower of Greyfriars exists, as it does today; the rest of their property is shown as an orchard. The gate itself survived until well into the nineteenth century, but has now completely gone. Sheriff's Orchard lies south-east of the gate and the fence also drawn on the map may be the rail or part of it already noted as the boundary between the Little Park and Great Park. The Ditch is all that remains to mark the city wall. Park Gate is built on the return of the wall to its originally intended line and is very close to the Park Street Gate, noted as in ruins, at the southern end of Little Park Street.

Within the Park there were still Park Hollow, no doubt the quarry already mentioned, the beginning of the ash avenue, the 'winning post' for what then must have been the new race track, the Old Windmill, and on the eastern boundary of the map, but whether within the park or beyond it is hard to tell, a large corn-mill on the Sherbourne River. Another mill is indicated by 'the Horse Way to the Mill' and 'the Foot Way to the Mill'.

Original house and solar

The sketch (Figure 177) shows the complete manor house laid out on a south-west to north-east axis with the hall and gatehouse facing each other across the courtyard. Their orientation is referred to as north, towards Coventry, and south towards the Park. The wings are thus 'east wing' and 'west wing'.

Much of the sketch is conjectural. The only certain buildings are the gatehouse, the two bays of each wing and the east wing extension, consisting of a building lying at right angles to the rest and six further bays southward comprising the solar. The pattern of the timber framing of this range, however, is not certain. The hall is most likely to have existed where it is shown, but nothing at all is known of its design. This is also true of the west wing, except for its two surviving bays. The sketch is not representative of a particular date as individual buildings were constantly being altered throughout the life of the manor. Nor are such details as chimneys shown.

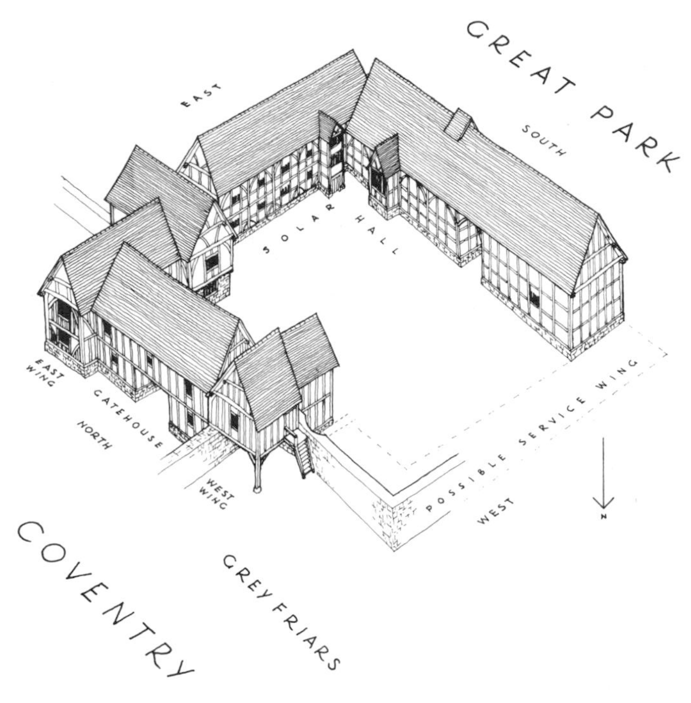

Figure 177 *Axonometric of complete manor house*

Cheylesmore Manor House is one of the rare cases in which almost as much can be learned about the building's history from documents as from the surviving structures themselves. Thus it is known from a document of 1421 that after its period of royal occupation in the fourteenth century, the manor house was already beginning to decay, but was repaired with timber from the park. Further deterioration had taken place by 1538–9 when the Commission's report on the adjacent Greyfriars property, at the time of its dissolution, mentioned 'an old manor of the king's called Chyldesmore' and added that the hall was down, but that the lodgings might be repaired with tiles from the friary. Leland, at about the same time, described Cheylesmore as 'a palace now somewhat in ruin'.

Townsend is said to have laid out much money in repairing the mansion between 1661 and 1685, but after it was acquired by the city in 1705 it did not remain for long in single occupation. In 1738 a weaver was proposing to make a tenement of what was then, probably wrongly, called the great hall. A hundred years later all the remaining buildings had either been divided into tenements or replaced by new weavers' houses.

At the end of the Second World War, there still stood the six-bay range of two storeys converted into top-shops. These were the characteristic dwellings of weavers; the top floor was either a shared workshop or a series of individual workshops, and at their north end was a much larger pair of top-shops of three storeys plus half-cellar. The latter building had encroached into the present east wing, curtailing its length by almost 3 feet.

In 1954 P. B. Chatwin[2] rediscovered the medieval roof of the long range and erroneously attributed it to the hall, no doubt the same building as the one referred to also as the hall in 1738. In 1955 the Royal Commission on Historical Monuments recorded it and identified it as the original solar wing.

Internally it consisted of a series of bays of unequal length, a short one alternating with a longer one. There were three upper chambers, each of two bays. The roof was a scissor-braced rafter-roof with crown-posts centred on each tie-beam and supporting a continuous collar-plate. The most decorative of the crown-posts was in the chamber nearest to the gatehouse (Figure 178). This was

Figure 178 *The solar roof, from a photograph by RCHM. Courtesy NMR*

Figure 179 *The north front before restoration*

Figure 180 *View from north-east*

an octagonal post with four curved and chamfered braces, two of them aligned longitudinally and jointed to the collar-plate, the other two transversely to the scissor-braces. The upper floor was also original with cross-beams and bridging beams with flat-section joists tenoned into them, their outer ends resting on the girdings. The cross-beams at each open truss were braced to the wall-posts of each side wall, as also were the tie-beams in the upper storey.

The Corporation demolished the whole of this building in 1955, at the height of the post-war redevelopment boom when many local authorities were acting with hardly less destructive zeal.

Before pre-restoration

If one approached from the city the group of buildings as it existed in 1965, the gatehouse was recognizable as of three bays, but the only sign of timber framing was within the carriageway in the central bay (Figure 179). Otherwise the entire exterior was nineteenth century in character, with gaunt square windows and the hardest of rendered surfaces. There was a stone-and-brick garden wall on the left, which protected a relatively large garden and a row of splendid limes.

East wing

The house to which the garden belonged was made up of the east bay of the gatehouse and the two surviving bays of the east wing. French windows opened into the garden from the north elevation. Round the corner, the east wall of the wing was of eighteenth- and nineteenth-century brick with domestic Georgian windows (Figure 180). A straight joint in the outer brick skin marked the division of the bays. An immensely tall chimney stack emerged from the roof on the same vertical line. The rear boundary of the garden was a range of lean-to outhouses built on to a brick and stone wall, marking the former solar range.

Figure 181 *View from south-west*

The front door of the house was in the west wall of the south bay opening on to a little yard, all that remained of the original courtyard of the manor house (Figure 181). This side of the house had been built out about 3 feet beyond the bay's original timber frame, and there was a further projection in the re-entrant angle with the south wall of the gatehouse, accommodating the staircase (Figure 182).

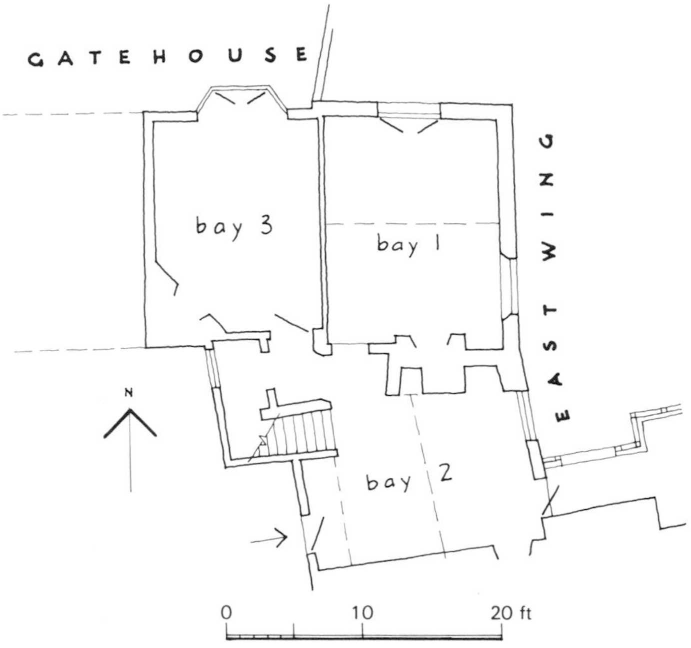

Within the house, the rooms and their decoration were, like the exterior, a mixture of late Georgian and Victorian, with some art nouveau in such details as fireplaces. It had been a comfortable small house of two living and three bedrooms, with as little evidence of timber framing inside as out. As usual the roof-spaces were the most revealing. The first glimpse of the roof of bay 1 showed a purlin roof of unusual design (Figure 183), while bay 2 had a rafter roof (Figure 184). The later survey, and further discoveries as building work progressed, showed that bay 1 was probably the oldest of the surviving structures. In the external north frame the truss has an exceptionally broad and cambered collar supported off a slightly narrower tie-beam by very wide struts. Beneath the tie-beam, the timbers are also broad on the exposed face but of no great depth. The collar is only 3 inches, the tie-beam 5 inches and the posts themselves not more than 7 inches. But the post jowls, which occur top and bottom, are 18 inches across the face, and the braces 15 inches. It is, however, the pairing of members that gives the whole frame its architectural quality. Not only are the two pairs of swept braces mirror-images but so also are the two corner posts, both heart-sawn from the same huge log (Figure 185).

The arch-braced truss II again has extremely broad but shallow collar and braces, the latter continuously tenoned, so closing the spandrel (Figure 186). There is no tie-beam

Figure 182 *Plan of Victorian house in east wing and bay 3 of gatehouse*

Figure 183 *Roof of east wing above inserted ceiling in bay 1 with intermediate arch-braced truss of unusual design*

Figure 184 *Rafter roof of bay 2 – after removal of ceiling. Longitudinal beam is the collar-plate, not an inserted ceiling beam*

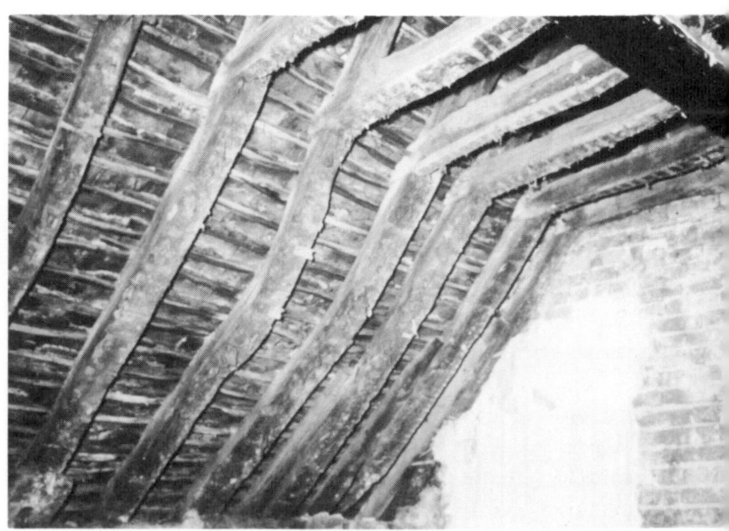

Figure 185 *Bay 1 north elevation as restored – all the main timbers, except the sill-beam, are original. Classic example of the scallop pattern*

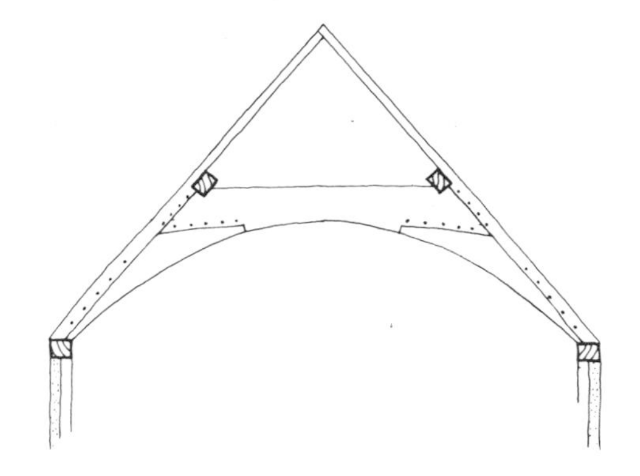

Figure 186 *Truss II – east wing, bay 1*

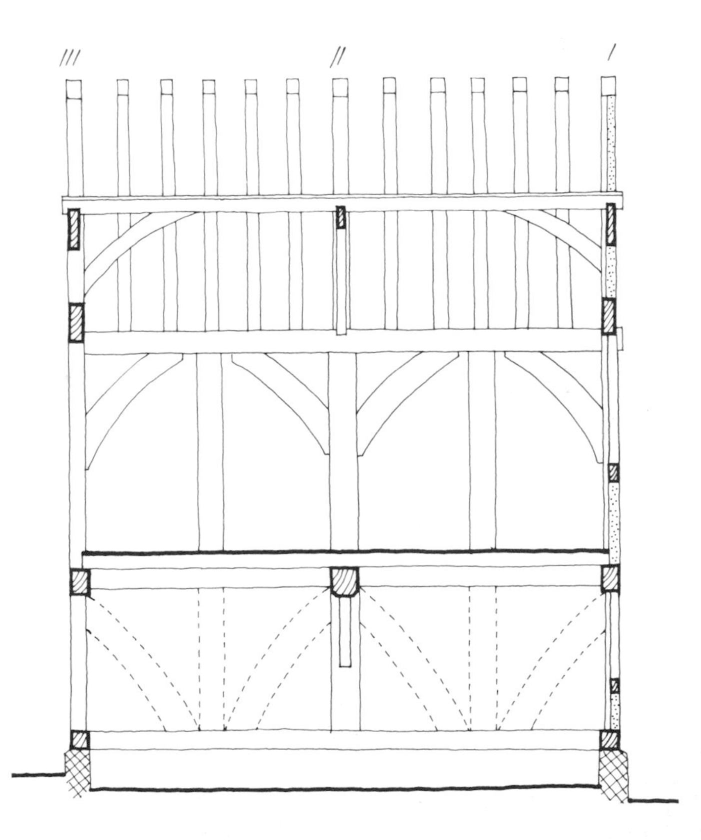

Figure 187 *Long section – east wing, bay 1*

and the posts, without jowls, are no deeper at the head than is the wall-plate but, like the other members, are extraordinarily wide across their face. Truss III is similar to T I, with tie-beam and knee-braces which have a longer sweep than those of T I but there are no lower braces. Lengthwise the components are no less unusual in their proportions (Figure 187). The rafters are broad and flat, resting on purlins of scarcely larger section than the rafters. The wind-braces are also of light section and narrow profile. As if to compensate for these, the wall-braces are of enormous width; so also are the girding-beams. The upper-floor cross-beam is 12 inches square and the mortices in its soffit for the knee-braces, missing but now replaced, are 2 feet in length. Only the floor beams, or joists, running longitudinally, are of the more or less standard dimensions. The timbers are of high quality, with chamfered edges but without moulding or carving. All this must point to an early date as well as to plentiful supply of timber for the royal carpenter.

Bay 2 was the most depleted of the entire group of structures. The most important survivor, except for the

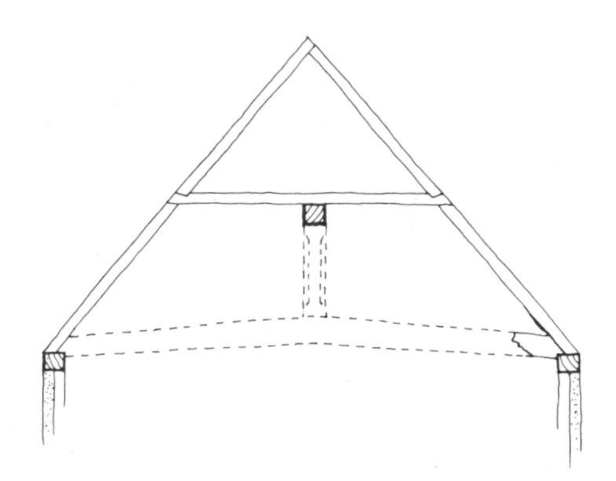

Figure 188 *Remains of intermediate truss of bay 2, east wing. Dotted lines show missing members*

Figure 190 *Stub-end of tie-beam of T II, being all that remained of the intermediate truss of bay 2, east wing*

Figure 189 *Vestiges of T I, bay 2, east wing. Sawn-off corner post, severed tie-beam and end of collar-plate and its brace to the crown-post are visible*

wall-plate and girding-beam of the west side wall, was the collar-plate that might have been mistaken for an inserted ceiling beam were it not for its mortice at midspan for the destroyed crown-post (Figure 188).

The bay is part of one of only four surviving buildings in Coventry with a rafter roof, linking it in style and probably date with the destroyed solar. But there was great difference in quality as compared with the latter, the rafters being only roughly squared and half-lapped to the collars instead of mortice-and-tenoned (Figure 184). The timbers, however, were substantial.

After stripping and taking down the chimney it was found that half of truss I had also survived complete with the end crown-post and brace to the collar-plate (Figure 189). The north-west corner post had been cut short about 6 feet above the ground floor, but its upper portion was still taking the weight of the remaining half truss and of a fair proportion of the roof. Its only support was the lath-and-plaster connecting it to the south-west post of bay 1. Of the intermediate truss nothing remained except a stub-end of the tie-beam on the west wall-plate (Figure 190), and T III had completely gone as the wing had lost 2 or 3 feet of its original length. The vestiges of the west side wall-frame have already been mentioned.

West wing

The west wing and two west bays of the gatehouse had been converted into bedsits occupied by a group of theological students. The wing had been extended to the south by a further brick-built bay and a lean-to along its east wall. The extension was a one-up/one-down with a diminutive scullery in the lean-to (Figure 191).

A larger westward addition contained a pair of three-storey top-shops entered from the south (Figure 192). The

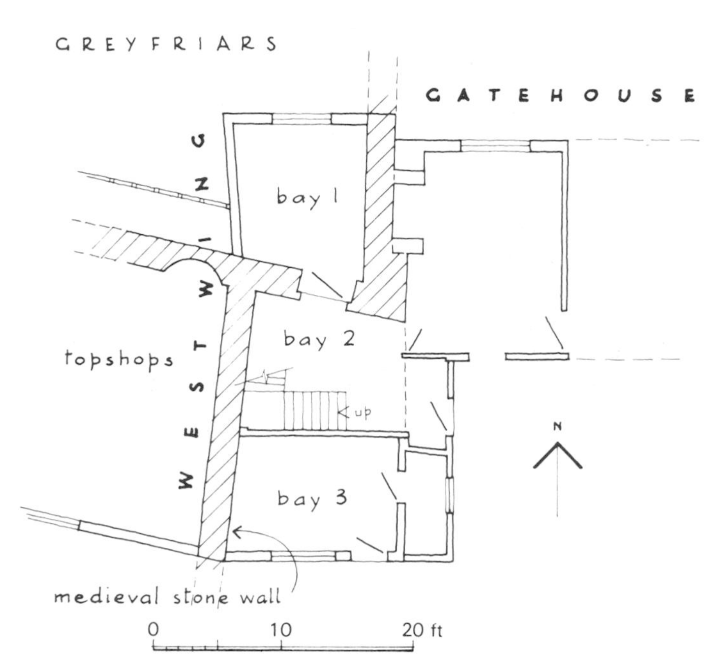

Figure 191 *Plan of west wing*

Figure 192 *View from south of west wing and attached top-shops*

north side of the wall up to first-floor level was of stone, nearly 4 feet in thickness. A curved recess on its south side within the building may have been formed for a spiral stair for access to the top of the wall when it was first built for the defence of the manor house. At its west end the wall turned southward, extending throughout the west gable of the top-shops. Most likely it had originally extended further southwards and may have been the outer flank wall of the service wing, corresponding to the solar across the courtyard (Figure 177).

Returning to the east end of the wall, it here turned acutely northward and became the party wall between the north bay of the wing and the gatehouse, supporting the former's upper-floor joists. It had also extended an unknown distance northwards as the boundary between the manor and Greyfriars monastery.

Its most remarkable feature, however, was a doorway in the east–west run just to the east of the curved recess (Figure 193). This had a huge ogee-arch on the manor side and a suitably more plain triangular head on the side of the friary. The jambs were rebated so that the original door, as the present one, opened into the latter. The discovery of this doorway not only confirmed the purpose of the wall but also solved the problem of the location of a gateway, already known from a document which said that the Black Prince had given permission in 1359 for

Figure 193 *Medieval doorway in stone wall*

the friars to have a postern from their hospital into the manor for carrying out their sick.

The design of the west wing (see Figures 175 and 177), completely disguised by eighteenth- and nineteenth-century alterations, shows how communication was made and the route by which the sick monks (sick is probably a euphemism) were carried out. On the monastic side, at ground level, was a *porte-cochère*, open to the north and west and probably fulfilling the purpose of a lychgate, with a guardroom over bay 1. This must have been reached by a ladder or wooden stair through a door in its west wall, of which the door-head still remains. The room has a fine stone fireplace and richly moulded timbering. In its south wall at the upper level, a four-centred arched doorway leads into bay 2, belonging to the manor (Figure 194). This bay is architecturally as imposing as bay 1, but structurally separate and of lesser span. This may have been because it was restricted by yet another stone wall that branched off the main one on the line of its side wall, perhaps enclosing one side of the old spiral stair, or it may have been the wall of another building.

The bay was, as it still is, a stair hall, and that is probably the best explanation of its smallness. An Elizabethan staircase, presumably succeeding an earlier one, had partially survived, and its splendid newel posts, string and handrail are now incorporated in the present stair (Figure 195). The landing has a doorway into the gatehouse and a range of windows in its south and east walls. The windows are, perhaps significantly, at high level so that the monks permitted on to the landing could not see into the manor courtyard (see Figure 216). The ground floor also had a door into the gatehouse.

Thus it seems that the monastic keeper would be admitted on to the landing to seek the permission of the guard

Figure 194 *West wing, bay 1 and on right doorway to bay 2*

Figure 195 *Newel and string of Elizabethan stair incorporated in new stair*

in the main upper chamber of the gatehouse to unlock the great door. This was then opened on to the friary side, and the sick monk could be carried into the ground floor of the south bay, but not into the courtyard, for the bay had no external door, and thence through the lower chamber of the gatehouse into the carriageway and out. All of this presumes that the gatehouse, as it then existed, had the same plan as the present one.

The framing of both bays of the west wing is more orthodox and certainly later than that of the east wing. The external walls are close-timbered throughout with elegant curved braces and a high degree of refinement. But the most enlightening detail is a small structural one. This is to be seen from the upper room on the monks' side, bay 1. Here two adjacent trusses are visible, the one being truss II of this bay, the other truss I of bay 2 on the manor side, divided from each other by the width of the stone wall beneath. That the span of the bays is different has already been noted but, as well as this, not only is each roof of different pitch from the other, but the trusses built by the friars' carpenter have the tenoned purlin system and the secular ones the clasped purlin system (Figure 196). Thus, though the two carpenters were no doubt working alongside each other, each followed his own method.

Lastly, concerning the west wing, its general design as well as details and proportions firmly date it to the mid fifteenth century, its fireplace being later.

Figure 196 *Adjacent trusses in the west wing, the nearer in the friars' room having tenoned-purlin, that belonging to the manor with clasped-purlin*

The gatehouse

The gatehouse is of one build of the mid sixteenth century (Figure 197(a), (b) and (c)). As we have seen there is plenty of information from the documents of decay and dilapidation but very little about rebuilding. We know that in 1538–9 the hall was down. If the original gatehouse was also dilapidated, the year 1549 when John Dudley, Earl of Warwick, became the tenant of Cheylesmore would be the most likely date, or shortly after, for its rebuilding. By this time, the medieval open halls were going out of fashion, except for festive or regal occasions, but for these the great halls of the castles were still available. Probably the solar was thus adequate for the duke's domestic needs, no doubt backed up by an extensive range of service accommodation. But gatehouses were to remain necessary, or at least fashionable, for a further hundred years at least and were generally well built.

This one is no exception; it is 44 feet 3 inches in length by 19 feet in width, and close-studded throughout. Its three bays, slightly varying in length from 15 feet 6 inches to 14 feet 3 inches, were erected from west to east, truss I being assembled against the already existing stone wall, shared with the west wing and older than both of them. There is a very large fireplace at the western end, back to back with that in the friars' room. A third fireplace on the ground floor was, unlike the others, extremely dilapidated. All three are probably contemporary with the gatehouse – certainly with each other.

Truss II is an open truss with tie-beam and pair of inclined struts and no collar-beam. The frame was complete and virtually undamaged. Its giant curved brace propping the carriage door-post seems slightly out of keeping with all the straight vertical framework of the rest of this building, but is no less visually right for its purpose. Its four-centred arched doorway into the carriageway is the successor of that through which the sick monks had passed.

Bay 3 had close-studded trusses, III and IV, similar to T I. There was a corner fireplace in each room (see Figure 175), clearly inserted and of stone but with wooden lintels and of inferior design to those of bay 1 and the west wing. It was intended to keep them, but they had to be dismantled to repair the timbers bedded in them, and they were not rebuilt.

The ground-floor wall-frame of bay 3 to the carriageway had been completely replaced by a 9 inch brick wall. Upstairs, however, the frame was not only complete but,

Figure 197
Gatehouse

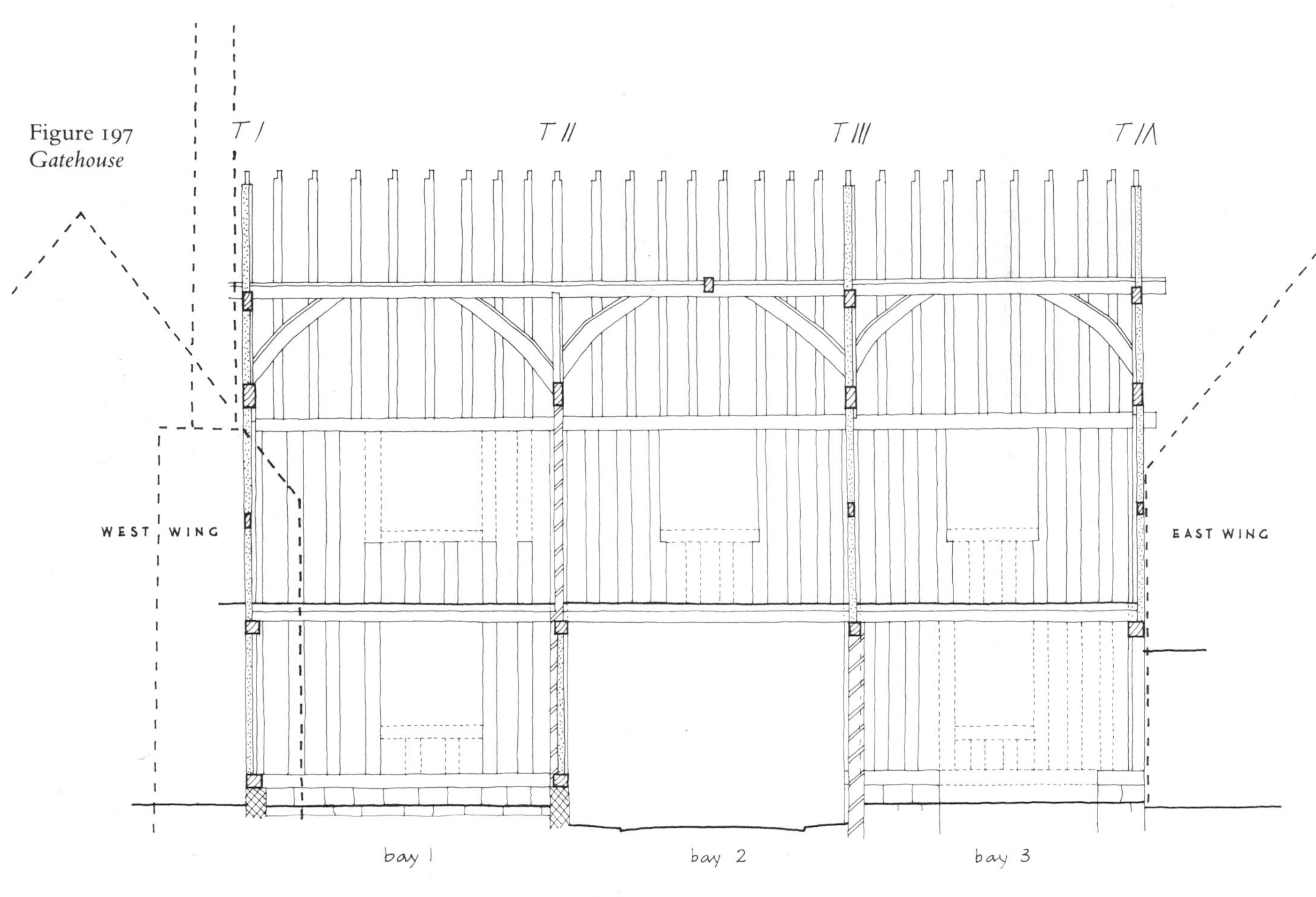

(a) Long section looking north – timbers shown in broken line had to be replaced

(b) Cross-section bay 1 showing T I

(c) Cross-section bay 2 showing T II

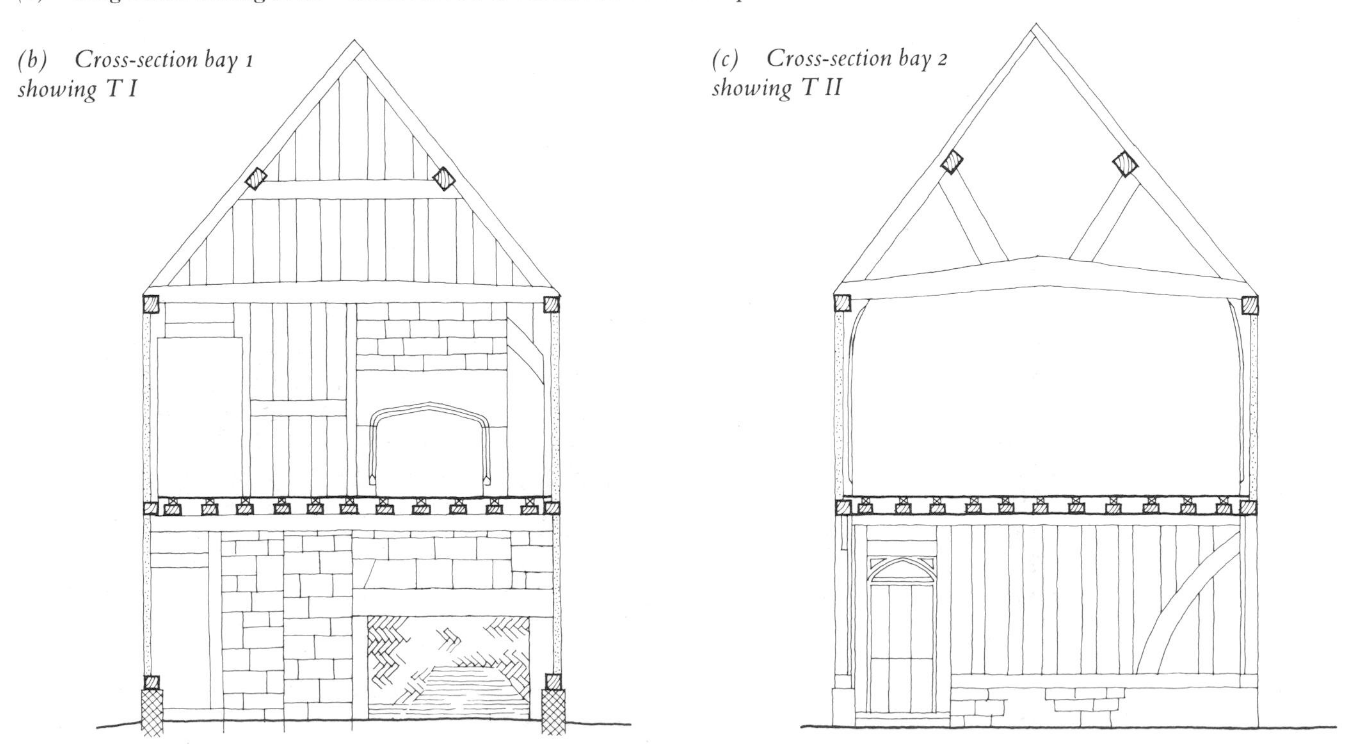

after stripping off layers of wallpaper and lath-and-plaster, its original panels facing the large room of bays 1 and 2 were found to have been decorated with typical Elizabethan figures. Unfortunately the painting was impossible to make out in detail as both the panels and timbers had been subsequently whitewashed, probably more than once. Only the slightest vestiges may still be discerned in three of the panels to the right of the door.

The south wall of bay 3 had a doorway against the south-east corner post and a window in the middle of the wall-frame at both floor levels (Figure 198). Despite evidence of alteration or reuse in the sill and stud on the east side of the upper window, they must be accepted as original. The mullions, all of which had survived in the top window, were square-set and as there was evidence of windows of the same type in other wall-frames, they provided the model for the new ones.

In this bay there was also a diminutive doorway in the north-east corner of the first-floor room through the end frame, IV. This was matched by an original door-head in the much older wall-frame of the east wing. Getting from one to the other involved a large step down into the east wing, ducking one's head at the same time. This of course suggests that the former gatehouse was an altogether lower structure.

Truss IV was also of interest as the panels immediately above the tie-beam had never been plastered on the outside, this side being protected by the extension of the gatehouse roof meeting that of the east wing. Here it was first found that the panels consisted of sandstone slithers slipped into wattle grooves and mortared together (Figure 199). Later it was confirmed that all the panels were of this construction, no doubt as a better defensive filling than wattle-and-daub, and they have since been found in many Coventry buildings. The rush-light burns on the tie-beam, as often found when attics have been used for sleeping in, were rather more difficult to account for, as the space beneath the link roof was severely confined. It must nevertheless have provided temporary refuge from military discipline for those who could negotiate a 12 inch opening through an unfilled panel! It might even have been a priest-hide.

As for general details of construction, sill-beams were laid on a red sandstone plinth about 2 feet high, the roof framing was of the clasped purlin type, and the purlins were bridle-scarfed immediately over the collars. The wind-braces were curved, tenoned into the principals and dovetail-lapped over the purlins. There was no ridge-piece.

All the bridled joints of the longitudinal members were

Figure 198 *Gatehouse south wall – the original window in bay 3, the stone fireplace wall to the left and the former house of the east wing in course of demolition on the right*

Figure 199 *Stone infill panels as first found in T IV of gatehouse in position protected by link roof to east wing, left foreground*

Figure 200 *Gatehouse T II from east showing roof details*

designed for erecting the frames in sequence from west to east. The quality of finish, like the carpentry, was uniformly high, the posts being shafted with chamfered edges, and the post-to-tie-beam knee-braces of the open truss II continued the profile of the shafts up to the tie-beam which was similarly moulded. A painted roundel still to be seen on the east face of truss II tie-beam has baffled all attempts at an explanation (Figure 200).

Two further details should be noted. First, the brackets at each end of the beam over the inner end of the carriage-way are Jacobean; thus they are one of the very few signs of work that could be attributed to Sir Robert Townsend, said to have 'laid out much money in repairing the mansion and died there in 1685'. He might also have inserted the fireplaces in bay 3 and done work to the other buildings which no longer exist. Second, the corresponding beam at the north entrance to the carriageway has a mortice about 3 feet from its east end, indicating a former post which carried the other leaf of the double doors and left space for a separate pedestrian door. A water colour, of extraordinarily 'primitive' style of *c.*1800, shows this post, but the doors have gone, and it stands on a plinth and sill-beam, thus precluding the possibility of a

Figure 201 *Watercolour of c.1800 of the gatehouse. Courtesy Birmingham City Librarian*

pedestrian door. The girding beam is painted white and has lettering 'CHILDSMORE PLACE'. The timbers are exposed and the windows roughly correspond with those before restoration but, clearly, they have been shaded in after the drawing had been completed, as the timbers are still visible beneath. It seems that *c.*1800 was thus the date of the final conversion of the gatehouse to the house and flats already described (Figure 201).

Restoration

The intentions set out in the preliminary survey report were supplemented in the introduction to the final specification. The main points were:

1. The aim is to preserve as much as possible of the existing structure.
2. Despite that, a large proportion of later brickwork and plaster will have to be removed and original timber-frames renewed.
3. Roof trusses will be exposed by removal of inserted ceilings.
4. Steel straps will be used to secure joints, particularly in roof trusses, rather than replacing the defective timbers.
5. Repairs will not be hidden or disguised.
6. New doors and windows will not be reproductions (this was before the discovery of the original window in the gatehouse).
7. Fireplaces will be retained, but their brick stacks, all of the present or last century, will be taken down.
8. The roof covering will be entirely renewed to provide insulation.
9. External timber framing will be exposed.
10. Infill panels of external walls will be replaced in modern materials but internal panels will be retained wherever they are capable of repair, some of the stone-filled panels being left unplastered.

There followed a key plan showing what was intended to be demolished and then elevations of every wall in the entire complex accompanied by notes of what was to be done to each of them. This was issued as the general specification accompanied by the bills of quantities.

At last, after almost two years of negotiation, surveying and preparation, the job book records the first site visit on 14 September 1966. The last entry was on 27 January 1969.

The contractor chose to tackle the job from west to east, but with considerable overlap from one structure to the next, especially towards the end of the work so that

Figure 202 *West wing*
(a) Uncovering of south bay timbering by partial demolition of brick extensions
(b) Six months later with south bay reframed and roofed but brick structure still not wholly demolished

trades such as flooring, glazing and services could be completed in one or at most two operations. Roofing by contrast was done immediately the structure concerned had been repaired. Nor was there any hurry to demolish the unwanted buildings, as they could be used for storage and even as scaffolding, as in the reframing of the south gable wall of the west wing (Figure 202). The top-shops at this end were not in fact demolished until the whole of the wing had been reroofed and the repairing operations had reached the east wing, the last structure to be tackled.

In the meantime, many more discoveries were made and problems met. One result was that the system of S and R drawings was initiated though not consistently carried through. Nor were the drawings done as one of the essential components of the tender documents, as in later contracts. Instead an S drawing was made after the frame concerned had been stripped, and generally a detail (D) was prepared to show what had to be done if this differed, as usually it did, from the instructions given in the specification.

The first of such a pair to be issued recorded the north elevation of the west wing – drawing S11 – and set out the repairs on D11, which included the window and a scissor-scarf detail (Figure 203). Many such D drawings

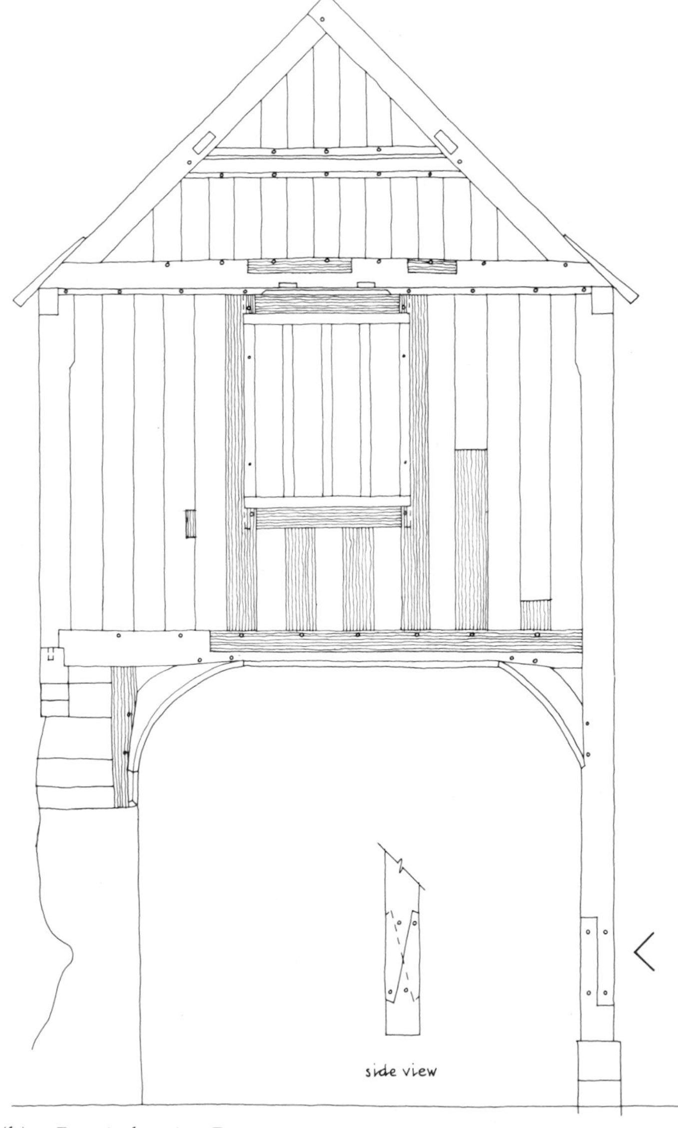

Figure 203 *West wing external truss, north elevation*
(a) Survey drawing S11
(b) Repair drawing R11

were full-blooded $\frac{1}{2}$ inch details with panels, plinths, roof-covering, gutters and so on, instead of being confined to timber repairs only. Nor were there any repair schedules.

This elevation contained the most important discovery, that it had been open on the ground floor forming the monks' porch to the ogee doorway. This had become clear as soon as the brickwork was taken down, leaving, incidentally, the immensely strong external stucco rendering not only still standing independently from floor to ceiling but supporting the window frame, entirely on its own strength. Taking up the wooden floor also revealed the original stone flag floor of the porch, as well as the plinth on which the north-west corner post stood. These were of course preserved and once again exposed, the post requiring a new lower length.

At the other end of the west wing, the tie-beam of the close-studded south truss revealed the evidence of the high-level windows. The tie-beam was chamfered on both lower edges, the chamfers being mitre-stopped at the positions of the mullions. In the soffit, and in the return wall-plate, the mullion mortices and the middle glazing-bar sockets were also discovered. The next disclosure was that the whole of the upper storey frame was an insertion. Presumably this reframing was done after the Dissolution, to get rid of the high-level windows and replace them by one at normal level.

Instead of the original close-studding, the panels were rectangular with relatively poor timbers. The only original stud was fortunately the one terminating the windows. A mortice in the side of this member established their sill level. It was also found that the bressummer, belonging to the original frame, and still preserved, was externally moulded.

The decision on whether to restore the earlier or later version of this wall (Figure 204(a) and (b)) was taken, as always, on practical rather than 'philosophical' grounds. The later one was structurally unsound and had half-lap joints where insertions had made impossible the use of mortice-and-tenons. It would have concealed the evidence of the original frame and its existence would most likely have been interpreted by future researchers as an aberration on our part. But the final argument was the architectural one, that the elevation required its close-studding and especially windows to restore the quality of the whole.

A different approach was adopted for the next problem. The fireplace in the ground-floor room of the gatehouse was found, when fully exposed, to have neither its original stone lintel nor one of its jambs. These had been replaced by crumbling brickwork and a sagging wooden lintel. The original stone breast, however, was necessary to support the hearth and fireplace above and the herring-bone brick fireback was worth preserving for its own sake. Should the jamb and lintel be replaced in stone, and if so what moulding should be carved on the lower edge of the latter – there existed the two contemporary upper-floor fireplaces as models – or should a purely structural answer be adopted?

In the event reinforced concrete with an exposed aggregate finish provided the most economical, honest and perhaps even architecturally impressive solution (Figure 205).

Figure 204 *West wing, south truss*

(a) The inserted wall-frame as found

(b) The original wall-frame as reconstructed

Figure 205 *Fireplace in gatehouse restored with reinforced concrete lintel*

Figure 206 *Gatehouse wall-plate, broken off at post II*

Figure 207 *Progress of gatehouse and west wing at April 1967*

But there is no substitute for the repair of timber frames. As usual with externally stuccoed buildings the appearance of Cheylesmore was deceptive. While stripping never resulted in the discovery of fewer original timbers than were expected, except of course in the case of the porch of bay 1, it invariably exposed far more, but all of them in dire need of repair if not replacement. In the gatehouse it was the wall-plates which had suffered the worst. Their joints at the posts had snapped and they had bowed outwards through each bay (Figure 206). Since wall-plates are trapped between the post-head tenon and tie-beam dovetail, their repair at this point or replacement, necessary in bays 1 and 2 of the south wall, required either lifting the tie-beam and truss or complete removal of the roof structure. More often it was the latter, as the purlins and wind-braces also had to be dismantled before the truss could be raised.

Nevertheless progress was steady and orderly through the gatehouse bays. It was being reconstructed in the sequence of the original erection. Thus longitudinal joints were correct for placing the next members into them. By the time reframing had reached bay 2, the west wing was ready for roofing (Figure 207).

Bay 3, with its inserted fireplaces and two brick walls replacing the lower storey cross-frames III and IV, required most attention. The sill-beams of these walls had of course gone and the south-west corner post, having been buried in the stone of the fireplace structure and plastered externally, was rotten. It was first decided to retimber frame III as all the evidence was there in the form of mortices in the girding-beam and corner posts.

This left only the east wall, frame IV, remaining as brick-built up to girding-beam level. This was a 9 inch wall, shared between bay 3 of the gatehouse and bay 1 of the east wing. It had been built beneath the girding-beam of the latter after it had already subsided so that its intermediate post, cut off slightly below the girding-beam, was about 8 inches lower than the outer corner post. The cross-beam, joists and floor had come down with it, and though the wall-plate spanning the two half-bays had not fractured it was severely stressed (Figure 208). It was decided therefore to reframe this wall in softwood studding with matchboarding on each face on a new brick plinth. The intermediate post was repaired with a new lower length and braced by the partition on either side. The frame of the gatehouse was left as it was, without restoring the close-studding of the lower storey even though the evidence of mortices in its girding-beam was complete.

It is worth noting that, in these juxtaposed structures, the later gatehouse was fully framed. Normally of course the newer structure is built on to the complete older one and so has a bare minimum of timbers. That the rule is reversed in this instance merely confirms that there was an earlier gatehouse antedating the present east wing, the latter thus being the one with the sparse framing.

The east wing was started in the late summer of 1967, when work had been going on elsewhere for about a year. The first intention had been to retain the whole east elevation of Georgian brickwork. Stripping, however, had uncovered several stages of building, alterations and patching, and the wall was not safe. Thus, the question of complete rebuilding in brick or reconstruction of the original frame had yet again to be faced. The former was impossible if the existing charm of the brickwork were to be preserved, and there was no point in rebuilding it if it were not. Since the three posts of bay 1 had survived within the brickwork, at least well enough to plot all their mortices (post I, rendered over its front face instead of bedded in brick, did not even require repairs) and the wall-plate was complete, there was no doubt about the design of the original frame of that bay.

There was only slightly less certainty about bay 2. Its north-east corner post was there even though, as we have already seen, half of the truss it had once supported had been cut off to make way for the chimney stack. This post, however, together with the wall-plate and the timbers already mentioned of the opposite wall, established the original framing system and the positions and size of the swept braces both from sill-beam to post and from post to wall-plate. The three surviving longitudinal members, including the collar-plate, had mortices close to their southern extremity. By plotting the curve of these braces according to those at the other end (that of the collar-plate still existed), the position and alignment of the former end wall-frame was also fixed.

So it was decided to reframe these bays as well. It resulted, amongst other things, in a great deal more new oak, to be finished and accurately jointed on the site. The former garden became an open-air workshop (Figure 209). It was still in use when the snows of January and February came in the following year. Meanwhile, bay 1 had been reframed and its roof covering completed by December.

Demolition of the last of the brickwork, the wall at the south end of bay 2, resulted in further discoveries of earlier structures. There was a massive stone wall about 6 feet long by 8 feet high, with a plinth offset on its north face, projecting from the east side wall of the bay (Figure 210). With further excavation it was found that this wall turned southwards where the original south-east corner post had stood, and which is now replaced. Further stone remains showed conclusively that there had been other buildings antedating even this earliest of the framed struc-

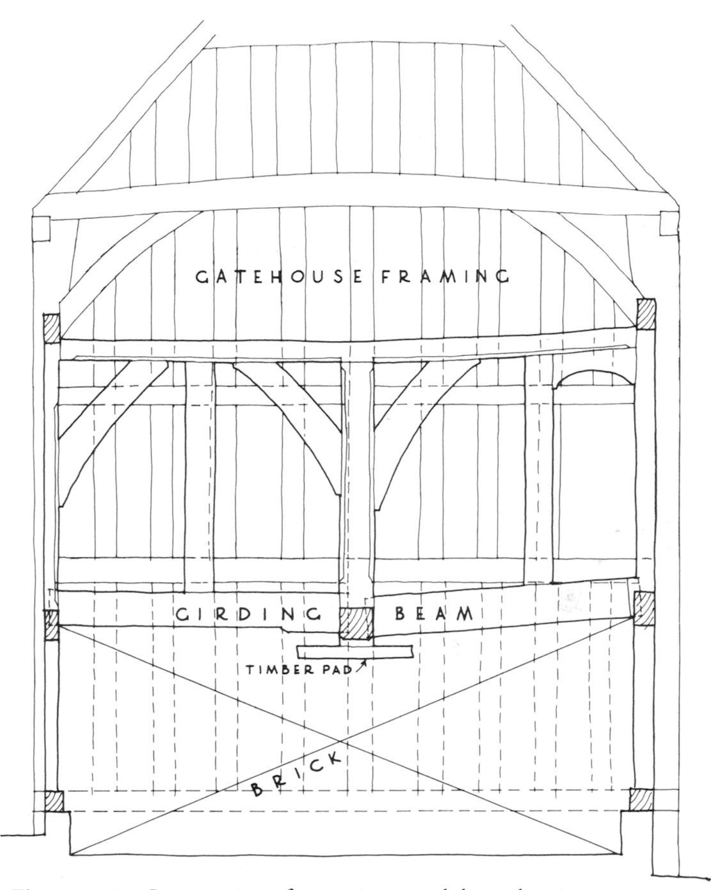

Figure 208 *Long section of east wing, north bay, showing subsidence of girding-beam, cut-off intermediate post and timbering of gatehouse (frame IV) built up against it*

Figure 209 *Timbers for reframing east wing*

Figure 210 *Remains of masonry uncovered at south-east corner of east wing*

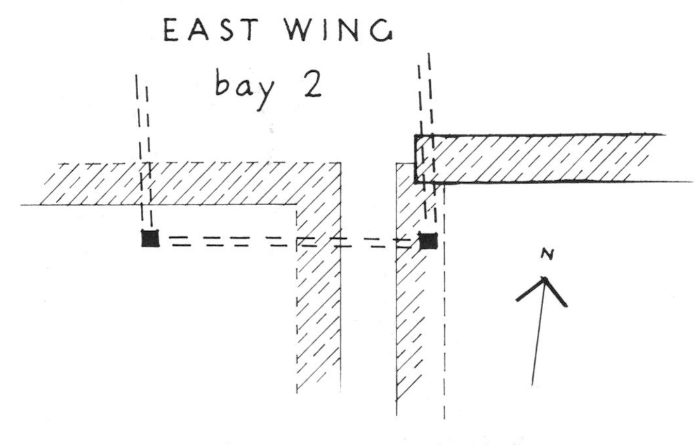

Figure 211 *Plan of stone remains at south end of east wing*

Figure 212 *Completed structure from south-east – panels still to be rendered and site to be cleared*

tures (Figure 211). Nor is this the only evidence in Coventry of masonry preceding timber framing.[3]

By May 1968 bay 2 had been framed, roofed and panelled, and the new stair turret built simultaneously. External rendering of the panels had just been started and the external drains and services were in. Internal trades, such as floors and ceilings, electric underfloor heating and plumbing were well advanced and the temporary strongroom had been built into bay 1 of the east wing. Nor even yet had all the brick buildings or garden walls been demolished (Figure 212).

The interior had been completed and furnished by the Corporation by November of that year, when the opening ceremony took place in the main marriage room, bays 1 and 2 of the gatehouse (Figure 213). In all the rooms the roofs had been re-exposed and the original structure left with as little 'finish' as could be reasonably accepted, even sawmarks being unplaned (Figure 214), and none of the wood, old or new, painted or varnished.

Meanwhile the sitework, designed mainly by the City Architect, was still going on, and there is no doubt that his plan of the surroundings of Cheylesmore, including his own Registrar's offices built in the following two years, has made Cheylesmore one of the supreme architectural surprises of modern Coventry. The whole group is screened from Union Street by a new range of offices.

Figure 213 *The completed marriage room, bays 1 and 2 of gatehouse. Courtesy Coventry Evening Telegraph*

Figure 214 *Bay 2 of east wing with restored tie-beam and crown-post. Note also the scarfed tie-beam of truss I of which half had been removed from the Victorian fireplace and chimney. Courtesy Coventry Evening Telegraph*

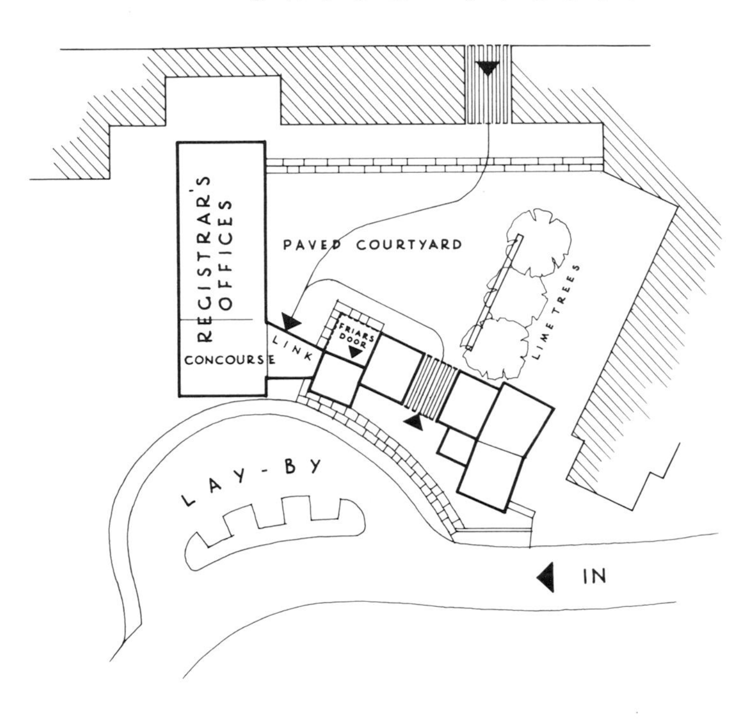

Figure 215 *Completed site layout by City Architect, including new Registrar's Office and commercial shops and offices along Union Street*

Figure 217 *The friars' door*

Figure 216 *The completed south front*

Figure 218 *Oak plaque to left of friars' door*

The approach is through another carriageway, reserved for pedestrians, beneath the upper floors of this building. The stone-paved courtyard is obliquely aligned to this entry, and Cheylesmore stands at the back of it at slightly lower level, still shaded by the limes and securely enclosed on both sides of the courtyard by an arm of the new offices on the left and the Registrar's offices on the right (Figure 215). This, with dark brick piers dividing full-height recessed windows, reflects not only the close-timbering of the gatehouse but also the deep colour of the natural sand render and the untouched oak.

The south side of the old building (Figure 216) is the traffic entrance, and parties go through the carriageway and turn left either to the great ogee doorway (Figure 217) beneath the friars' bay of the west wing or into the main entrance in the modern link. The not least important detail of all restorations done for the Corporation is the plaque giving the original plan and description of the building (Figure 218).

CHAPTER NINE

MUCH WENLOCK GUILDHALL

History[1]

The Minute Books for the Royal Burgh of Wenlock begin in 1461, their most often quoted entry being:

> Upon the 23 and 24 days of this Monethe of September 1577, was reared the howse over the prison house, Mr Thomas Ludlowe beinge baylif of this town and franches.

This building, apart from later alterations, could be the present committee room, the gaol under it then having two cells and occupying its full length, and it *could* have been reared in two days (Figure 219(a), (b) and (c)). The gaol is, of course, the older building, but an earlier reference of 1538–41 refers to an 'arch roof . . . of the Chamber of Great Wenlock' and to 'roofing the belfry with lead'. Whether this was an earlier structure over the gaol or a different building altogether is unclear. If the former, the explanation may be that the original guildhall had been of two storeys and built entirely of stone, and that it was then taken down to the present level of the gaol walls, its upper storey being replaced in timber. As we have seen, this would not be the only example of timber replacing stone; and one reason for the change might be the substantial increase of floor area as a result of substituting framed walls only 4 inches in thickness for the 2 to 3 feet thickness of stone.

The present larger courtroom of three bays standing on five ground-floor bays must have been built fairly soon afterwards (Figure 220(a), (b) and (c)). It is not dated in the records but '1589' is inscribed on the coat of arms still to be seen inside the building above the present judges' bench. More convincing evidence that it is of this period, if not the exact year, is the building's construction, especially the design of its two external roof trusses and its wall-frames. The former have crossover bracing, typical of Shropshire in this period (Figure 221), while the latter with close-studding and intermediate rails are typical of a wider area and probably longer period. Moreover, in both of the buildings the wall-frames are so similar, not only in general design but in timber sizes and spacing, that there could hardly be more than ten to twenty years between them. The only difference is in each building's respective floor structure, but the different nature of their substructures, one on solid walls, the other on posts and beams, would account for that. The two buildings are thus roughly contemporary.

It is important to note that the entire guildhall was severely plain at this stage (Figure 222(a) and (b)). The present pair of gabled windows over the street did not exist. One of the original windows may still be identified (there were, doubtless, similar ones in the other two bays). In the rear elevation of the courtroom there seems to have been only one window, that of which the blocked opening may still be seen.

In 1674

> It was agreed . . . shall be allowed . . . such money as he shall lay out for pointinge of the Court Howse and boarding over the inner room . . . whyteninge the hall and greeninge the Cort posts . . . Latting and Plasterynge the common gaole overhead to keepe out the smoake and nesty smell out of the Election howse for pavinge the street over against the cort howse beinge in the harte of the Market.

'Pointinge' and 'boarding-over' are no doubt painting and panelling. The 'Court Howse' must be the present courtroom as the committee room is named further on as the Election howse. Thus in the first part of the quotation we have some indication of the courtroom plan. It contained a hall and inner room. Since it is of three bays these clearly occupied two of them. The third must have been the entry or anteroom with a stair to the market and serving both buildings. This is bay 1, and its stair no doubt succeeded the earlier outside stair in the same position, but which of course gave access only to the committee room – the one structure at that time built.

The second part of the quotation then refers to the gaol and committee room. Lathing and plastering was a fairly new treatment – clearly a great boon to Wenlock's councillors, sitting immediately over their prisoners. Lastly, the description of the courthouse 'in the harte of the market' is of interest in suggesting that the guildhall was on an island site, surrounded by the market, the butter market being only that part of it beneath the building.

The year 1678 appears both in the building, again above the bench, and in the documents. The inscription reads,

a

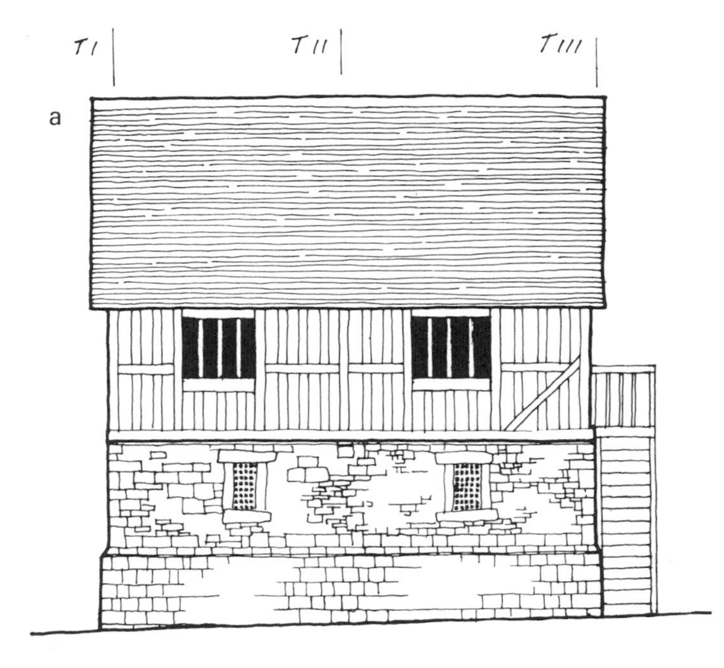

b

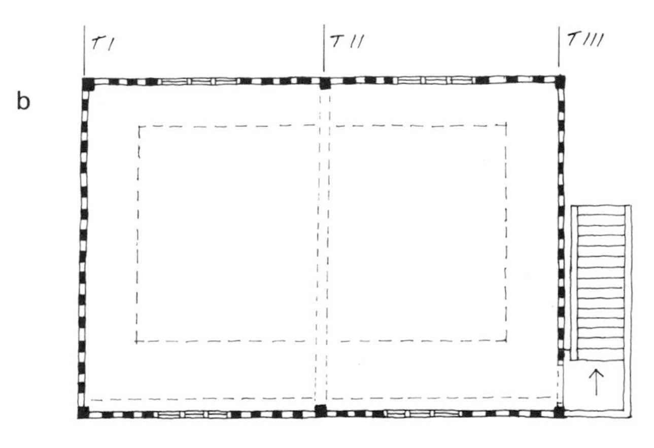

c

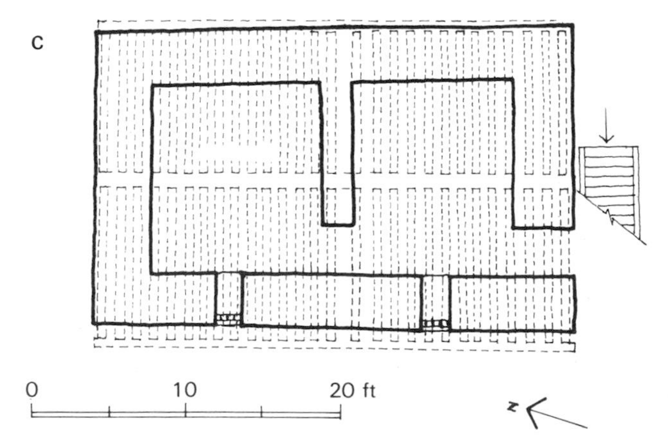

Figure 219 *The courthouse of 1577 reared on original stone gaol*
(*a*) *The gaol at ground floor*
(*b*) *Courthouse, present committee room, at first floor*
(*c*) *Elevation*

'This work was done in the time of Master Launcelot Stephens, Bailiffe', and may refer to the document which reads,

> It is Constituted and agreed unto by and with ye consent of the said Bayliffe Bayliffs Peers & Comonalty that pte of the Townehall which is below ye next cros beame to ye table shall be sepated and devided in such sort as ye said Mr Bayliffe shall appoint and that lower Pte to be imployed for a Butter market & to be from time to time kept cleane by ye pson who shall rent ye tole of ye markett being allowed for ye same out of his Rent five shillings and ye charge of makeing ye foresaid partition to be deducted out of ye interest of ye money given by Sr Thomas Littleton.

This pretty well establishes the butter market as occupying bays 2, 3, 4 and 5 of the ground floor of the courtroom, bay 1 being partitioned off and probably still being the staircase bay. There is no longer any sign of either this partition or the stair.

Hardly less important, it is also documented in the following year,

> there shall be a Pentice erected and built on the east side of the Corne Markett for the defence of the same from stormes and tempesteouse weather and to preserve the timber of the Hall from wetting and perishing.

At about the same period, though unfortunately not documented, there were also substantial alterations to the upper floor. First, the tall windows of the street elevation were introduced (Figure 223). Second, bay 1 was thrown into the main part of the building by removing the partition along truss II, thus creating the present three-bay courtroom. The stair must now have been removed to the further end, but was probably still an outside stair. Third, and most important, the roof was almost entirely reconstructed. Only the external cross-frames, trusses I and IV, remained as before.

For trusses II and III are quite different from these, though identical with each other, and their design strongly suggests the late seventeenth or early eighteenth century (Figure 224). The struts, instead of supporting the collar as in earlier trusses, are jointed into the principals, with the collar set immediately above this joint. It is difficult to say whether the decorative braces from struts to collar are contemporary with the trusses or Victorian embellishments.

The dating evidence of the longitudinal structure of the roof is still more convincing. There are two equally spaced tiers of purlins on the east side, and three unequally spaced on the other. The lowest purlin on this west side spans

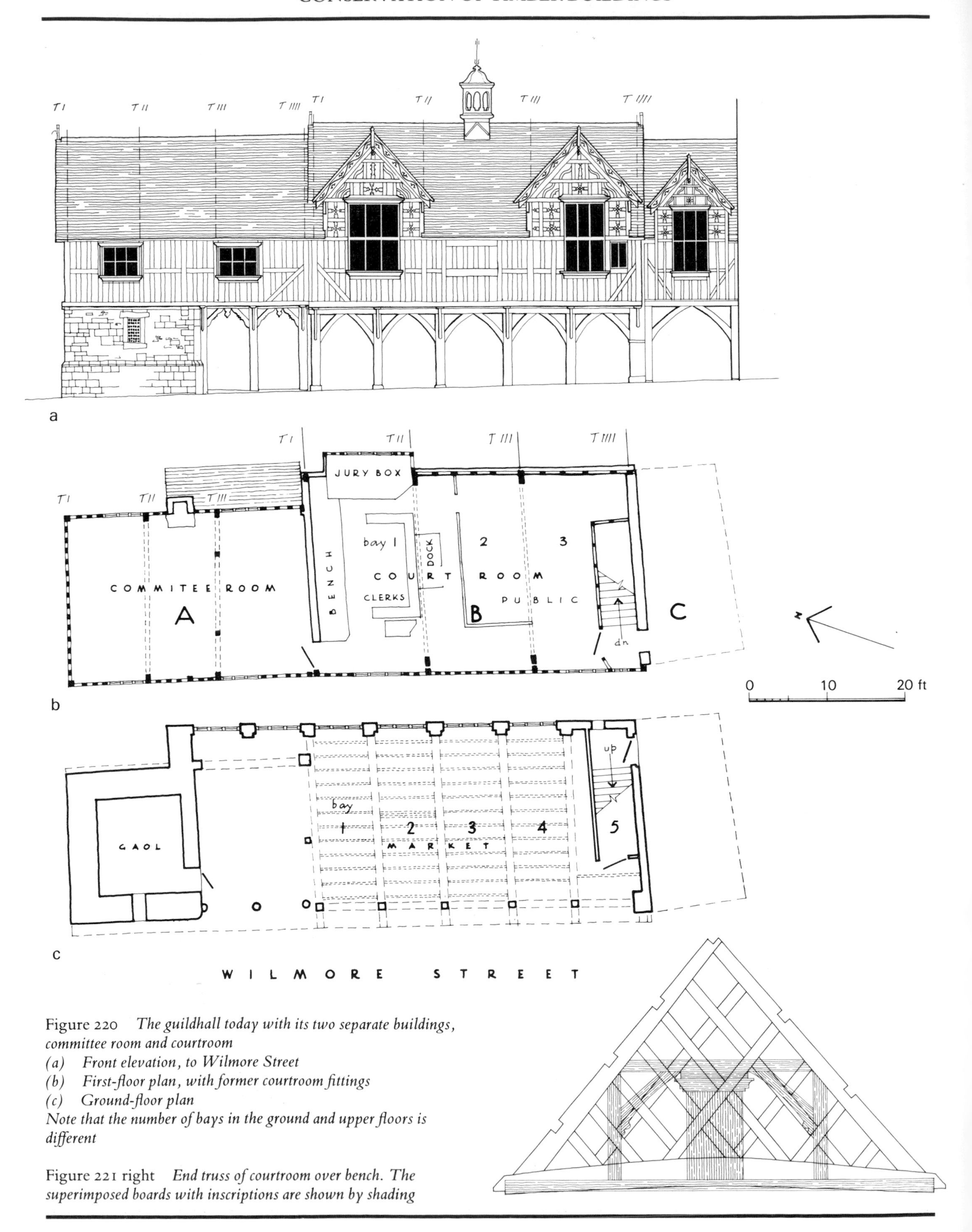

Figure 220 *The guildhall today with its two separate buildings, committee room and courtroom*
(a) Front elevation, to Wilmore Street
(b) First-floor plan, with former courtroom fittings
(c) Ground-floor plan
Note that the number of bays in the ground and upper floors is different

Figure 221 right *End truss of courtroom over bench. The superimposed boards with inscriptions are shown by shading*

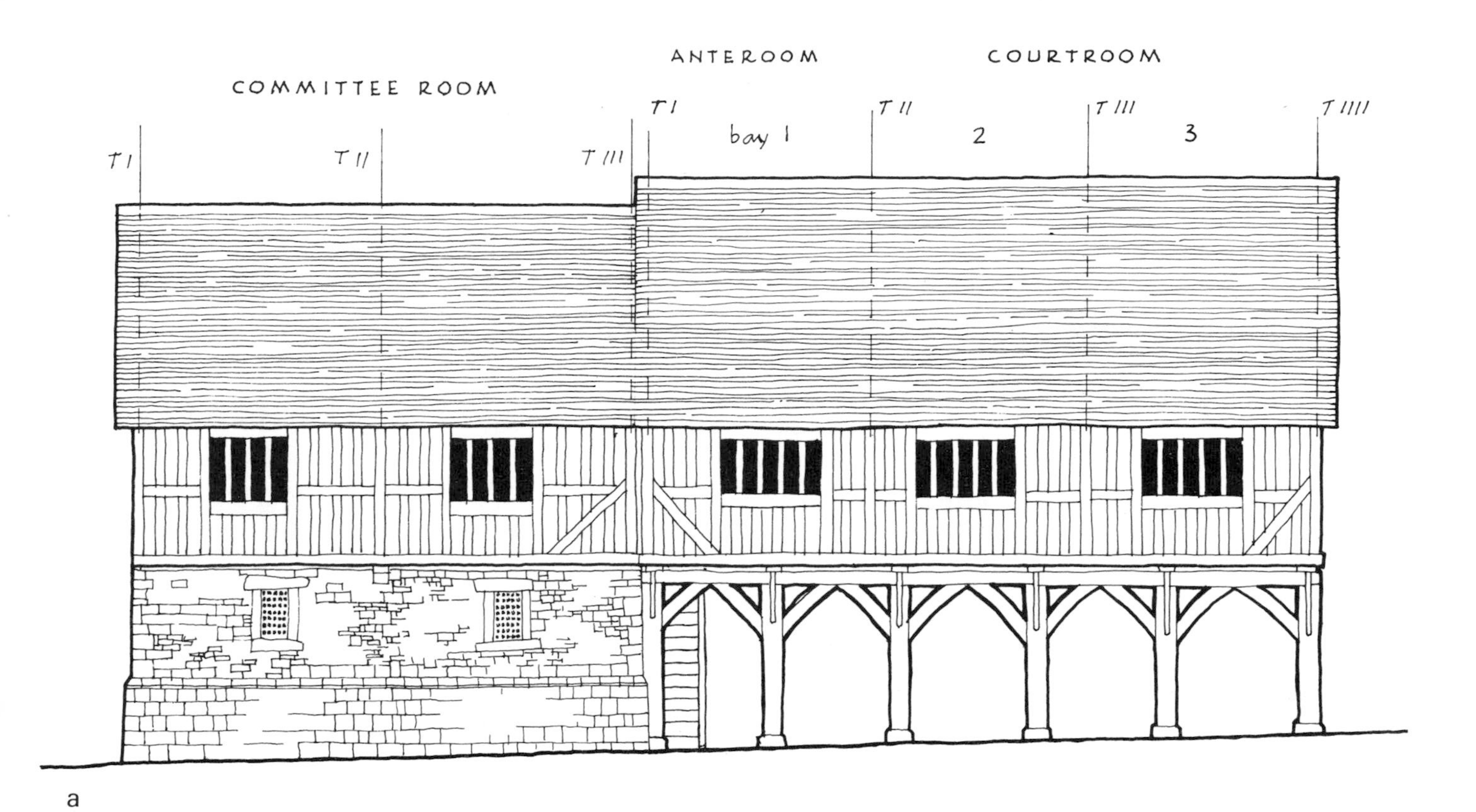

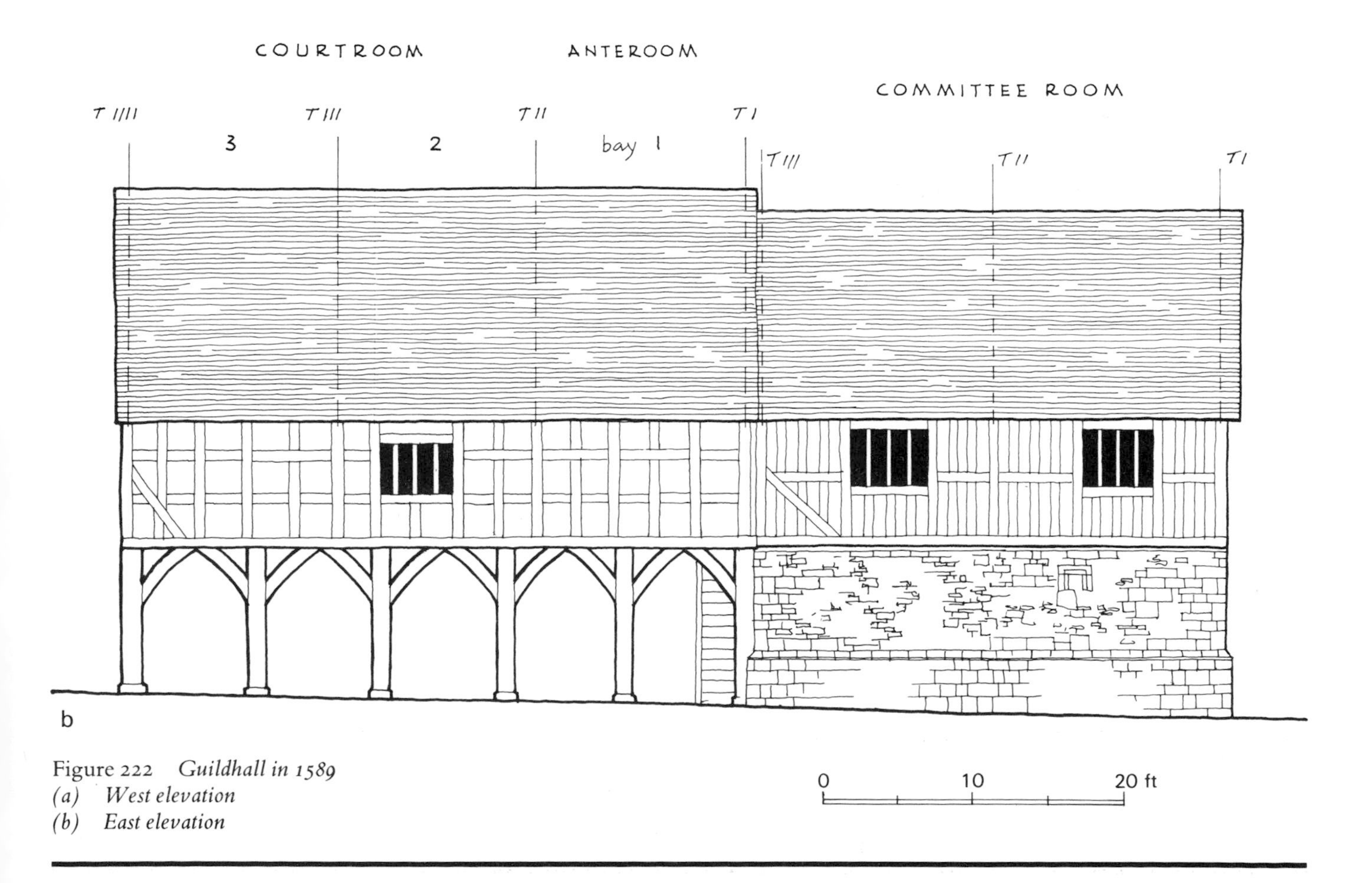

Figure 222 *Guildhall in 1589*
(a) West elevation
(b) East elevation

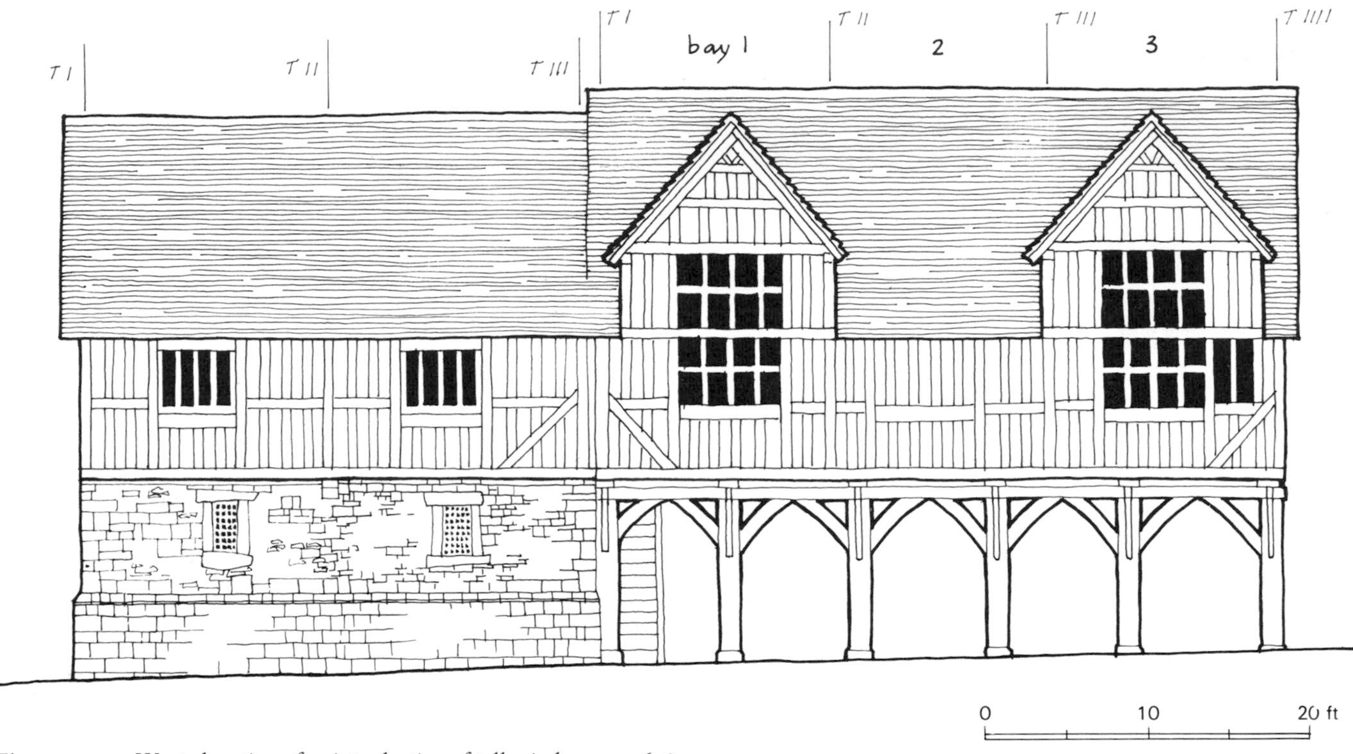

Figure 223 *West elevation after introduction of tall windows, c.1678*

bay 2 only, and there is no sign that it was ever of full length and then cut out. This side of the roof seems therefore to have been designed to incorporate the tall windows. But as well as the difference in the number of purlins, there are also different types of scarf-joints as between one side and the other (Figure 225). The bridled scarf of the purlins on the east side is normal for the sixteenth and seventeenth centuries; these therefore are probably the purlins of the original roof reused. The tapered scarfs on the west side, however, are exceptional and must be of the later date. Moreover, the principals have been trenched to fit the purlins exactly, so that there is no question that they could have been put in later.

Lastly, the roof has no wind-braces. Wind-braces began to be superseded in the late seventeenth century by bay-length diagonal braces notched over the rafters. Such braces would have left no trace had they been removed when this roof was later close-boarded, as it still is. On the other hand, the earlier kind of wind-brace, triangulating the principal-to-purlin joint, would have left clear evidence. But the complete proof that truss II is later than the former partition between bays 1 and 2 is that the truss has no sign of having been a closed truss. Only the posts are morticed for the former partition's rails.

There were further alterations in the eighteenth century. Builders' accounts of 1719–20 show that brick was being used, and included an item of £2 4s. 2d. for building a chimney, no doubt the predecessor of the Victorian stack on the east wall of the committee room. But the greatest change was in the reconstruction in brick of the east elevation (Figure 226). Not only were the original posts and braces removed (perhaps the pentice of 1678 had failed to preserve them) and the arcaded wall built, but the upper wall-framing, including the bressummer, was completely encased in brick. The south gable wall was similarly bricked and all the timbers below the truss removed except the upper corner posts and perhaps the lower post of the street front. This, if it still exists, is now also encased within the present brick pier, into which the jetty bracket incongruously disappears. Lastly, the elegant cupola and weather-vane are also of the eighteenth century (see Figure 220).

The nineteenth century was to see the most drastic alterations of all. Beginning with the committee room, there is an entry of 1869, 'Resolved that the 2 cells under the Town Hall be taken down and thrown into the Butter-market.' In the event, the north cell was retained. All the timberwork of the substructure, replacing the south cell,

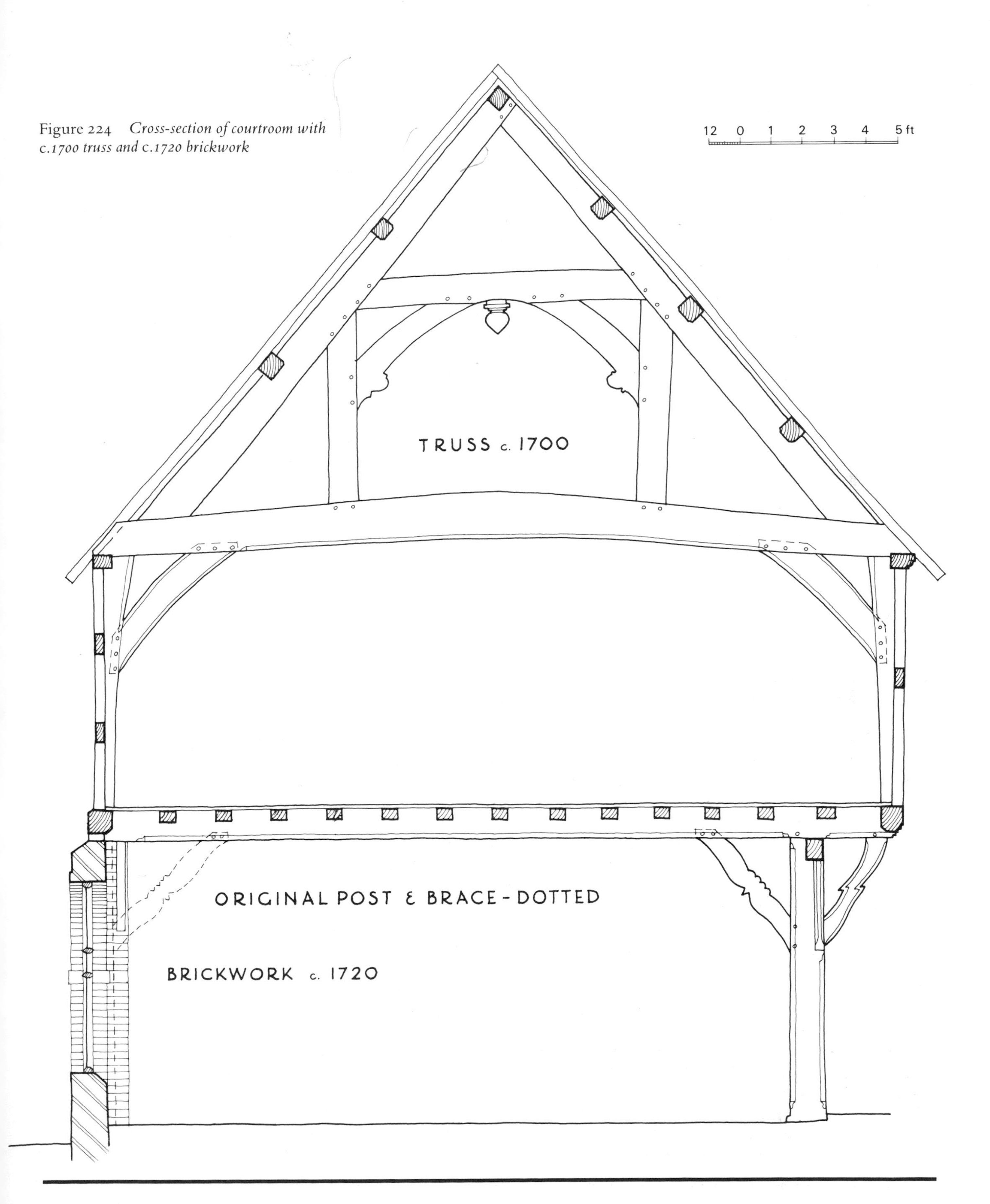

Figure 224 *Cross-section of courtroom with c.1700 truss and c.1720 brickwork*

Figure 225 *Purlin joints of courtroom roof*
(a) Bridled-scarf on east side
(b) Tapered-scarf on west side

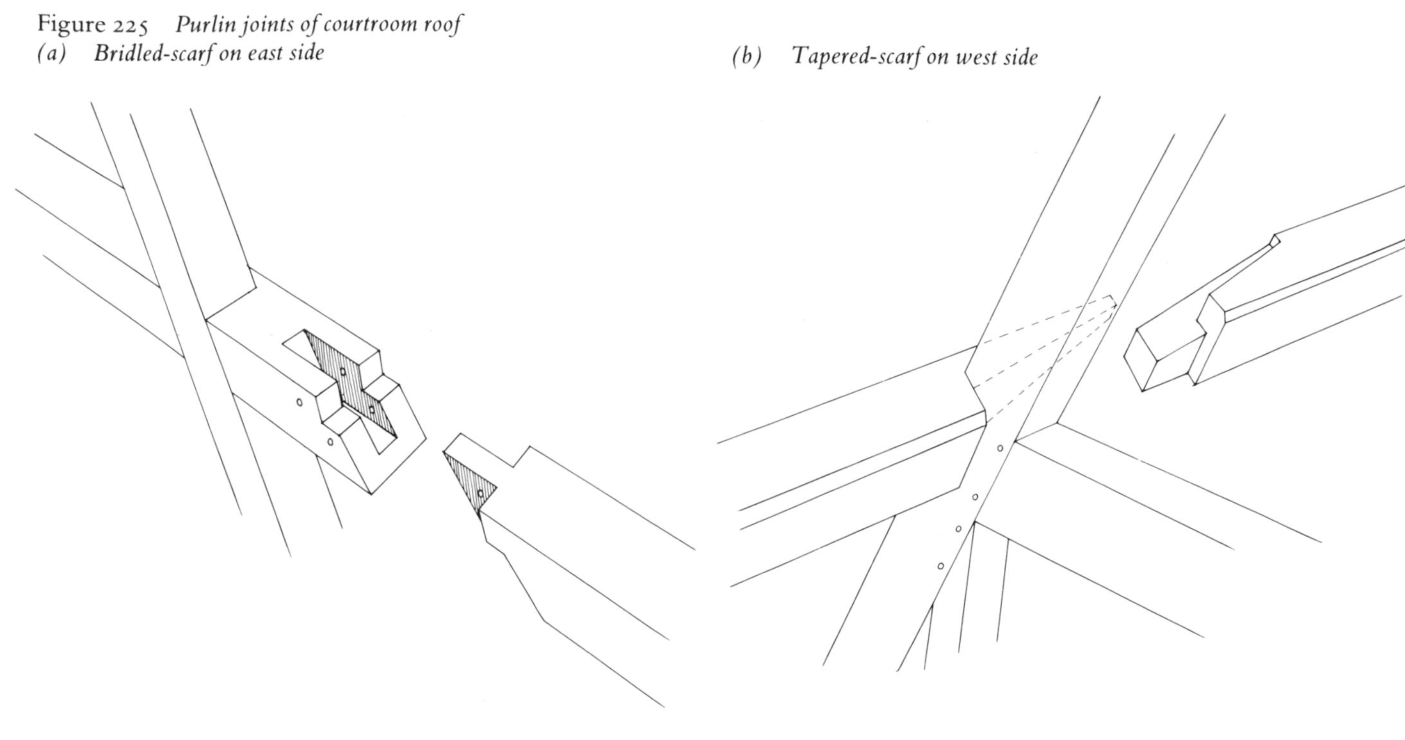

Figure 226 *East elevation c.1720*

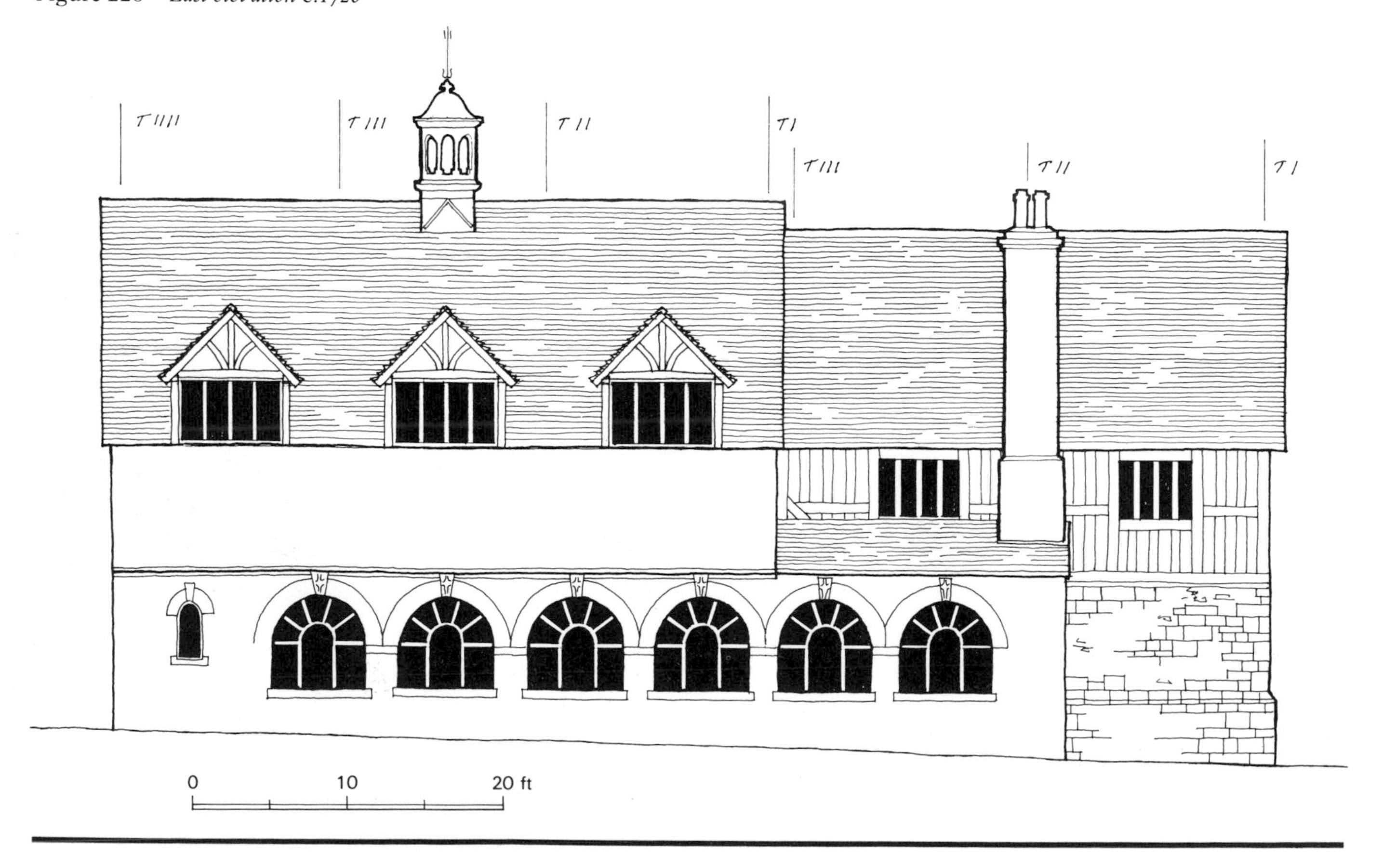

can easily be recognized as Victorian in its ornateness, emulating Elizabethan or Jacobean. The jetty brackets, for instance, support false beam-ends and nothing else. Little otherwise was done to the exterior, except to replace the original windows, probably simply ovolo-moulded, with the more elaborate versions that still exist throughout the guildhall.

As for the interior, the original committee room had one intermediate truss, placed centrally between the windows as indicated by the wall-framing. This was now removed and the two existing trusses inserted. There are no posts in the wall-frame to support them; thus there are no knee-braces to their tie-beams (Figure 227). They take their bearing off the wall-plates only, each on the line of one of the front window jambs. The tie-beams are boxed, the boxing of one of them possibly concealing an original tie-beam, the other probably a softwood or compound beam. Two inner posts set 5 feet in from the side walls, each with highly ornamental and unfunctional brackets, give some intermediate support. The trusses have slender queen-posts and splayed braces. The purlins, of which there are two to each slope, are also slender and chamfered. The entire roof structure is thus Victorian. So also are the rich panelling and 'Elizabethan' chairs ranged along the north wall.

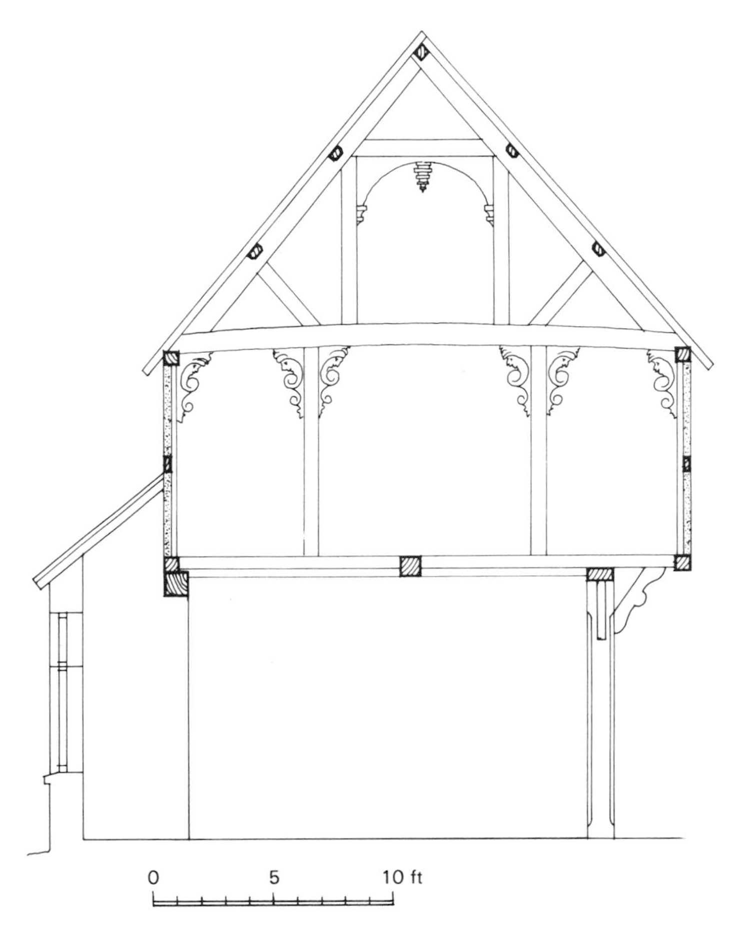

Figure 227 *Victorianized committee room – truss III*

As for the courtroom, first is the jury box which was constructed on the east side of bay 1. This gives some architectural distinction to the garden elevation, even though it was designed with little regard either to traditional framing technique or to its effect on the original structure (Figure 228). It is entirely of softwood, with timbers like matchsticks in comparison with customary oak members; moreover, to provide headroom within it, the original wall-plate across the bay was cut out, thus removing lateral restraint between the posts of trusses I and II.

Second, this room was fitted out as a law court, with raised dais and judges' bench, witness box, prisoners' dock, clerks' enclosure, and public rail; it even contained a set of stocks on wheels. The introduction of these fittings involved certain structural alterations, of which the most serious was the removal of the large swept knee-braces of the two internal trusses (see Figure 224). This must have been found necessary as the braces restricted headroom along the side walls, while the fittings prevented circulation in the middle. In place of the braces, additional posts were bolted to the inner edge of the existing posts, some with diminutive knee-braces (Figure 229). Tenoned into the tie-beam at their head, the posts were no doubt intended to restrain transverse racking. It is tempting to think that two of these posts may have come from the

Figure 228 *East elevation before restoration*

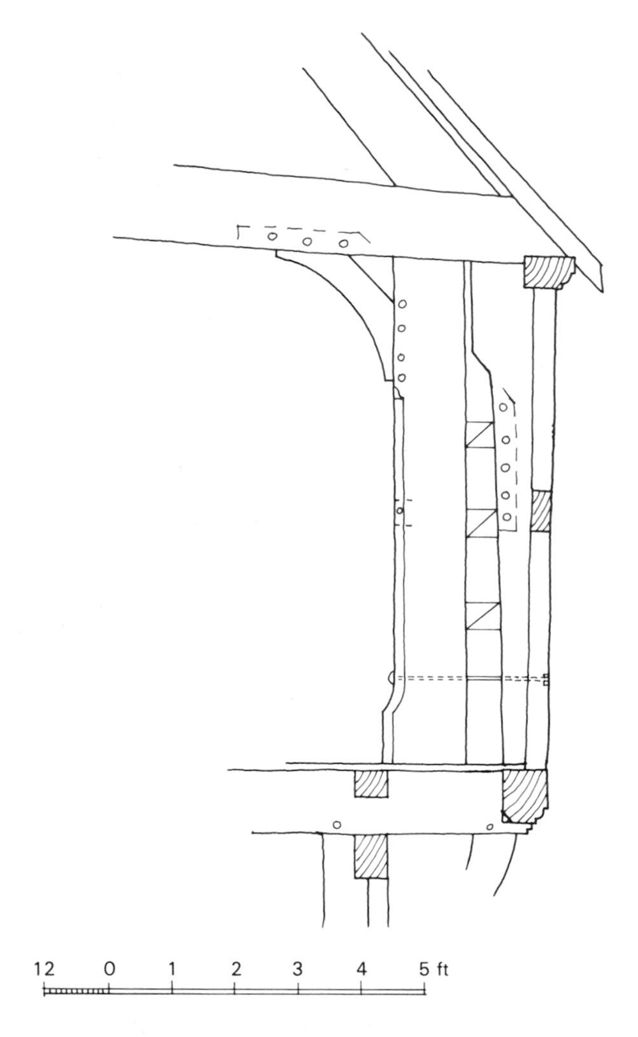

Figure 229 *Typical inserted post and knee-brace*

Figure 230 *Foot of inserted post before restoration*

committee room when its roof was reconstructed, especially since, with jowels top and bottom, they are very similar to the north-east external post of this building. The design of the stopped chamfers is also appropriate for the building's date. This theory must nevertheless be abandoned as all four posts had clearly been within one and the same building. Thus they were obtained from another building, probably in the locality and also being 'restored'! On the street side these inserted posts had nothing to stand on except where the toe of the jowled foot caught the edge of the joist, with the result shown in Figure 230. Finally the wall-plates were severed across the tall windows, as at the jury box. This may have been done when the windows were first constructed but it is more likely that in the earlier version the wall-plate continued across the windows as their transomes.

The last major alteration was the construction of the extreme south bay, set slightly obliquely to the rest of the street front (Figure 231). This provides the way into the present door and staircase, of the same date, and, on the upper floor, the cloakroom and w.c. The front wall is again framed in softwood as the jury box but with decorative panels copied from those of the window gables, slightly refined. This new bay can be dated to 1868 from the minutes, which read

> Proposed . . . that a retiring Room for the use of the Magistrates and a water closet be erected according to the Plan laid before this meeting by Mr Nevett at the south end of the Town Hall the cost of same together with the enclosure of the passage leading into the Church Yard not to exceed the sum of one hundred and seven pounds and carried *nem. con.*

This was in February; then in August 1868, we have the entry

> Ordered that the Councillor of the Wenlock Ward examine the works in progress and if satisfied pay the sum of £80 on account of the contract.

And again, in 1870

> (Resolved) that thanks of this Council be given to Messrs Cooper Riveton & Sons for their handsome donation of four windows for the Butter Market under the Town Hall.

No doubt the eighteenth-century arcading on the east side had been open until this date.

In all of this the first mention of a name is that of Nevett. Though builder rather than architect, Nevett could very well have designed and carried out all the Victorian alterations. There is, however, a distinct likeness between those parts of the guildhall and the museum built on the opposite side of the road in 1878, and the latter was by an architect, Smith. Perhaps he also had a hand in the former, even though the minutes do not seem to note him.

The architect Lloyd Oswell is also mentioned, but rather too late for the bulk of the work. The entry for July 1891 reads

> The Town Clerk reported that the Committee had met and had instructed Mr Lloyd Oswell Architect Shrewsbury to prepare plans for the proposed repair and improvements at the Guildhall and that Mr Lloyd Oswell had prepared a preliminary plan of what he recommended but that no estimates of the cost had yet been obtained.

and September 1891

> The Mayor read the report of the Committee recommending the acceptance of the tender received from Mr Bowdler of Shrewsbury subject to the deductions suggested by Mr Lloyd Oswell the Architect reducing the estimate to £470.
>
> With reference to the heating of the building, the Town Clerk explained the differences between the high and low pressure systems of hot water heating and stated that Mr Oswell recommended the acceptance of Renton Gibbs estimate for the high pressure system of £50.

Thus to summarize this history:

1541 — Gaol already in existence with 'an arch roof or office of the Chamber of Great Wenlock' over it, and 'Richard Dawley paid £13 6s. 8d. (and other costs) including 14s. 4d. for roofing the belfry with lead'. Shingles previously mentioned (1539), also laying a carpet (1540).

Figure 231 *View from Wilmore Street with 1868 extension (nearest bay)*

1577 — This building removed or destroyed and the committee room, still called Town Hall, reared in two days on top of the gaol – of two cells.

1589 — Courtroom built at south end, subdivided into separate rooms according to its three bays, with stair in bay 1 to serve both buildings.

1624 — Mostly interior alterations, but of interest in the use of paint and lath-and-plaster.

1700 — Courtroom substantially altered – partitions taken down, tall windows added and roof reconstructed.

1720 — Whole of timberwork, including pentice, on east side replaced by brick; cupola erected.

1860–1890 — Extensive alterations, inside and out, without regard to traditional timber-frame system; south bay added.

Survey

There were two main sources of trouble – traffic in Wilmore Street and the alterations of the last century. The combination of these would in a few years' time have seen the collapse of the front of the courtroom on to the road. In one sense the Victorians were even responsible for the problems caused by modern traffic. For with the erection of the south bay the guildhall, instead of presenting an island in the town plan of Much Wenlock, became a street frontage. However attractive may be the present secluded lawn and trees at the back of the building, the traditional plan would have enabled even today's traffic to circulate – and with room to spare. Nor would the essential urbanism of an ancient market-place, the more impressive at Much Wenlock with the church forming its northern boundary, have been lost.

But that is another question; the more material result is the damage to the guildhall. Wilmore Street is 21 feet 10 inches wide at its best and 17 feet at worst. Taking the jetty into account the actual carriageway was only 15 feet down to 10 feet in width, and the upper floor of both buildings actually oversailed the kerb of the pavement (Figure 232(a) and (b)). No wonder their bressummers had been replaced and refaced with boards that were constantly being grazed and splintered by traffic. On one of the first of our survey sheets is noted, 'Braces II, V, VI are already replacements – only VI resembles the original. Braces III and IV are original – IV has been fractured and splintered, III less seriously damaged.'

There was another cause for concern. Within the jury box ominous fungi were appearing. There was also serious damp in the w.c. of the south bay, again with suspected dry rot of the window lintel and ceiling joists. Apart from these obvious defects, the arrangement of the fittings had over the years become inconvenient for the county court still held there, the interior decoration was threadbare, and the heating system and all other services were out-of-date and inefficient.

At that stage the Borough Council called us in. The survey was begun in October 1968. Its outcome was a specification for stage 1, issued in December 1969. But not all that time had been taken in surveying. That there was a lot to be done and more to be examined before the work could be specified was obvious within a fortnight. A meeting with the councillors at the Guildhall resulted in the town clerk, A. G. Matthews, being authorized to take us on and enquire into the possibility of grants, for which a global estimate was required. £16,000 seemed to be a safe figure, based on the time that would probably have to be spent on each operation. Even so this was hardly more than a guess, as we still had had insufficient experience to apply rates for building elements – roof, walls, floors – according to past work. The estimate was at least adequate for the purpose of grant aid.

In this respect, Much Wenlock borough was in the enviable position of possessing a nationally important building but the minimum of administrative responsibility. It was within the district of Bridgnorth, and over

Figure 232 *Wilmore Street widths*

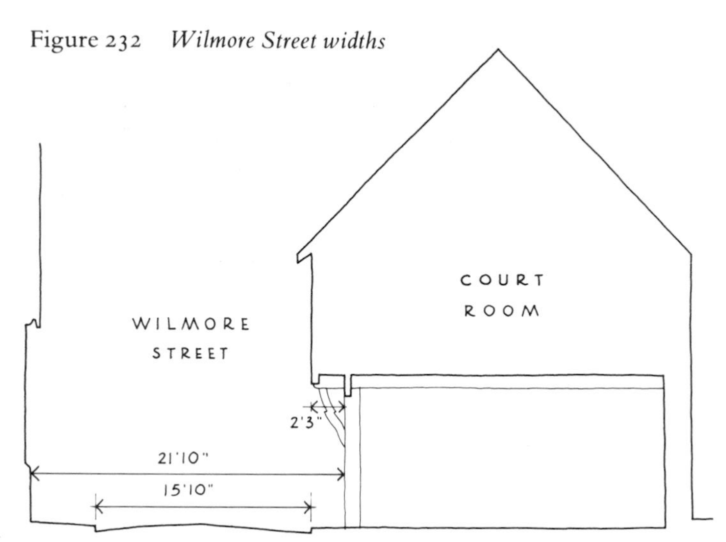

(a) At its widest towards south end of guildhall

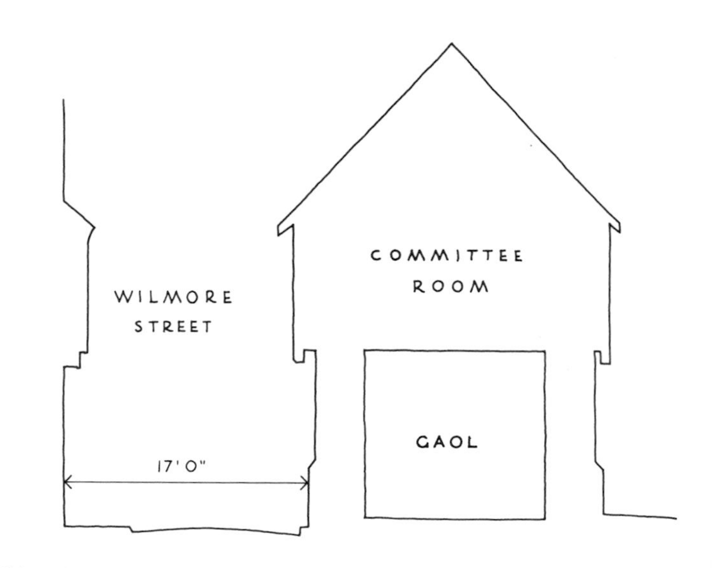

(b) At its narrowest at north end

Bridgnorth council was the County. With the addition of Historic Buildings Council, there were no less than three sources for financial aid, resulting, thanks to the town clerk's skill in negotiation, in a 25 per cent contribution from each. Thus Much Wenlock had to find a mere £4000.

The more relevant parts of the stage 1 specification are reproduced below. That it was considered necessary to prepare such a document emphasizes the degree of detail that had to be gone into to obtain a fixed price for the main, stage 2, contract. It also affords a convenient description of the work which the survey up to that stage had shown to be necessary. This is summarized, as it was for the councillors, in the interim report of 16 February 1970, extracts of which are also given later. The less technical reader may now like to turn to these extracts, skipping the next few technical pages. Those who like specifications may read on, noting that the various structures are referred to as blocks A, B and C, being respectively the committee room, courtroom and south bay. The buttermarket is the complete covered area beneath the guildhall.

Note that references to drawing numbers in the specification do not apply to the figures in this book.

Specification – stage 1

General

The Guildhall, Much Wenlock, is to be reconditioned. There are two stages of work. This specification covers stage 1 only. The eighth-scale drawings outline both stages.

The builder's work in stage 1 involves dismantling, taking down, storing, scaffolding, propping and protection, as a preliminary to stage 2. During stage 1 the architect will carry out a detailed investigation of the opened-up parts of Block B so that fixed prices can be obtained for the reconditioning.

The programme of work is as follows:

Stage 1 December 1969 to February 1970
Stage 2 February 1970 for twelve months.

Scaffold

1 The builder must provide all scaffold. . . . The way through block C must be kept open. . . .

2 Scaffolding will be required for the stage 1 stripping and propping and for access to parts of the building which require detailed survey work for stage 2. The props will also be required for stage 2 to hold up wall-plates, tie-beams, bressummers etc. Arrangements must be made for these to be retained in position during stage 2, even though the same builder may not be employed in both stages of the work.

3 *West wall* – see drawings no. 5, 6 and 11.

Block A: Access required to wall-plate level and bressummer level.

Block B: Access to apex of window gables, all areas of the wall-framing to wall-plate height, front and back (including rear dormer windows) and concealed junction in gables between blocks A and B.

Block C: Access to top of front gable and all other parts of wall-frame. In addition access is required over the full height of the concealed bargeboard under the gable between blocks B and C.

East wall – see drawings no. 7 and 8.

Block A: Return scaffold around junction of block A and B giving access to concealed bargeboards.

Block B: Access to wall-plate and to apex of bay window including side cheeks.

Block C: Access to wall-plate and the concealed bargeboard between blocks B and C.

End wall – north elevation

Return west wall scaffold around the corner to give access to the first-floor framing of the exposed gable and the apex of the bargeboards.

4 *Block B – internally*

Scaffolding and planking will be required for access to strip the walls and ceilings and for detailed survey of the roof and wall-frames. Some support may be obtained between the trusses but the bulk of any super-imposed scaffold load should be taken down to the ground through the first floor, and not propped off the first floor without the consent of the architect. Access will be required to inspect the whole of the roof timbers, dormer timbers and trusses.

5 Buttermarket – further staging must be provided at approximately 6 feet 0 inches below first-floor level for a detailed inspection of all four bays of the floor.

6 It is necessary to retain the whole of the scaffold in position on site until stage 2 of the contract is let. This stage should follow on quickly behind stage 1. The scaffold will be maintained by the stage 1 builder until 1 February 1970 or as agreed on site.

7 At ground-floor level a continuous hoarding must be provided.

Fittings

8 Move the following unfixed furniture to store in adjoining committee room, or ground floor of assembly hall opposite:

13 pew benches, 2 tables (one with red cloth and desk), 2 chests, 1 round table, stocks on rollers, barrow on table, books, gun, breastplate, 5 large chairs.

9 Carefully dismantle the following furniture and move to store (as above). See record drawing no. 10.

(a) Public rail (softwood) in 2 lengths total 28 feet

long, 4 feet 3 inches high including posts and rails, uprights, floor strap, hinged gate.

(b) Dock – unfasten and dismantle 2 posts, rails, balustrade, seat, gate hinges and straps to floor.

(c) Clerk's bench – unfasten from floor and dismantle, corner posts, boarding, rails and seats with matchboarded backs.

(d) Witness box – unfasten from floor and remove in one piece.

(e) Main bench (raised area) – take up floorboards, joists, framing and fastenings (see heaters under services). Take out steps at each end including risers, nosings, and carriages. Unfasten four main posts with bookrest and scroll carvings, including panelling, to leave original floor line and vertical wall surface free of all framing and fixtures.

(f) Jury seats – dismantle 2 lengths of seats and bookrests. Take up the floorboards, joists, framing, together with treads and risers. Take out seat ends and backs.

10 Carefully dismantle or take down and store the following fittings and linings from walls and floor:

(a) *West wall*
Bay 1 (stairs end): carefully remove the picture (Dr Brooks) and panelled area behind, including the dado rail and matchboarding under, skirting, and all battens, grounds and fixings.
Bay 2: 'Bailiffs' and Mayors'' board including matchboarded panel under with the moulding, skirting and all battens, framings and grounds.
Bay 3: the moulded panels on either side of the window and the panelled area below the window including the skirting, dado rail, and all framing behind to the face of the plaster.

(b) *North wall*
Protect and leave this wall intact, except as specified elsewhere under taking down.

(c) *East wall*
Bay 3: bay window – take down all the stained panelling at each side and below the mullioned window including battens, back to the main softwood framing of the bay. Take down the panelling in bay 1, not within the bay window, adjacent to the north wall.
Bay 2: take down the 'Members of Parliament' board with fixings, and battens. Take down the matchboarded area under including the dado rail, skirting and all battens and grounds.
Bay 1: carefully take out the tall bookcase with cupboard doors, shelves and cupboards under. The books and contents will have been removed by others.

(d) *South wall*
Staircase glazed screen and canopy. Protect and leave intact.

West wall

11 Carefully probe the inside wall lining for the studs and rails visible externally. The extent of this work should be agreed with the architect on site. Carefully remove plaster and lath. Protect any original wattle-and-daub infill panels, and any decoration or painting found underneath.

Expose the wall-plate and bressummer and all construction under.

12 Open up the original window by removing the four brick infill panels as shown on drawing no. 6.

13 Allow for opening up forty-two infill panels of brickwork. This work includes the removal of both finished faces of plaster, the removal of brickwork, carting away of materials and protection by polythene afterwards.

It is possible that the infill panels mentioned in the clause above are original. If this is found to be the case then the infill must be carefully protected on both sides, and retained intact. Instructions will be issued on site at that stage of the work.

East wall

14 Remove square sections of the brickwork for needling below the wall-plate. Needles to be supported from ground level not from the first-floor joists.

As a preliminary to taking down the external brick skin, the internal lath-and-plaster should be removed to ascertain the extent and condition of the original timber frame.

When this preliminary investigation is complete, instructions will be issued for the removal of the brickwork.

Take down the whole of the brickwork in bays 2 and 3, and between block A and the projecting bay 1 of block B – in all 28 feet long, 8 feet high, approx 9 inch thickness. The existing brickwork string course above the window arches must stay and mark the lower limit of brickwork removal. All brickwork should be removed above together with fascia and r.w. gutters to expose the wall-plate and all framing. Nothing must be taken down beyond the straight joint between blocks B and C.

The original framing should not be regarded as load bearing in its uncovered condition. The props must be kept in position through stage 1 and arrangements must be made with the scaffold hire firm and the builder for stage 2 to retain the scaffold in position and hand over on completion of stage 2.

15 Jury window area – take out the eight brick infill panels between the brackets above the brick string course. See other notes for opening up in buttermarket and courtroom.

Take all necessary precautions for propping, shoring the window structure in its existing position.

16 The extent of taking down the brickwork of the return of the east wall at the north end of block B will be decided on site when the corner post I has been located and exposed.

17 Protect all new wall openings with 1000 gauge polythene, battened to the outside of the timber frames.

General

18 Take down all lath-and-plaster work above wall-plates, including the main ceiling and false ceiling in bays 2 and 3. Carefully remove the boarded and panelled area of ceiling in bay 1 over the bench and clerk's table. In all three bays the ceiling joists (or collars) should be left in position until stage 2. (See note about removal of electrical wiring.)

Take down all areas of plaster within the gabled and dormer windows so that the whole of the structural framing is open to view and inspection. This includes ceilings, cheeks and spandrels, and vertical faces on each side of the windows.

19 The floor has been frequently repaired and is now a mixture of hardwood and softwood, in plain and jointed boarding in various widths and lengths. The whole of the boarded floor should be taken up and set aside for possible reuse, to be decided at an early stage. Allow for carefully taking up and storage only.

20 Take down the plaster ceilings of the buttermarket including all laths, battens, packing, infill material. Only the ceiling area under block B to be taken out, i.e. bays 1 to 4 only. All the floor joists should be exposed, with the main beams, and all later props, insertions, straps, trimmers, and spacers retained, but exposed for detailed inspection. All services will have been removed as a preliminary.

21 Remove matchboarded and framed duct, 25 feet long by 3 feet 4 inches by 1 foot 7 inches as preliminary to dismantling heating pipes and removal of floorboards and ceilings.

22 See specification relating to services. In addition remove all wires and pipes which are not in use.

23 Remove inserted posts of T II and T III, both sides, but first prop the wall-plates and tie-beams from the ground floor. Remove the oak pegs securing the post heads to the underside of the tie-beams. Remove the bolts and carefully lower the posts of the west wall through the floor to the buttermarket. The posts of the east wall must be lowered to release the tenons and then brought up to the first floor.

24 Remove half-brick infill in the framework of the tall windows, including the brick cheeks. Allow for protection of all openings.

25 Remove the boards covering the bressummers of both blocks A and B. The feet of the studs, beam-bearings and weatherings can then be inspected.

This specification went out to tender to four contractors who had been previously interviewed at their offices. All, as usual, said they had experience of restoration, though not all thought that preserving ancient buildings was worth while. Each was told that a larger contract would follow and that the winner of the first, if his work was satisfactory, would be given the opportunity to negotiate with our quantity surveyor the cost of the second, and so complete both contracts.

Galliers of Shrewsbury submitted the lowest tender of £1705 for stage 1. Slightly more than a decade later, it is difficult to credit such a price for the work set out in the specification. Nor has by any means all of it been included; a major section on dismantling and disposing of the entire central heating system has been omitted.

Interim report

The interim report was submitted when stage 1 contract was well under way. Though issued to the contractor to give him some advance notice of the work to be done in the second contract, its main purpose was to inform the council and the contributing authorities including the Historic Buildings Council. The first part set out the guildhall's history, already described and so omitted here, then the building's construction and condition. This is summarized as follows, without however repeating defects already noted.

Roof

The roof covering consists of tiles, laths and one-inch boarding. The roof-covering rests on rafters most of which are original; all of them, however, have been severely mutilated and their positions altered, and few are now of one piece from wall-plate to ridge. Many are reinforced by new sides nailed on to obtain a level surface for the boarding. On the east slope the upper purlin has been fractured in bay 3. All the purlins have now to some extent sagged and been packed and reinforced on their upper face; one of them, the purlin already noted, has an ovolomoulded mullion from an earlier window acting in this capacity. The lower purlin of bays 2 and 3, east, have been removed and a beam inserted, spanning between the trusses to carry a Victorian ceiling over the dormer.

Collars, about 3 inches by 2 inches in section, have been nailed to the rafters to carry the main ceiling.

The trusses and main frames
T I has not been altered except for the cutting of one of the cross-braces to gain access into the roof space of block A. T IV has been bricked up on both sides, but some if not all of the timbers have survived probably now in a decayed state.

Window structures
The construction of the two tall windows contains several reused timbers. Their valley rafters seem to have been the cross-braces of an earlier truss. Their gable purlins are also reused timbers, but only that on the north slope of the bay 3 window can be identified as a section of a wall-plate on which are still the joints for a main post, studs, and tie-beam – all in apparently new condition. The windows themselves are Victorian. So is the jury box, but the valley trimming of the rafters is of earlier date, suggesting a former upper window in this position. The whole of the new structure is in softwood, including 7 inch by 4 inch purlins, 4 inch by 2 inch rafters and ceiling joists, and external framing of 5 inch by 4 inch softwood. The window bay projects about 2 feet beyond the east wall face and is supported on seven compound cantilevered brackets. Each bracket is composed of two sections about 8 inches by 4 inches, laid one on top of the other. Only two of the brackets are bolted through to an original floor joist.

The other two high-level windows on the east slope are lightly framed in softwood and sit on the wall-plate of the brickwork.

Wall-frames
On the west side practically the whole of the original framework survives with the important exception of the wall-plate in bays 1 and 3, removed for the Victorian windows. The bressummer, except for a short length of the original in bay 3 of the substructure, has been replaced; the replacement sections of unknown age extending across bays 1 and 2 and 4 and 5 of the substructure are not morticed for the wall-framing. Pegs of the original section are out of position in relation to the studs. Each stud and even the two intermediate main posts are skew-nailed from the front on to the top surface of the bressummer. The post of T II has already slipped 2 inches outwards at the base and all the nails have corroded and the timber through which they are driven decayed. Each of the wall-frames has shifted outwards at wall-plate level. The outward inclination at the centre of bay 2 is about 8 inches. In bays 1 and 3 the Victorian windows which are of oak and well constructed have tended to hold the wall-frame vertical. But each has shifted outwards about 6 inches at plate level and then leans somewhat inwards from that level to ridge.

The timbering of the east wall, including the bressummer, is wholly decayed owing to its contact with the eighteenth-century brickwork. The original plate still partially survives but another one has been placed on the brickwork outside it. It is the latter on which the rafters now rest. There is no wall-framing in bay 1, replaced by the jury box.

Conclusions
The following conclusions are still subject to modification in the light of further stripping, but a minimal scheme for restoration must take them into account:

1. A continuous wall-plate must be reinstated in each side wall.
2. The bressummer of each side wall must also be renewed, or extensively renewed with mortices for the wall-posts and frames.
3. The west wall frame will require repairs and reinstatement of the tenons of each stud now nailed to the existing bressummer.
4. To make these repairs possible the entire roof covering must be removed. This would in any case be necessary to bring back the transverse frames into the vertical and would also be desirable for provision of roof insulation.
5. At least six out of the original twelve purlin lengths will have to be replaced, also 50 per cent or more of the rafters.
6. The inserted posts are doing no work; they must be removed and the knee-braces as in the original building reinstated.
7. The tall windows of the west elevation must be taken out and adapted, with minimum alteration, to the

reinstated wall-plate.

8 The jury box will be treated by drilling and chemical impregnation for dry rot and woodworm, and its two brackets and side framing, infected by dry rot, replaced.
9 The east wall framing of bays 2 and 3 must be wholly renewed as originally and all the main posts taken out for repair.

These are the essential structural repairs of block B. Its present external appearance will be altered only in such detail as will not be noticeable. Internally the reinstated knee-braces will inevitably reduce headroom on either side of the court. Otherwise reinstatement will be identical with the interior before it was stripped.

A contract should be let for these repairs while a further contract (stage 3) to make the building usable is being prepared. Stage 2 will thus be confined to structure, windows and insulated cladding, so that the thermal insulation factor becomes satisfactory for the floor as well as the roof and walls.

This contract should start consecutively with completion of the current stage 1 contract and the cost should be negotiated with the present contractor, Frank Galliers.

There will be a further report on stage 3 covering such matters as heating and other services, and repairs and decoration of block C. Block A has not been surveyed in the same detail as block B as it has not been stripped. In view of the findings in block B, it will be closely examined involving a further amount of internal stripping.

Not all of the work set out was done. Block A was giving no cause for alarm and so was not touched, except for superficial repairs and installation of new services. There was no stage 3, the whole of the work being done in stage 2. Nor was the entire roof covering removed. Only the roof tiles and battens were taken off, leaving the boarding. This was repaired and finally painted white on the underside and exposed with the rest of the restored roof structure. The tiles were replaced on new battens, felt and counter-battens, nailed through the boards to the rafters, with 2 inch glass-fibre quilt laid between them. Reinstatement of the courtroom fittings also did not take place, the legal authorities preferring to have only the judges' bench, slightly less elevated than before, and tables and chairs for the rest of the court personnel.

Surveying was continued throughout stage 1 and it was invaluable to have the contractor's labour available for opening up inaccessible parts of the structure and also to be able to use his ladders and scaffold to get up to them in safety. He completed the contract, as programmed, at the end of February and the second specification and set of details were ready for negotiating the second contract soon afterwards.

Restoration

Galliers' itemized prices on the specification came to £10,900 and, on the strength of that, work began on 6 April 1970 even before the contract had been signed. The first operation was to strip the roof of block B of tiles and battens. It was then found that the cupola was rotten. Nor was the full extent of the work to be done to the east wall-frame known until its brick casing had been taken down to bressummer level, exposing both sides of the timbers. Not only were the wall members, as anticipated, too rotten for reuse, but the amount of decay of posts, bressummer and wall-plate was worse than expected, post I and the wall-plate having to be renewed entirely (Figure 233).

Making allowance for these and a few other items and adding provisional sums for electricity and heating brought the final contract sum up to £12,500.

There was a full-scale meeting for signing the contract on 28 April. As well as the signatories – town clerk and contractor – there were the carpenter/foreman, representatives of subcontractors and of the timber suppliers – Venables of Stafford. The specification was studied clause by clause, that for timber repairs – coming in for most

Figure 233 *The east wall framework exposed by stripping*

of the discussion, particularly on the practicability of cleaving timbers. This was eventually abandoned.

Meanwhile it had been learnt that the whole of Wilmore Street was to be taken up to lay the town's new sewer. One consequence was that most of the scaffold on the street, erected only a month or two earlier had to be taken down; this caused a great row between Bridgnorth Council and Much Wenlock. The former maintained that our work should not have been started, the latter that they had waited long enough and then not been informed when the sewer work was imminent. Possibly the town clerk did know. At any rate he was able to turn what at first seemed a disaster to his council's advantage. Not only was Bridgnorth obliged to pay for taking down and re-erecting the scaffold, but the period during which the street would be closed and traffic diverted, essential for the sewer but not considered necessary for the guildhall, was extended for the duration of our contract as well, giving the contractor invaluable working and storage space. But, of greatest importance, it was possible at the end of the contract to widen the pavement at the expense of the carriageway and protect the building by erecting bollards so massive that the outcry was turned against them instead of the reduced street width. They still stand; so also does an undamaged guildhall.

Despite its extremely thorough preparation the contract turned out to be difficult and lengthy. It was not finally completed until December 1971, taking a year and eight months; even after that the additional work required by the council, and maintenance defects, took us to the end of the following year.

One reason was our own and the contractor's inexperience. For instance, the specification for cleft oak was impracticable, the result of a misunderstanding of traditional methods. It caused delays in delivery of timber for the east wall-frame at the beginning of the contract, when, owing to the absence of scaffolding on Wilmore Street, this was the only part of the structure that could be started. When the timbers at last arrived in mid-June, to a revised specification allowing half-sawn members, they were taken to the contractor's yard for framing. This too was a mistake as they came back, three weeks later, with several studs and rails framed so that their heart-face was to the inside. Still worse, the plain surface of each bay frame, made up, as the original, of timbers not always the same thickness, was the wrong way round, so that the thicker members would have projected to the outside. The whole frame had to be dismantled, mortices altered and framing started all over again.

The bressummer and wall-plate of this wall also caused trouble. Having set these up on the job together with the repaired posts while the wall-frames were in the yard, it was clear that the relative heights of the bressummer to wall-plate at the post positions were quite different from the surveyed measurements recorded before the brickwork had been taken down. Also the bressummer had been damaged by trying to get it into place with the aid of a sledge hammer. Thus all these members had to be taken out, and the work restarted. It is easy to see after the event that the complete wall-framing, including bressummer, wall-plate, posts and the wall members, should have been laid out on the ground and exactly fitted before erection, and then erected altogether.

Finally, concerning this ill-fated wall, long before the end of the maintenance period all the completed panels had to be internally stripped, leaving the subframe and external skin of rendered woodwool. These subframes were not, as in our later jobs, double-frames and they had been very badly fixed. Their repair was a major operation. But the real cause of trouble was that the amount of shrinkage of the green oak had been underestimated.

The west wall, when at length work could start on that, fared hardly better. The specification and details called for an internal subframe of 3 inch steel angles to secure the studs to the bressummer, the wall-frame to the posts and the top member to the wall-plate, restraining, if not fully

Figure 234 *The west wall framework showing sill and head of original window and outward lean of frame*

correcting, its outward bow. By the end of May it was found that the extent of its leaning over the street was increasing as a result of the disturbance to the opposite wall (Figure 234). The contractor therefore took the precaution of tying it back with a wire rope looped round the wall-plate and, passing between the joists of the upper floor, anchored near ground level by a baulk of timber set across the outside of a window opening.

Two weeks later, on Tuesday 23 June, my assistant reported in the job book,

> On Thursday an attempt will be made to pull back the building to its original position.

In some consternation, I wrote up the meeting that took place that Thursday as follows:

> Situation of job was that steel angles and 3 in × 2 in timbers had been screwed to the frame to hold the panels. The belly of the wall-plate of bay 2 had been reduced the previous day by the same means, the rafters having been lifted out of the wall-plate bird's mouths. Wire ropes with Stillson pulley-blocks had been secured to the new bressummer in the opposite wall.
>
> It was intended by the contractor to force the entire structure of the roof and west wall over to the east. This would be dangerous and impracticable.

The job book continues,

> The fixing of the steel work as found yesterday in no respect conforms with instructions, nor was it ever anticipated that forcing the structure back to the vertical would be attempted. This is the completely wrong approach. The proper operation is temporary suspension of the roof and west wall together, but excluding the bressummer, and propping the frame by needling beneath the wall-plate in such a way that the natural tendency will be partial or complete reinstatement to the vertical position. The separation of the rafters from the wall-plate is wrong and the work already done must be put back; the work not done, i.e. stripping the west wall, should already have been carried out.

A fortnight later the wall had been stripped of all its brick panels (three or four original wattle-and-daub panels being retained), the top steel angles had been fitted along the wall-plate of bay 2 and across the windows of bays 1 and 3, and the lightened framework was suspended a few inches above the bressummer. It had however been impossible to force back the rafters into their bird's mouths along the wall-plate. The inclined needles were inserted and the wire ropes were retained in case of disaster but not tensioned. Three weeks later the sceptical foreman phoned to say that the wall-plate had slipped inwards a distance of 4 inches. At 6 inches the wall-frame would be vertical; this soon happened, and so the steel subframe could be completed except for its screwing to the bressummer.

Troubles were, however, far from over. Only a length of 5 feet of the original bressummer was usable. This was $10\frac{1}{2}$ inches on the face by $8\frac{1}{2}$ inches in width, and the moulding on the lower external edge was quite different from that on the inner edge. The bressummer delivered by Venables in two lengths on 4 August was 10 inches by 10 inches and the moulding of each edge was the same. At least it seemed to be correctly morticed for the studs and posts, of which the latter had already been fitted *in situ* with slip-tenons.

The half-inch difference in depth could easily be overcome by setting the bressummer on the cross-beams that much higher, but both of the timbers still had to be returned to the sawmill to correct their mouldings. In all later jobs the timber merchant's responsibility was confined to delivering unworked, square-edged timbers only, the contractor having to apply chamfers and mouldings and, as before, cut the mortices-and-tenons and all the other joints.

The bressummer was redelivered a fortnight later and the three pieces were jointed on the scaffolding, the original 5 feet length being between the two new ones. Slipping the complete 45 foot bressummer sideways on to the cross-beam tenons was certainly the trickiest job the contractor had to undertake (Figure 235). It was done (as always when problems are likely to be encountered) when the architect was *not* present. The wall-frame was then lowered on to the bressummer. The tenons exactly fitted the mortices of the new bressummers, but they landed exactly *between* the mortices of the old!

The instruction to the contractor had been that, as soon as the wall-frame had been lowered, the steel angles already connecting the feet of the studs should be coach-screwed to the bressummer so that there should be no possibility of movement between the wall and the bressummer, and that in addition steel flats should tie the angles securely back to the cross-beams. Never was an instruction carried out with such unnecessary speed in case the architect should demand that this operation too should be done all over again, merely to correct the positions of the old mortices. But the error was not even noted. Nor was it noticed, until several years later, that the new bressummers had no pegs and those of the old one had not even been withdrawn. So, they remain (Figure 236).

Of lesser problems, all windows needed more repair

Figure 235 *Jetty detail showing bearing of bressummer (new timber) on cross-beams*

Figure 237 *Replacement knee-brace – glass-fibre insulation for the west wall, instead of insulated panels, also shown*

Figure 236 *Detail of west elevation, showing new and old bressummer*

Figure 238 *View of courtroom as completed looking south*

than had been anticipated, both of the east dormers having to be scrapped and reconstructed in oak. It would also have been better to get rid of the jury box. For, less than a decade after the contract was finished, and despite the fact that all its softwood framing was chemically treated and that considerably more than its affected members were replaced, rot again appeared. Fortunately it was only wet rot, at the foot of the north-east corner post and along the bressummer.

As for the rest of the structure, the replacement knee-braces as delivered were much too big (Figure 237). This was the result of playing safe by the timber supplier after the experience with the bressummer. Eventually, as worked and fitted by the contractor, all such braces were made as architecturally right as they are structurally necessary (Figure 238).

Insulation of the building consisted of double-skin panels in the east wall, as already noted, and an internal layer of glass-fibre for the west wall (see Figure 237). The reason for the different treatment was that several original panels were preserved in the latter, while the former, as we have seen, was entirely new (Figure 239). Also the west wall was to be internally faced in oak panelling, bearing various rolls of honour of mayors and other dignitaries. In the event, it was finished in plasterboard. For the roof the existing boards were preserved, taking the place of the usual fibreboard but otherwise the standard detail was followed. This meant that the Victorian openwork barge-

Figure 240 *Victorian bargeboards replaced on solid planks and restored cupola*

Figure 239 *East elevation as restored*

Figure 241 *Completed jury room with lowered dais, painted panelling, restored wall-plate across jury box and new tables*

boards could only be reused by mounting them on solid planks to mask the edge of the insulation. This too was the only means of preserving them without a considerable amount of repair and replacement (Figure 240).

Lastly, the opportunity for the Council to have everything new, from defective structure down to fittings and switches, was too good to miss. The new oak strip floor would have been damaged by having the old fittings screwed to it. Thus all except the dais, slightly lowered (Figure 241), were abandoned or consigned to the butter-market or museum. This led to our designing new tables and chairs to take their place. Double windows to improve sound insulation on the street side were another innovation. The final item was a display case placed in the entrance passage which contained a plan of Much Wenlock and an illustrated history and description of the guildhall.

The final cost was £12,348 16s. 8d. – a saving of precisely £151 3s. 4d.

CHAPTER TEN

WELLINGTON INN

Manchester

The building

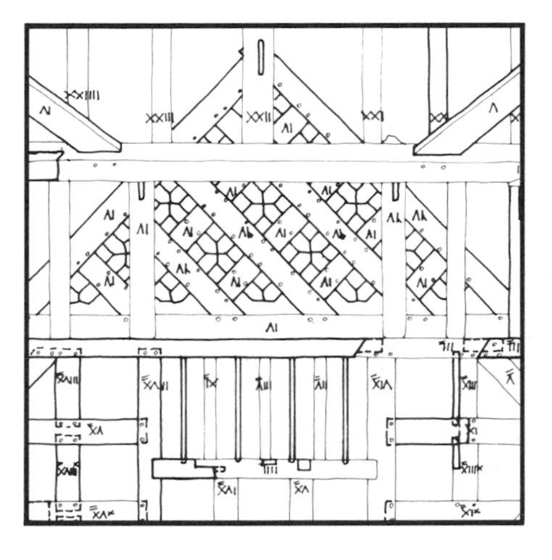

The Old Wellington (Figure 242) is Manchester's only reminder of her former timber-framed city centre (Figure 243). The devastation of ancient buildings, both from bombs and redevelopment plans, was practically complete even before the Royal Exchange corner tower had been repaired. The nearby building 'Deakins Entire' (Figure 244) was the last to go before the Ministry took action and scheduled the Wellington as an ancient monument, even though it was architecturally of rather less value than Deakins. Sinclair's Oyster Bar was also designated 'of historic interest', though it hardly even deserved to be listed. The interest of the two together, however, was that they marked the whereabouts of the Old Shambles.

The Corporation and its developer (Arndale's Central Western District, or CWT), undaunted by this obstacle, made them the focus of their £10 million scheme and undertook, with the aid of Ove Arup as engineers and Pynford as contractor, the unique operation of casting a reinforced concrete raft under both buildings and jacking them up 4 feet 9 inches on massive concrete piers (Figure 245). This was not only to make headroom for the underground car park and storage for the future shops, but also to bring their ground floors up to the new pedestrian precinct level. The operation has been fully written up by John Charge[1] of Ove Arup, the latter remaining the consultant engineers for our restoration above ground.

Bass Charrington were the clients. We began in December 1971 with a stripping schedule, made more difficult than usual by the steel framework that had been inserted from top to bottom of every bay to stiffen the structure for its lifting (Figure 246) and by the external

Figure 242 *East and north elevation as restored*

Figure 243 *The market-place and Shambles in 1810 with the east gable end of the Wellington on left. Courtesy Manchester Central Library*

scaffolding and corrugated sheeting for its protection while the new development went on all round it (Figure 247). The survey thus had to be done amidst not only an entanglement of steel but also trailing wires of the temporary lighting and heating system. At least the building was warm.

The schedule was set out floor by floor and bay by bay, beginning at the top (Figure 248). It was issued in May 1972. Nothing happened, however, for two years, except endless negotiations about Bass taking over the building from the developers who were in the thick of labour troubles, about grants from the Department of the Environment and Manchester Corporation, about insurances, forms of contract, the programme and so on.

At last on 18 March 1974 the stripping began and it was possible to survey the structure and write the 'Survey report and proposals for restoration'. The building had at first been a typical late medieval two-storey jettied structure (Figure 249(a) and (b)), at the end of a row which included the predecessor of Sinclairs. In the next century it had been enlarged by another floor with two diaper-framed side gables as well as the end gable (Figure 250(a), (b) and (c)). The windows were changed, the jetty underbuilt, presumably to enlarge the rooms of the ground floor, and the first floor construction of both bays 1 and 2 was altered, the original dragon-beam being reused in the former as one of the bridging beams. A fireplace and stair were also inserted in bay 2, taking up about 4 feet of its width. The original trusses were reused at the higher level. Except for the fireplaces and staircase, practically all of this has survived.

In the nineteenth century the Wellington underwent a series of changes. The earliest of the photographs (Figure 251) shows the lowest storey already brick; this must have replaced the framing probably at the beginning of the century. It also shows the east gable elevation, facing the market, with new and much larger windows behind a gigantic pair of spectacles. The ground floor was the Wellington Inn, under the name of 'Samuel Kenyon'. On the elevation facing the Shambles, the gable at its west end had already disappeared.

By 1897, the upper storeys had become Ye Olde Fysh-

Figure 244 *The last city centre timber building, except for the Wellington, awaiting demolition. Courtesy Manchester Central Library*

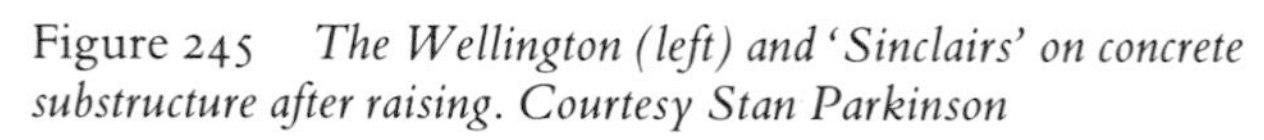

Figure 245 *The Wellington (left) and 'Sinclairs' on concrete substructure after raising. Courtesy Stan Parkinson*

Figure 246 *Internal condition of Wellington at the time of survey. Courtesy Martin Charles*

Figure 247 *The Wellington concealed by sheeting at time of survey. Courtesy Martin Charles*

Figure 248 *Stripping schedules*

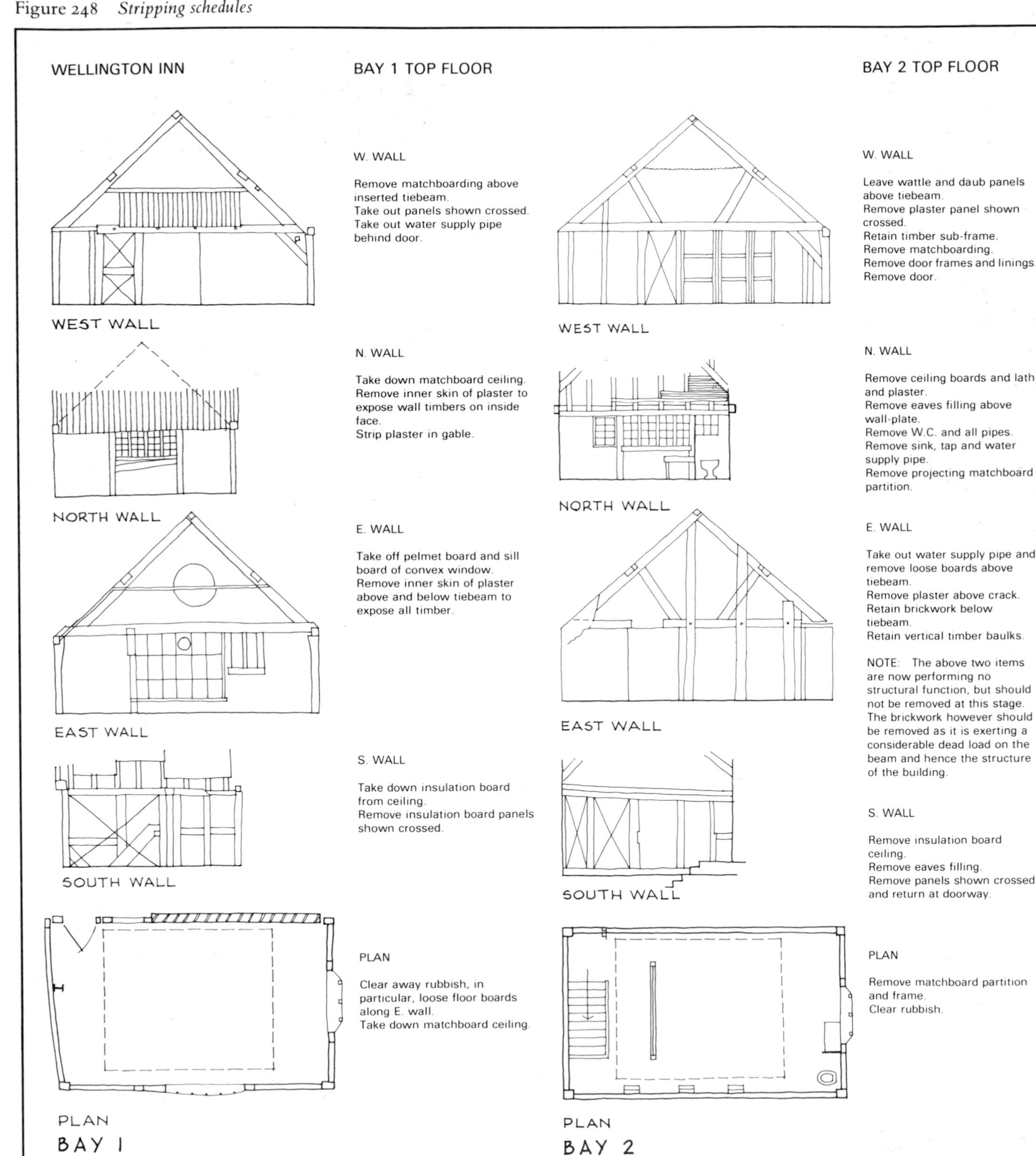

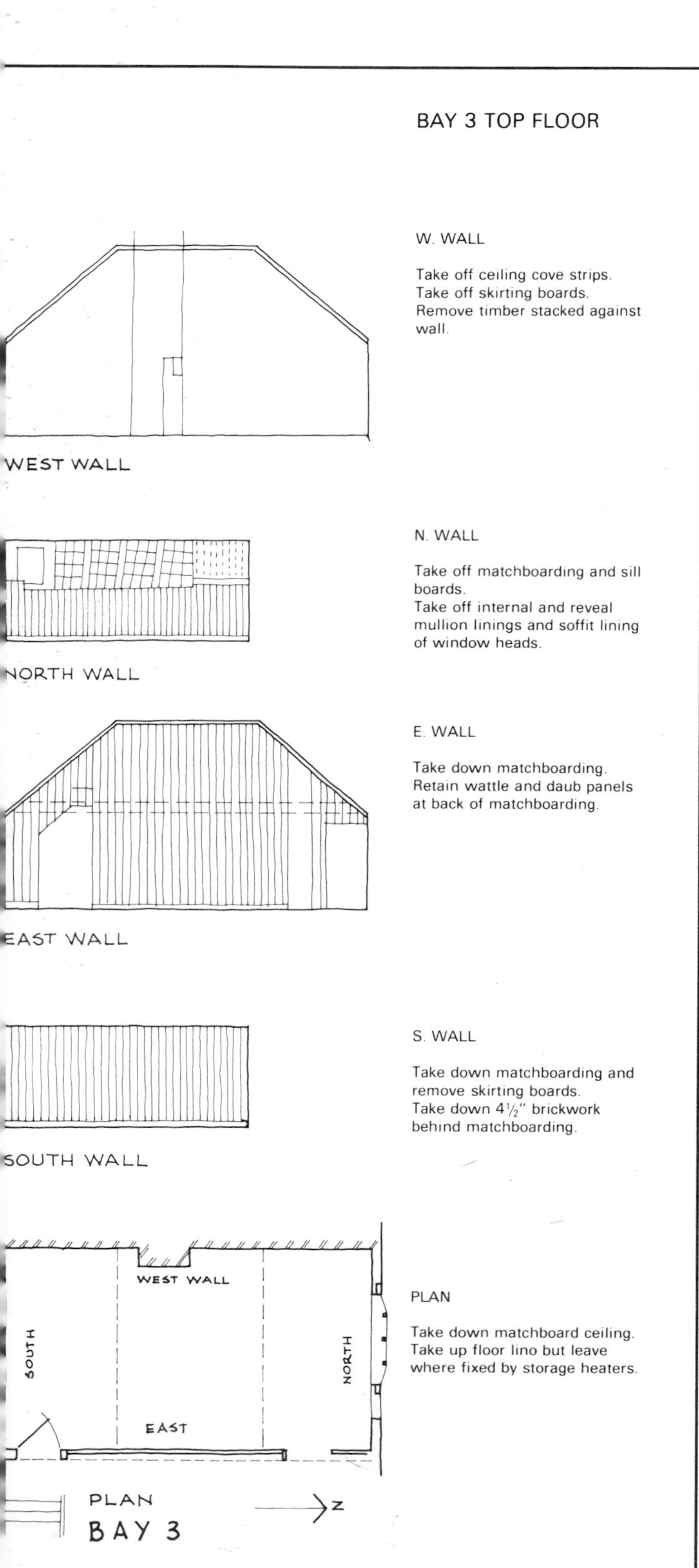

Figure 249 *Probable first design of Wellington*

(*b*) *View from north-east*

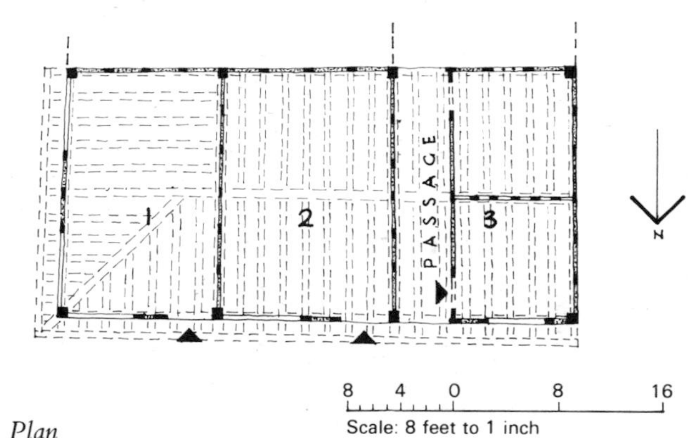

(*a*) *Plan*

ing Tackle Shoppe with a large clock in the main gable, but otherwise, except for the removal of some of the external bric-a-brac, there was no change, and Kenyon was still in business.

It cannot have been much later when, at the further end of the side elevation, a 14 inch brick party wall was erected between the Wellington and Sinclairs, encroaching on the Wellington by a distance of 2 feet into bay 3. This resulted in the whole of the end frame and roof truss being removed, leaving nothing to tie together the opposite wall-frames; the horizontal wall-frame members were simply 'bonded in' to Sinclair's brickwork. Why the party wall was not built on Sinclair's side of the truss has never been discovered (Figure 252). Our first proposal to overcome this problem was to move the whole Wellington the same distance to the east and reinstate it (Figure 253). This however was too much for the authorities and,

Figure 250 *The Wellington as altered and heightened in seventeenth century*

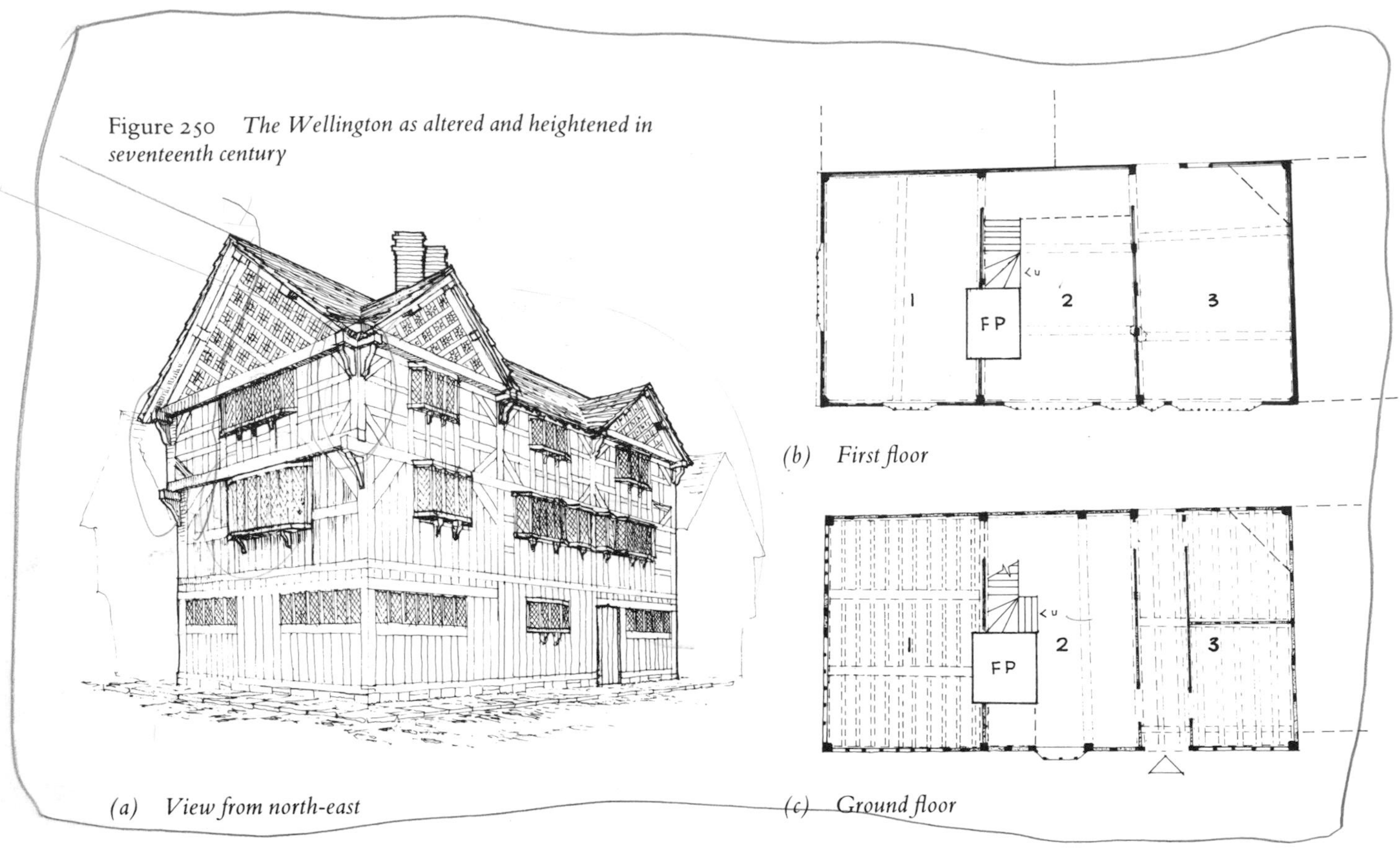

(a) *View from north-east*

(b) *First floor*

(c) *Ground floor*

Figure 251 *The Wellington in 1870. Courtesy Manchester Art Galleries*

Figure 252 *Plan of building before restoration. Note brick party wall encroaching into bay 3*

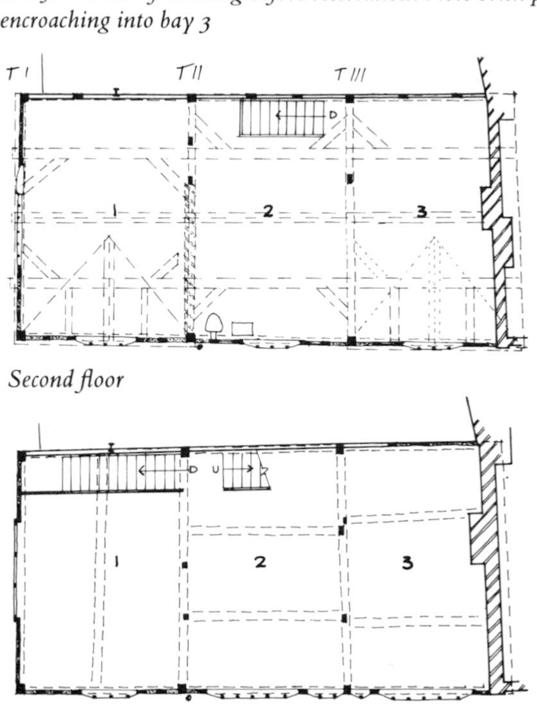

(*a*) *Second floor*

(*b*) *First floor*

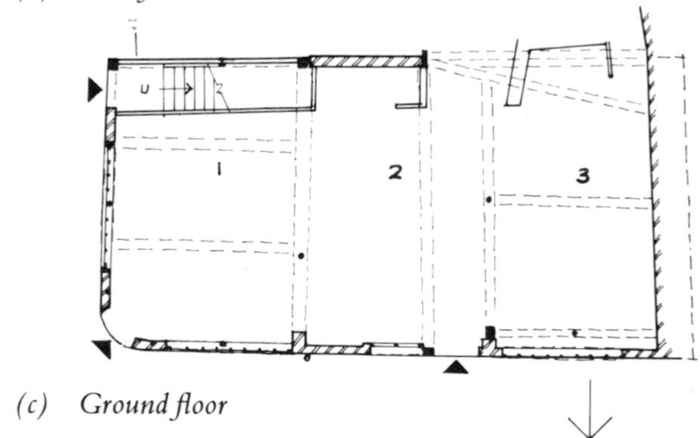

(*c*) *Ground floor*

Figure 253 *First proposal – moving the building 2 feet to east to restore T IV*

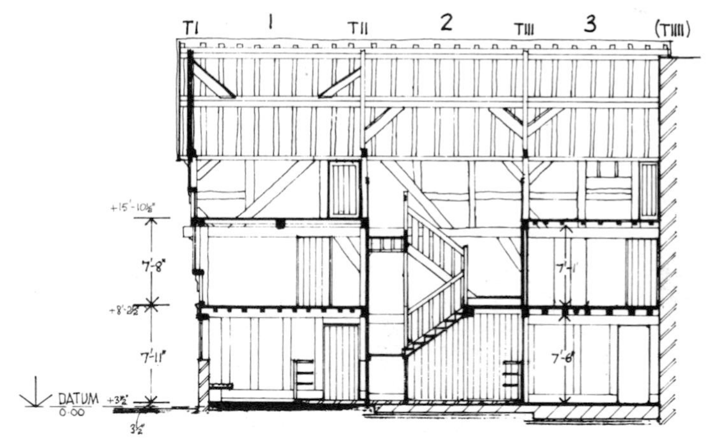

Figure 254 *Long section showing rear wall-frame with large braces indicating former rear building and proposed staircase etc.*

as it would also have involved additional cross-beams beneath the raft, the idea was dropped. Alternatively, the ground-storey structure could have been preserved as it existed and the first floor and upper framing moved, so restoring the jetty of the east elevation. But this also was rejected as too radical, and the projection beyond the building line, even though it did not occupy any pavement space, promised to raise intractable problems of land ownership and ground rent. Clearly the law has become more complicated since the days when every town building was jettied.

Thus we were left with the technical problems of tying back the wall-frame and restoring the gable without reconstructing the end truss. A vertical steel channel, held back by steel rods bedded in the party wall, was erected on the face of the wall to master the ends of the timbers. It was still possible to restore the side gable as its tie-beam had been jettied on brackets and so could oversail Sinclair's brickwork. There was no problem of restraint of the rear wall-frame as the new development, with service accommodation for the Wellington, was built hard up against it.

The timbers of the rear frame were few but clearly indicated the roof line of a former building, and there was probably a little courtyard to its west, accessible through the Wellington (Figure 254). Thus protected from

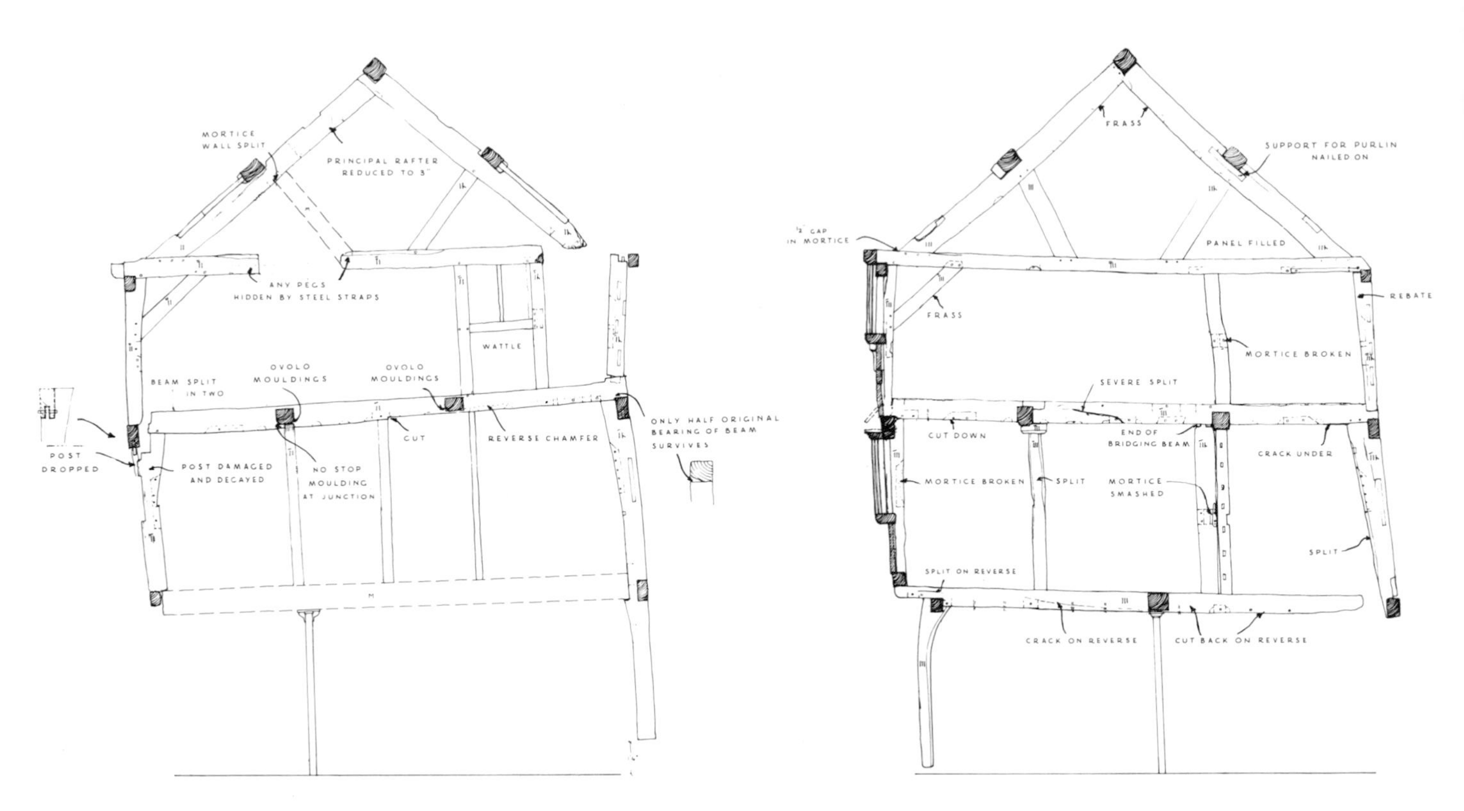

Figure 255 *S drawing T II*

Figure 256 *S drawing T III*

weather, the whole of this side of the building was in better condition than the rest.

Elsewhere, trusses II, III and the long side frame were the worst (Figures 255, 256 and 257). The insertion of the former fireplace of bay 2 had caused the tie-beam of truss II and the upper cross-beam to be cut out where they would have penetrated the flues. The first-floor beam had been entirely replaced by a softwood member, intermediately supported by a cast iron column. Constant leaking of a soil pipe had rotted the post and adjacent timbers, causing them to subside some 12 inches.

Truss III was hardly better, but this frame still had a vestige of the original jetty and the roof truss was more complete. But the real problems were to be seen in the north side wall (Figure 257). First, the horizontal framing of bay 1 had subsided with post II and pulled the east corner post towards it. Second, it was found that the order of erection had been reversed when the top floor was added. The bressummer and second-storey girding-beam (original wall-plate) had been erected from west to east, but the whole of the upper structure, including its wall-plate and purlins, had directional jointing from east to west. Moreover, the insertion of the long ranges of windows at the same date had interfered with the first-storey framing; the peg-holes (as if setting a precedent for the anomaly at Much Wenlock, already noted on page 189) only intermittently conformed with the studs. The bressummer had been fractured and the girding-beam and brace joints pulled apart. Add to this the mutilation of the main roof structure and wall-frame when truss IV was removed, and it may be seen that reframing on the ground followed by piece by piece re-erection would be the only practicable answer.

But as well as the subsidence of post II, the building had been declining gradually from east to west, and this fall, which also amounted to nearly 12 inches, had been, as it were, perpetuated by incorporating in the raft two steps, each on the line of the intermediate trusses. For the completed building these changes of level would have to be eliminated as incompatible with a workable public bar. The floor levels of bays 2 and 3 thus had to be raised; while it was easy, in fact necessary, in realigning the frame of the old building to raise the upper floors, this of course could not be done with the concrete slabs of the rear building. The result was even tighter planning of the latter than its already severe space limitations demanded.

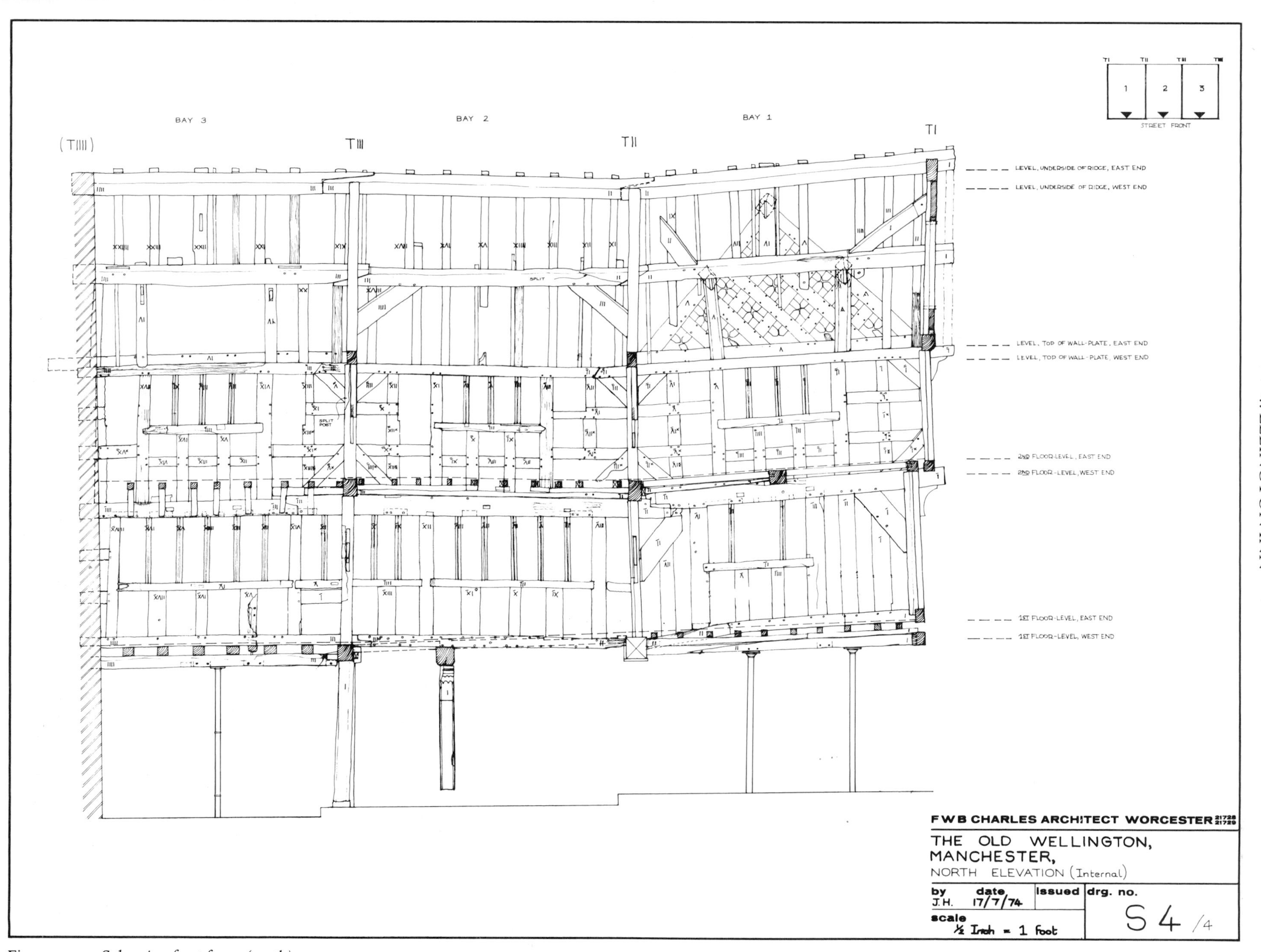

Figure 257 *S drawing front frame (north)*

The scheme

Bass required the use of the entire Wellington instead of, as hitherto, the ground floor only (Figure 258). The ground floor would of course remain the public bar as it always had been, if without the traditionally mixed company of the Old Shambles. The counter of those days was a massive mahogany affair, still available in store. So also were the cast-iron tables with mahogany tops. The latter are again in use, but the restored building had to have, amongst other things not thought of in the old, a fire-protected and regulation-size staircase. The space this would occupy made reuse of the counter impossible. For the new one, Bass's architects were in favour of veneers and artificial finishes on the grounds that pubs had to keep up with fashion and be changed every few years. We insisted on solid wood and designing for permanency, as the building itself. They gave us our way, and it was designed in baulks of iroko.

The first floor was for lunches. Fixed furniture, if only for reasons of economy of space, was necessary, and here we gave way to the Bass architects' preference for settles of traditional design rather than more simple modern benches. The top floor was for special lunches and dinners. The dominance of the re-exposed roof, typical of the north in its crude strength and weight of timbers (Figure 259) made choice of furniture relatively unimportant, so long as it did not try to compete.

The conservative approach to the design of the ground floor was carried to its exterior. The brick walls, windows and continuous fascia would be retained. The only difference was that the new fascia would have to be considerably shallower than the old one so that the girding-beam could be exposed, instead of enclosed and consequently rotting behind it. The existing doors were also right for this scheme and needed only repair and regraining. The brickwork, however, having moved and loosened as a result of the subsidence, had to be rebuilt and the windows entirely remade.

The design stage included several meetings with the DoE as they had some say in the restoration of a scheduled ancient monument. Characteristically, they suggested that the fabric and structure should be preserved exactly as it stood by means of inserting a permanent steel frame. This idea, however, was dropped when the problems not only of the state of the structure but the lack of space for meeting the client's requirements were pointed out. There were also meetings with officers of the City Departments, in particular the City Engineer and Fire Officer. Both were concerned only to be helpful. The former remarked at one of our earlier discussions, 'If the building has to be taken down and rebuilt, the regulations will apply; if you are only repairing it, there is nothing I can insist on except in new work, such as the staircase.' Treatment of the frame for fire resistance was the last thing either of them would have asked for, particularly as in the City Engineer's

Figure 258 *Plans, including rear accommodation, according to final proposals*

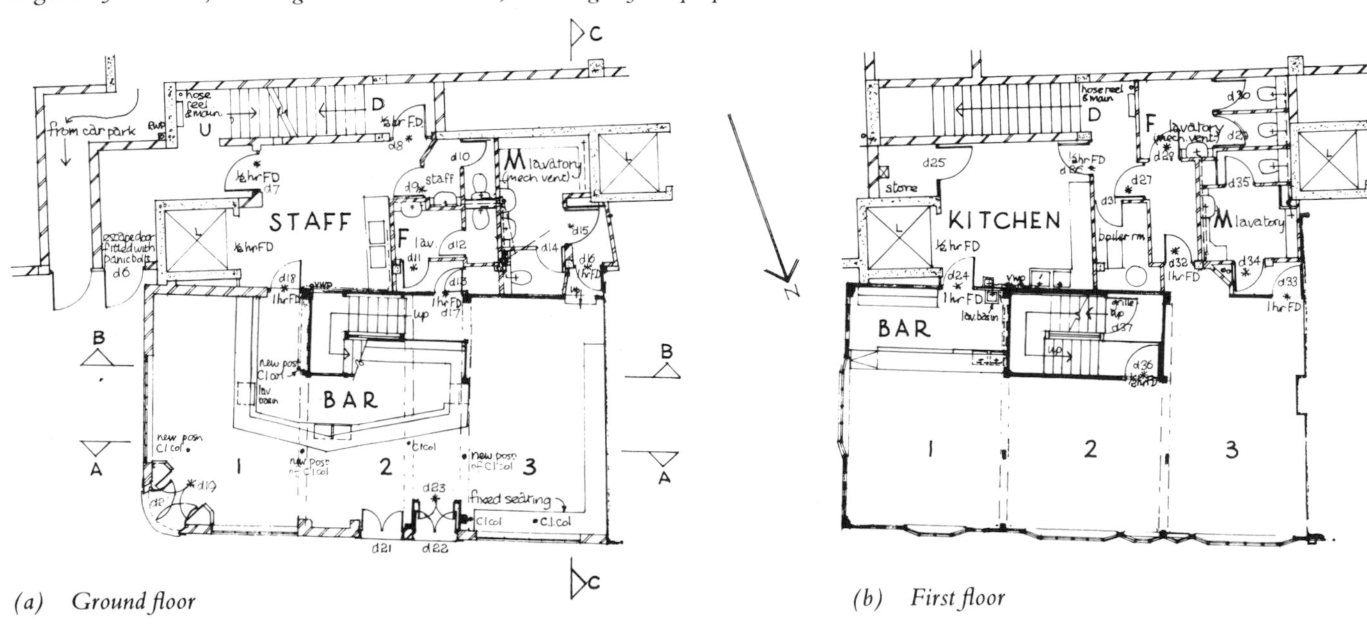

(a) Ground floor *(b) First floor*

experience, the fibres of timber could be affected by such treatment. They both agreed that oak was for all practical purposes non-combustible and usually safer in fire than reinforced concrete or steel. There only had to be alternative means of escape. The top floor thus has a way out on to the flat roof of the rear accommodation and the new staircase had to be enclosed with a half-hour fire-resistant door at first floor, the stair being directly accessible at ground floor and top floor.

Only with the developer and his architects were there problems, for big money was involved. The developer was responsible for compensation to Bass for any damage caused by the lifting or later. He was also under contract to hand the building over exactly as it had been before Bass moved out. This would have involved the developer in expense for reinstatement; Bass of course did not want this, but they did want compensation for doing it themselves, albeit in a different form. The main compensation however was for loss of profit while Bass were excluded from the building. Bass's concern therefore was not to take it over in a manner that might be construed in court as terminating their compensation. 'Take-over' and 'hand-over', 'date for possession' and so on, had to be very carefully defined, nor did Bass wish to be left with a building legally usable but without access by the public. Dates for these stages of progress were thus planned with extreme caution. No 'starting date' was acceptable *before* 16 January 1976, four years after our appointment, but the 'completion date' had to be 22 April 1977, giving only fifteen months for the real work!

The contract

With the time for preparation set by the leisurely pace of the law, not only bills of quantities, one-eighth scales, S & R drawings and schedules were ready for tendering, but also a full set of engineer's and our own half-inch details. Only heating, ventilating and fittings were left to be covered by provisional sums.

The firm of Temple Somerville won the contract against three other invited contractors; all of them had been interviewed. Their estimate and contract sum was £118,920, on which the final account showed a saving of £5,326.30. The contract included for the whole fitting out of the new rear accommodation already built by the developer. Thus our first task when work started was to prepare a schedule of defects of their work, so that they would not have to be put right within Bass's contract.

The contractor's first operations were the removal of the external scaffold and sheeting, stripping the roof and stacking the heavy stone slates. Then they erected a new sheeted roof with enough space over the trusses for hoisting and lowering the big roof timbers. This work was also complicated by the steel frames, which Ove Arup refused to allow us to move until each bay had been secured by new scaffolding, for which there was hardly

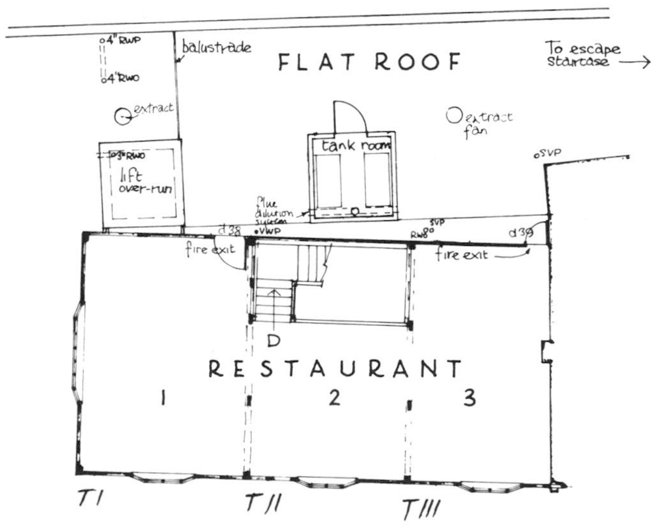

(c) *Second floor*

Figure 259 *Restored top floor*

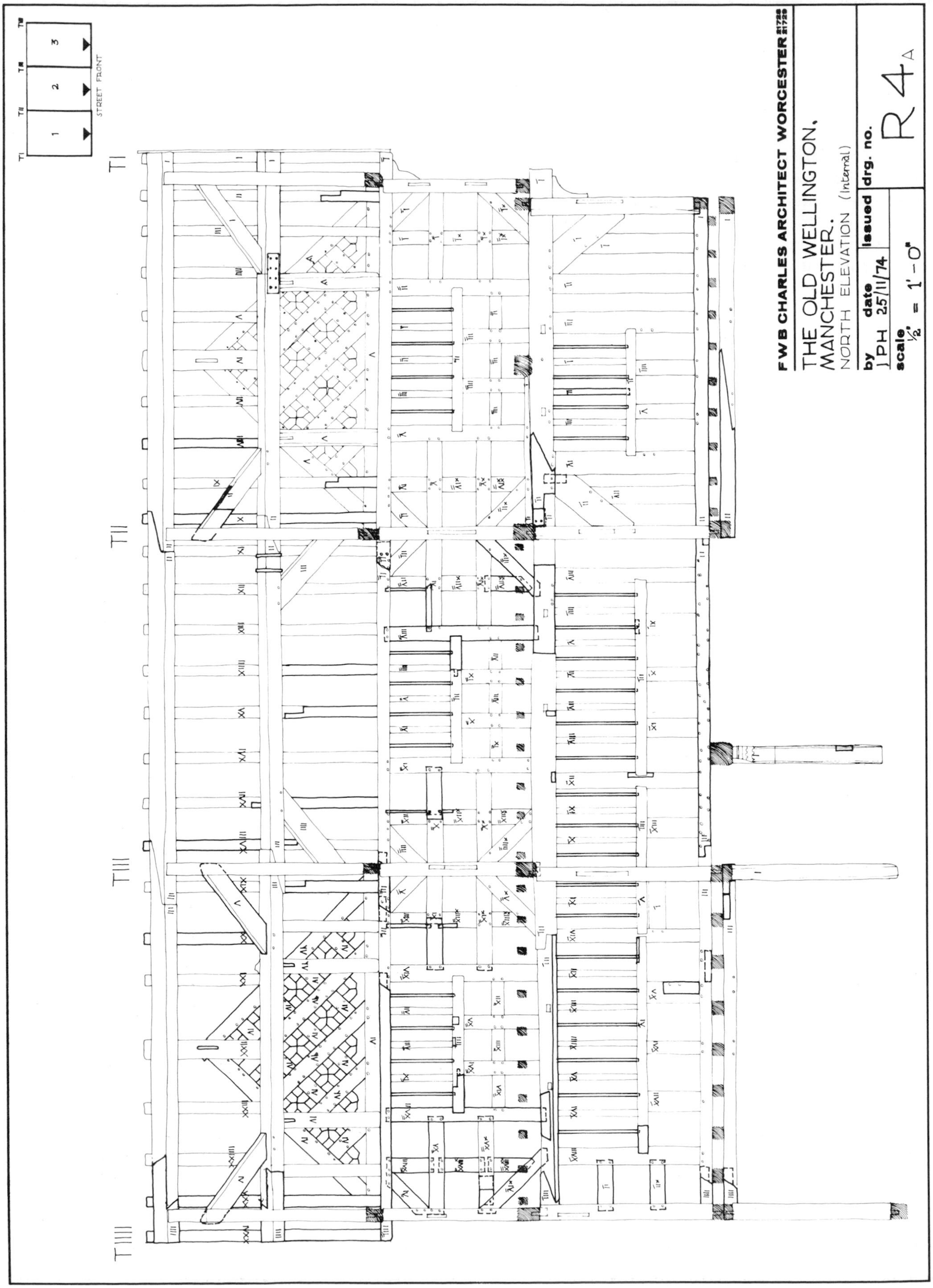

FWB CHARLES ARCHITECT WORCESTER 21728 21729

THE OLD WELLINGTON, MANCHESTER.

NORTH ELEVATION (Internal)

by IPH date 25/11/74 issued drg. no. R4A

scale 1/2" = 1'-0"

STREET FRONT

TI TII TIII TIIII

Figure 260 *Reframing of north wall*

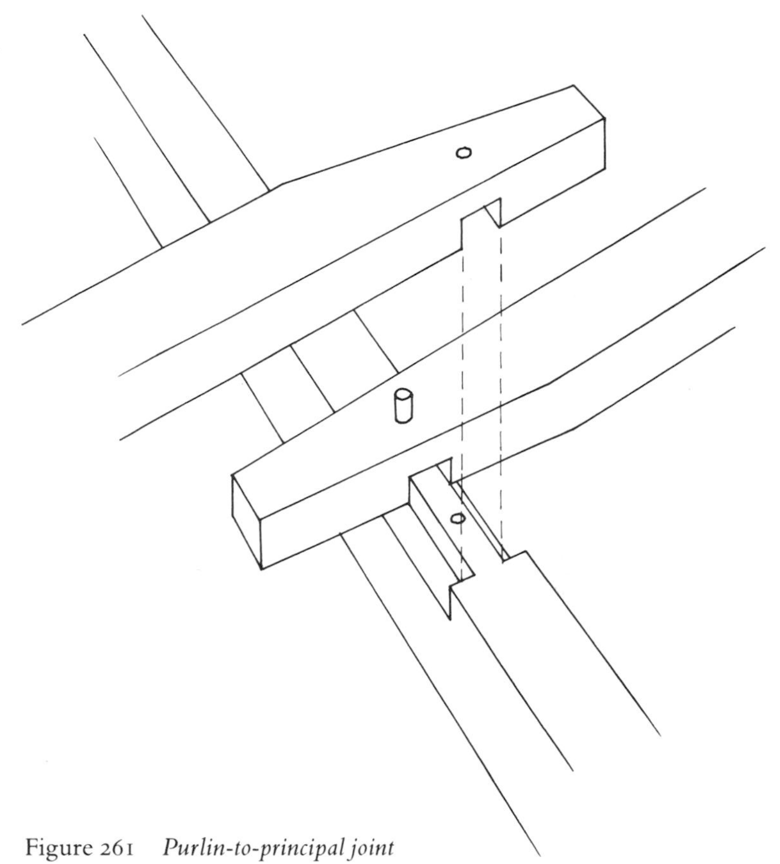

Figure 261 *Purlin-to-principal joint*

enough room for erection, or until the structural timberwork had been repaired. The panels were also taken out, and all the timbers numbered with the usual plywood plaques.

In the numbering of timbers, the contradiction in the erection sequence as between the original building and its later superstructure was disregarded, and it was assumed that east was the upper face throughout and that the erection sequence for both phases of construction had been from east to west. This sequence was correct for first dismantling the roof and top floor of bays 3 and then 2, and for re-erecting them in reverse order. Bay 1 would remain standing, except for truss II and a few wall-frame timbers that had to be taken down for repair. Site restrictions, however, made it impossible to dismantle more than one truss at a time. There were also weight limits on the surrounding concrete slab. The amount of timber that could be stacked on it was determined by load-bearing capacity within a working space that was hopelessly inadequate for laying out a frame. New or repaired timbers had to be worked according to measurements taken on the scaffolding and final adjustments made in the air instead of on the ground. The amount of propping was enormously increased as compared with what would normally have been necessary.

The relevelling of the frames (Figure 260) posed few problems in re-engaging the joints of the purlins and wall-plates – the former because the joints were simple trenchings over cogged principals, typical of the north west (Figure 261), and the latter because the joints were the usual bridle-scarfs and the morticed member could be offered up to the relevelled member.

But the double-bridle joints of the girding-beam were much more difficult, and a new piece with splayed scarf and simple top bridle had to be let in. This replaced the rotten part of the timber and enabled the brace to be secured with a slip-tenon. The bressummer rejointing, even though the timbers had to be renewed through bays 1 and 2, proved impossible without introducing a reverse joint; this, a simple half-lap scarf, had to be made on the 'wrong' side of post II, in bay 2 instead of bay 1. But despite this falsification of the jointing, the peg-holes, already noted as mostly out of register with the studs, were faithfully reproduced in the new timbers – a degree of archaeological fidelity that can perhaps be overdone.

Another result of having to repair and re-erect the members in limited lengths of framing was that progress tended to appear to be better than it was. Normally dismantling is followed by a long period of work on the bench and reframing on the ground; there is virtually no progress to be seen on the building. Then re-erection takes place in a matter of days. In this case, bay 3 was being re-erected only three months after the contract had started, and dismantling of T III began in the following month. Final adjustment, however, had to await the reframing of T II, several months later. Work thus alternated from frame to frame, with none completed until all of them were repaired. The job book entry of 12 April reads, 'As much of bay 3 framing has been erected as is possible before the overlapping bay 2 timbers have been relevelled. Therefore bay 3 must be temporarily braced, as agreed by Ove Arup. On 26 April, 'dismantling bay 2 now almost complete', but on 7 June the entry reads, 'bay 1 should now be repaired instead of completing bays 2 and 3.' The normal course of completion bay by bay was thus impossible.

There were also minor troubles with an inexperienced contractor. The job book reads, 'It is again emphasized that all pegs in construction joints are to be left projecting. This allows joints to be tightened up. Repair joints *only* to be glued.' And again, 'All oak must be used in accordance with specification – i.e. *heart to outside*.' Or yet

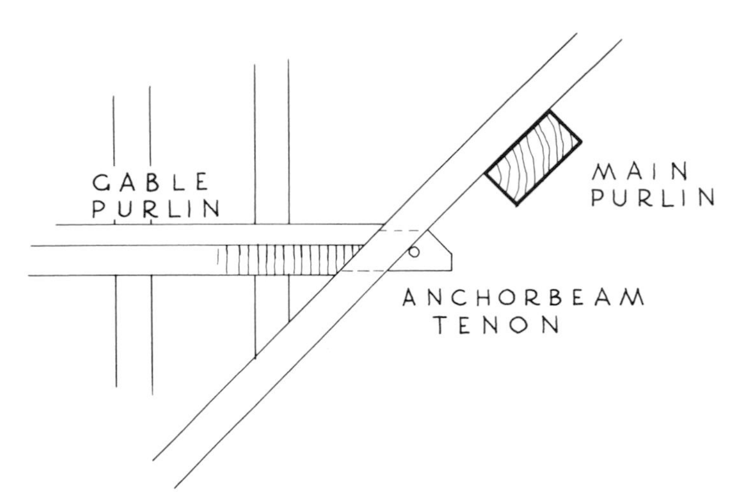

Figure 262 *Anchor-beam or tongued-tenon system of jointing gable purlin*

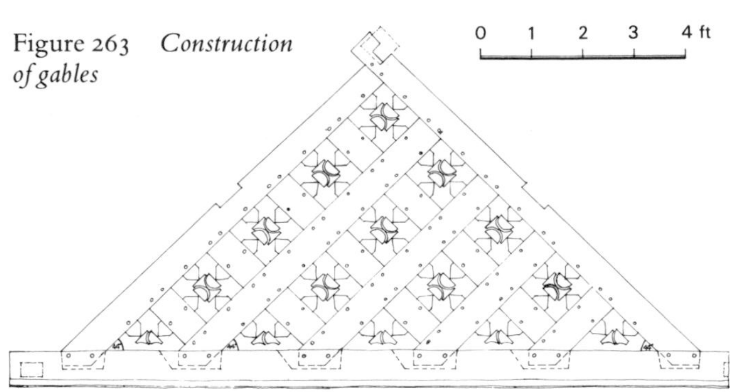

Figure 263 *Construction of gables*

(*a*) *Elevation. Note enlarged mortices to ease assembly*

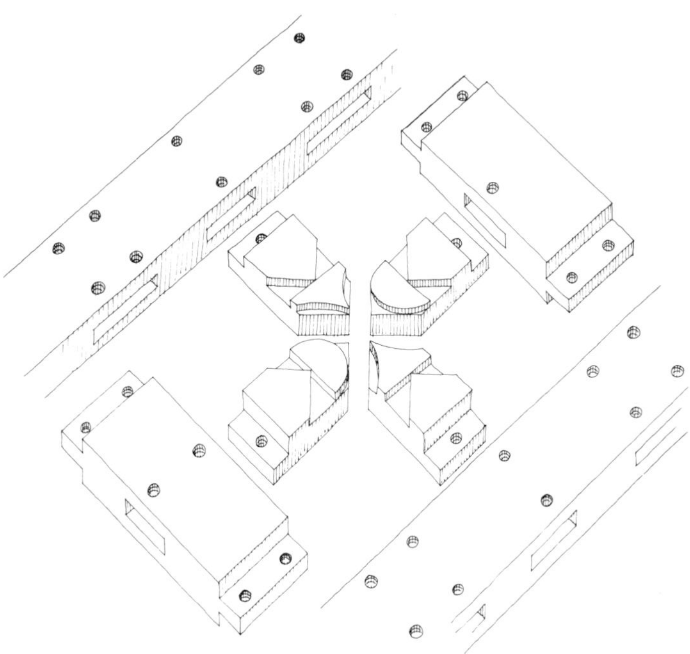

(*b*) *View of exploded infill panel; the recessed surface is painted to match the plaster*

again, 'Mouldings must be cut properly, *not* to reproduce wear and tear.'

There were also additional repairs despite all the preparations. These arose mainly out of the engineer's assessment of the condition of the timbers when they could be examined on the floor for the first time. While the quantity of plating and bolting sometimes seemed to us excessive, the special conditions of the contract, especially the tight programme, precluded the possibility of working out an alternative. The strengthening of the first floor was certainly necessary and there was no better way than following the Victorian tradition of propping the beams with cast iron columns. There were already two of them, obstructing the limited floor area of the bar, but these had to be increased to five, all of them being obtained from Ironbridge Gorge Museum. The bar, at least in that respect, is thus now more Victorian than the Victorians left it.

Reconstruction to wall-plate level had been completed by 4 October, and the roof structure was ready for rafters by the end of the month. As well as the jointing of the purlins over the principals, the method of tying back the purlins of the north wall gables (Figure 262) and their diaper-timbering are details typical of this region (Figure 263).

Lastly, to confirm the date of the second build of the Wellington, a Scottish coin of 1602 was found in the brace mortice at the head of the top rear post III.

In the interior, iroko was used extensively, not only because it is half the price of oak (or one-quarter when air-dried oak is required as for joinery) but because it is a handsome and very stable wood, obtainable in wide planks and boards. The floorboards for the Wellington were specified as '$\frac{3}{4}$ inch (full) ploughed-and-tongued boarding in 8 inch face width with $\frac{3}{4}$ inch by $\frac{1}{4}$ inch plywood tongue, well cramped up and secret nailed at every bearing'. The counter-top and front were also iroko, the former 18 inches by 2 inches, the latter in horizontal boards 7 inches by 3 inches (Figure 264). The first-floor bar was of the same wood with vertical board construction similar to the staircase enclosure (Figure 265). The staircase, also in iroko, was being fitted by 10 January 1977 (Figure 266).

Figure 264 *Iroko counter in bar, also showing one of the cast iron columns from Ironbridge Gorge Museum*

Figure 266 *New staircase*

Figure 265 *Upstairs bar*

Figure 267 *Ground-floor bar showing additional cast iron columns and photo-mural on end wall*

By the end of January fittings and finishes of all kinds were being installed. On 14 March the first-floor boarding, left until last, was almost complete, the ground-floor bar was 'in course of fixing', and the party wall, originally intended to be left as brick except at the ground floor, where a photomural (Figure 267) of the history of the Wellington was to be mounted, was plastered to full height. By the beginning of April electric fittings were in progress and the heating system had been tested.

On 9 May 1977 the building was opened to Bass staff, the press, television and those who had been mainly responsible for its restoration. From then on the Wellington Inn resumed its role as Manchester's oldest pub.

PART FOUR

BUILDINGS

Figure 268 *Siddington Barn, Gloucestershire*

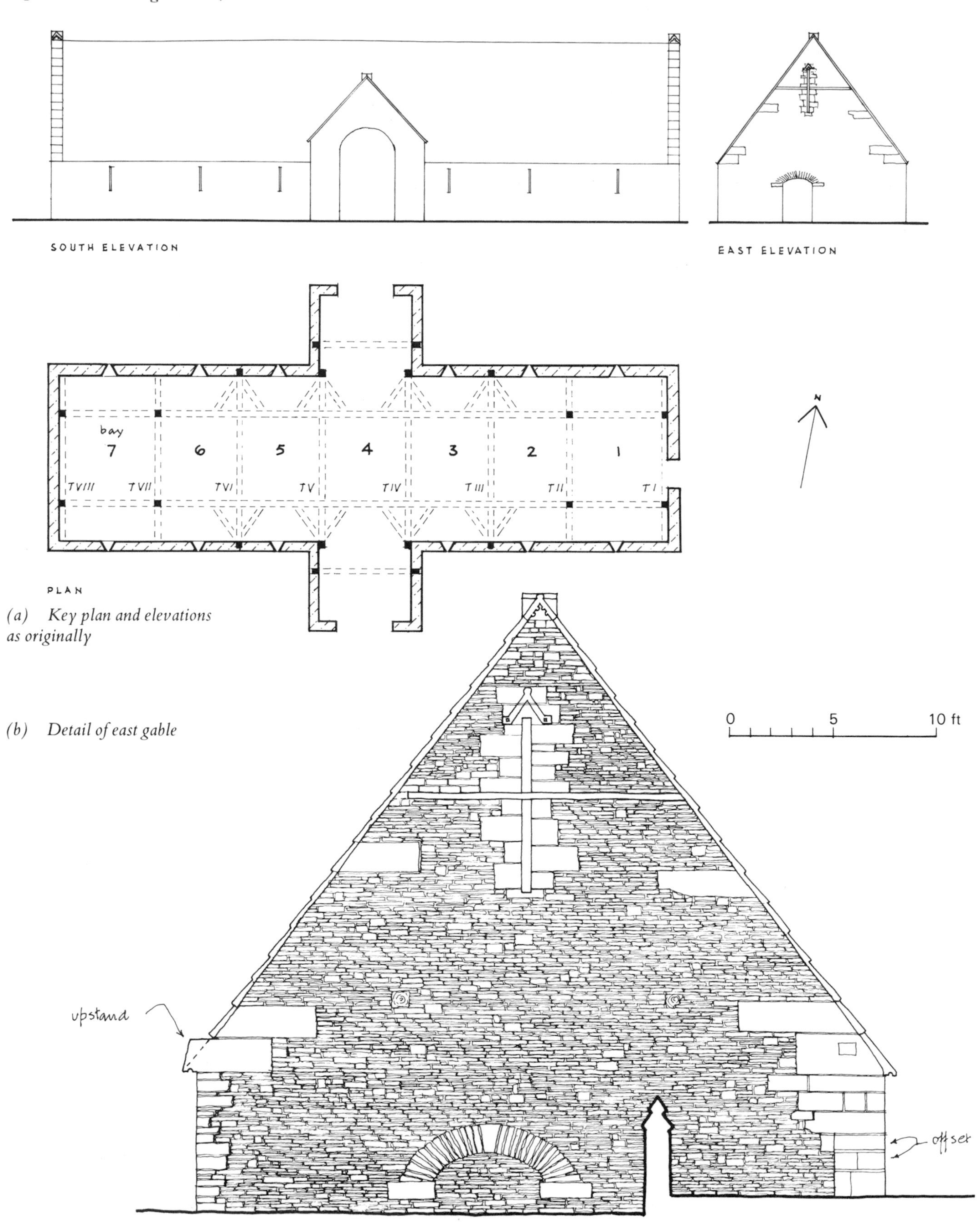

(a) Key plan and elevations as originally

(b) Detail of east gable

CHAPTER ELEVEN

BUILDINGS

The following are some of the buildings that have been surveyed and to which reference has been made in earlier chapters. Most of them were the subjects of reports such as that of the Old Crown, summarized in Chapter 5. Some of them were also restored, though not always by us. They are set out in date order. The descriptions are intended to supplement the details already mentioned in order to present them, however sketchily, as buildings; in one instance – the Bailiff's House, Bewdley – this is done by reproducing the actual survey sketches.

Siddington Tithe Barn

Twelfth century. Near Cirencester, Gloucestershire. Figure 268(a) and (b).

This building may have the oldest timber frame in England.[1] It has all the features of lap-joints and scissor-braces and heavy substructure characteristic of the earliest buildings. In addition the stonework of the east gable has certain details which can be ascribed to the Anglo-Saxon period. A very tall slit window is divided half-way up by a horizontal stone 'weatherboard'. Its hood mould is also reminiscent of wood. The masonry reveals several periods of construction and has many other differences from that of medieval tithe barns as, for instance, absence of buttresses. Lastly, it is sited in relation to the church, which is Norman, in such a way that the obvious explanation is that it was built before the church.

There were originally seven bays and eight trusses of which T II and T VII were aisled, and the rest base-crucks.

In its restoration – urgently needed – it will be necessary to use epoxy resin, despite its dangers (pages 126–7), for this is the only means of consolidating the feet of the cruck-blades buried in the masonry, short of taking down and rebuilding large areas of the stone walls.

Eastington Hall

Thirteenth century. Upton upon Severn, Hereford and Worcester. Figure 269(a) and (b).
(See also Figure 32.)

This is probably the oldest base-cruck hall in Hereford and Worcester, with huge Gothic door, leading into the through-passage, a spere-truss and screen of later date. The central arch has a heavy boss and scissor-braced upper roof. An ordinary farmhouse until the beginning of this century, it was more than doubled in size in 1911 in half-timber construction, reminiscent of Lutyens's work in timber.[2] The hall was given a further bay joining it to the solar range, the two having originally been separate buildings (page 77).

The hall floor was taken up for relaying in 1979 and beneath the layer of sand on which its tiles were laid was the original mud floor, consolidated with lime and still showing indentations of heavy furniture. It had been a virgin site until the hall was built. There was no sign of a central hearth, but the considerable disturbance in the corner in front of the passage and against the side wall had signs of burning and this conformed with the most usual site for the hearth in the lesser medieval halls.

Martley Rectory

Martley, Hereford and Worcester. Figure 270.
(See also Figure 70(a), (b) and (c).)

Martley Rectory is the most 'archaic' of the post-and-truss houses in the design of its timber frame. It now has a Georgian front of considerable charm, a mostly Victorian entrance front, and a fine Elizabethan staircase. The original plan and much of the structure have, however, survived. That it was an important priest-house, with a spere truss and three-bay solar, is suggested by its closeness to the parish church. The hall roof is remarkable; the roof-pitch was originally 57 degrees, and the pattern of swept braces in the central truss is reminiscent of a giant cusped arch, interrupted by the tie-beam. The spere truss is hardly less odd, with a bite out of its massive tie-beam to complete its Gothic arch defined by the knee-braces.

In plan, as already noted (see page 77), there is a distance of about 5 feet between the upper end of the hall and the solar. This and the design of the hall, which is so different from that of the undoubtedly fifteenth-century solar, suggests that it must be very much older, perhaps even of the thirteenth century.

Middle Littleton Tithe Barn

c.1300. Near Evesham, Hereford and Worcester. Figures 271 and 272.

This great barn was restored for the National Trust from 1975 to 1977. Its plan, like Siddington, has an aisled bay at each end and base-crucks intermediately.

It has been finally dated to *c.*1300 by the archaeologist James Bond.[3] He discovered from his study of the Chronicles of the Abbey of Evesham that Abbot John de Brokehampton, 1282–1316, built a grange, 'a very fine

Figure 269 *Eastington Hall*
(a) *Key plan showing medieval roofs and later building in outline*

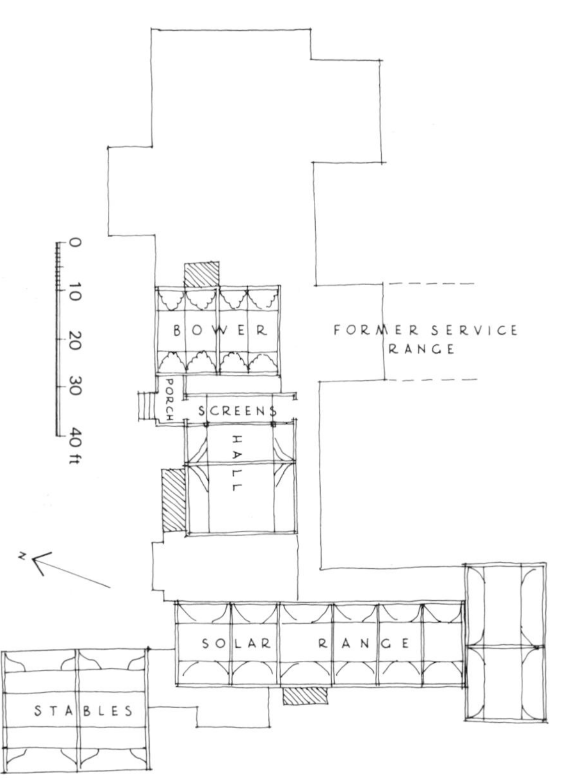

Figure 270 *Martley Rectory. Key plan showing medieval roofs and later building in outline*

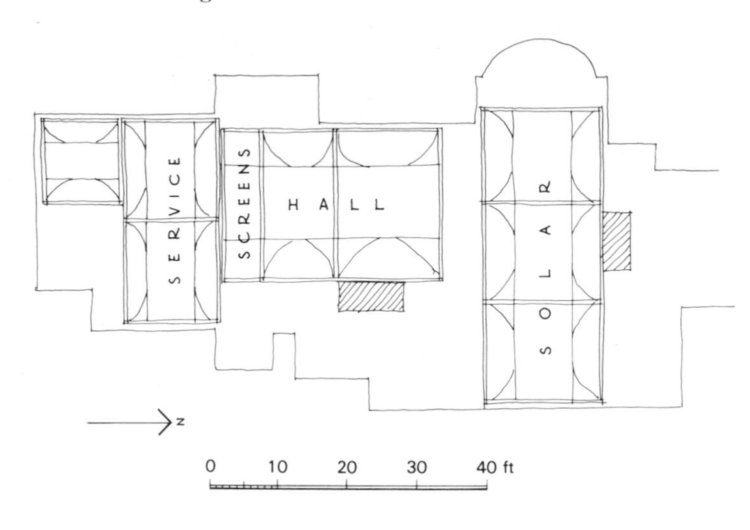

(b) *Hall truss*

Figure 271 *Middle Littleton Tithe Barn. Plan, section, elevation*

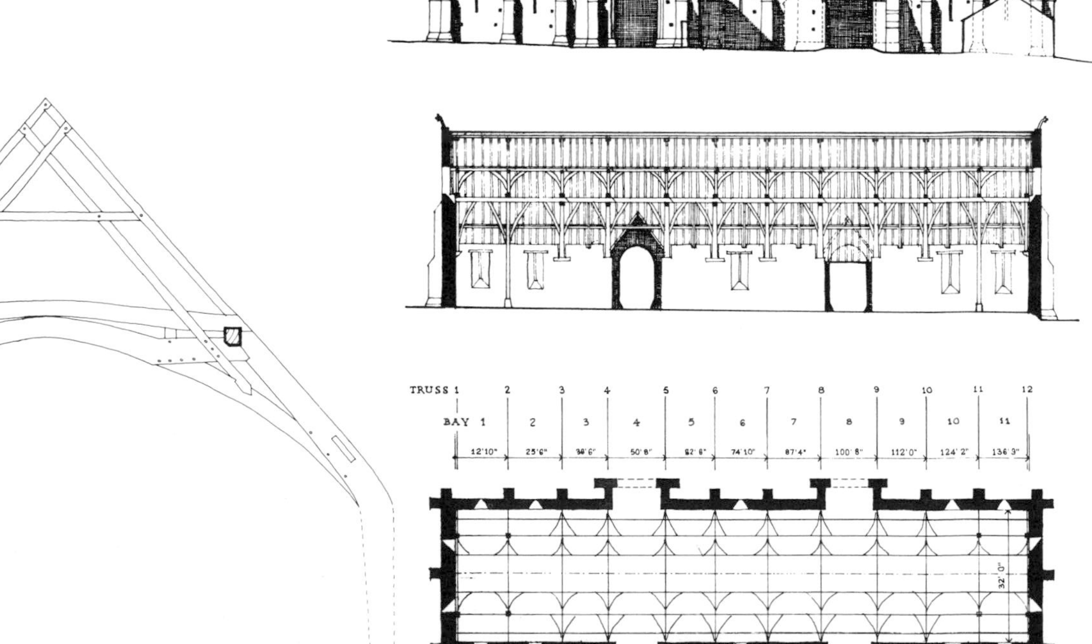

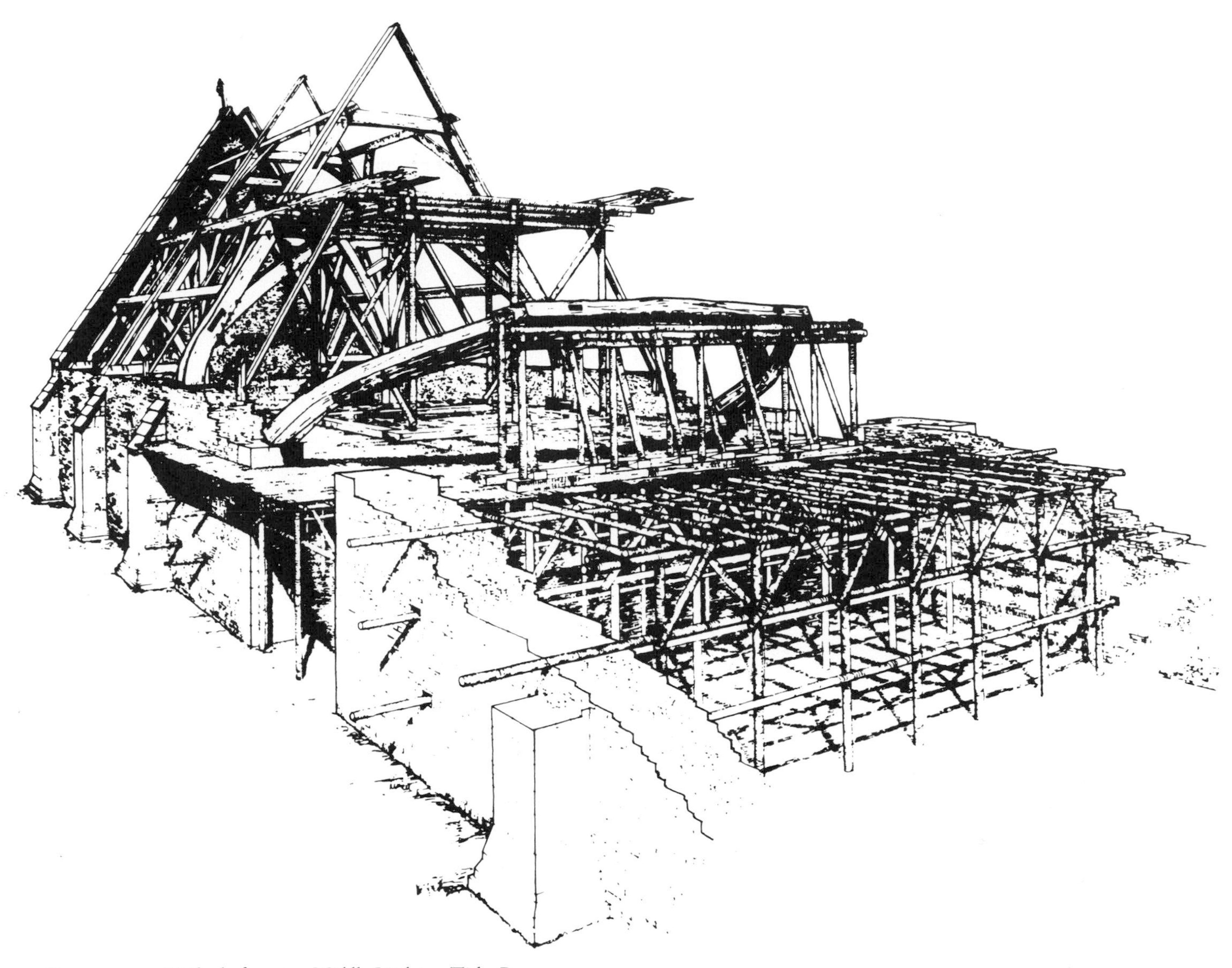

Figure 272 *Method of rearing Middle Littleton Tithe Barn*

one', at *Littleton*. Previously it had been ascribed to Abbot Ombersley, 1367–79, but this barn was located at *North* Littleton. Since Middle Littleton has the parish church it is clearly the village more likely to be written as Littleton. The argument is settled, however, by a description of the barn as later subdivided between two owners, giving the number of bays and other details which exactly conform with this building.

It is structurally complete except for the second wagon porch destroyed at the end of the last century; this had the result that the crucks, no longer buttressed, had spread dangerously, and the roof was in a state of collapse, propped by inserted posts and cross-beams. The scissor-scarf of the aisle-posts has already been mentioned (page 124). Its early features, besides the style of the timberwork, are its main entrance, an almost half-round Gothic arch, and its remarkable stone finials reminiscent of Norwegian stave churches. Though these cannot be dated they suggest an earlier origin than the conventional finials of many other barns and churches.

Rectory Farm

Fourteenth century. Grafton Flyford, Hereford and Worcester. Figure 273.

This now vestigial base-cruck hall could be dated to the previous century. The only early feature it lacks is scissor-braces, as at Eastington Hall and West Bromwich Manor House.[4] Instead, the central truss has a short crown-post with four-way braces. The upper tie-beam on which the crown-post is set and the base-cruck-collar are together well over 3 feet deep at midspan, and the arcade-plate, elbowed wind-braces, arch-braces and cruck-blade are in

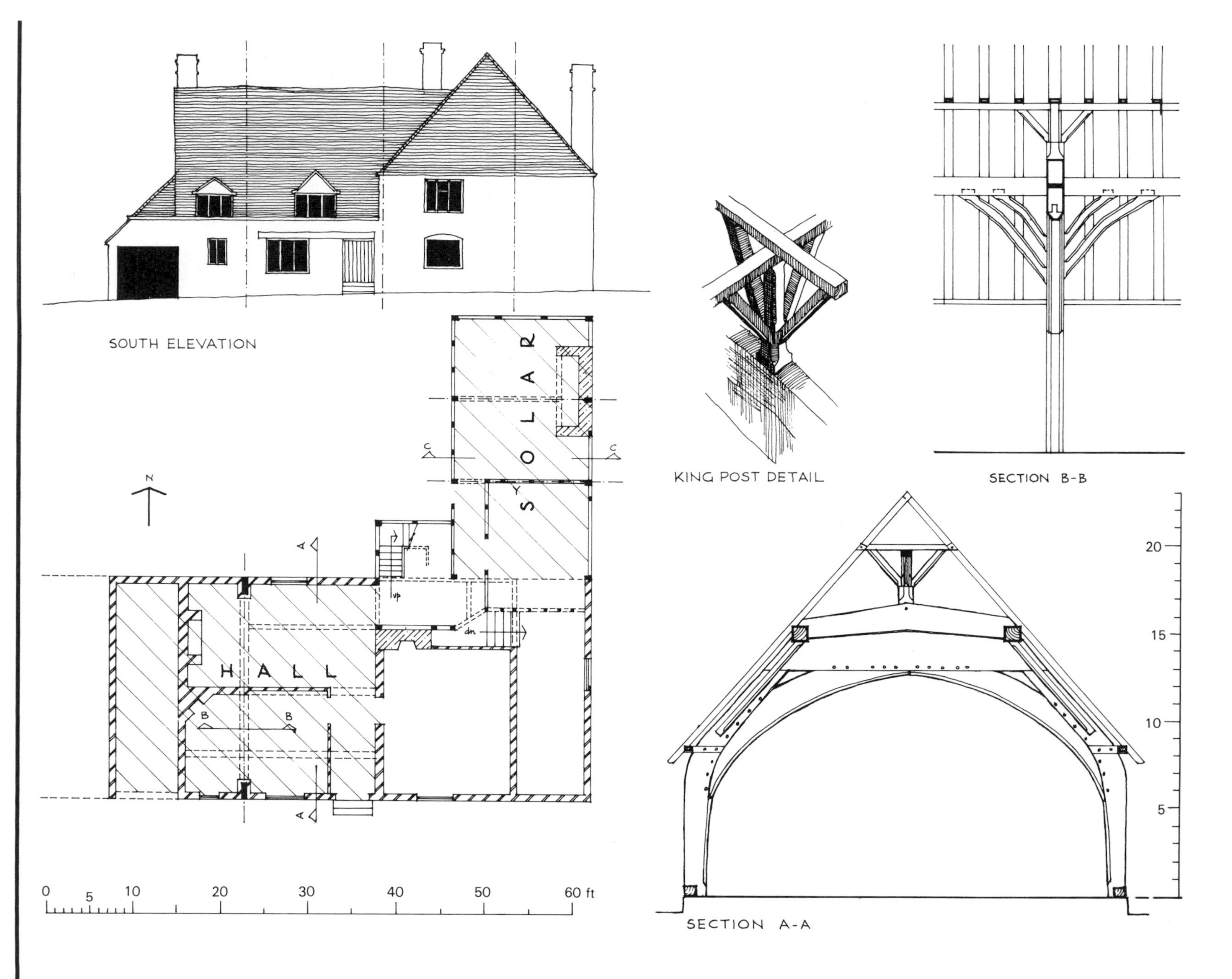

Figure 273 *Rectory Farm, Grafton Flyford, near Droitwich. Plan, elevation and details of building as existing. Shaded area shows original hall and solar*

proportion. Little more of the hall structure has survived, but the solar, of three bays with elegantly moulded timbers and probably of the mid fifteenth century, is untypically related to the hall, originally with its own solar bay. It may be another example of the disconnection of the two buildings.

Amberley Court

Fourteenth century. Hereford and Worcester. Figure 274.
(See also Figures 31 and 72.)

In contrast with Rectory Farm and the other base-crucks, the superstructure of this hall is a purlin roof. The purlins, however, as already noted in Chapter 1, do not support the rafters but are part of the longitudinal bracing system, acting as distance pieces for the trusses. Also there is no upper tie-beam. The arcade-plate, instead of resting on the collar, is tenoned into the face of the blade. The superstructure has both main and intermediate trusses, both with large cusping. A detail worth noting is that the chamfer on the lower edge of the cusps is deeper than anywhere else. As seen from the floor of the hall, the depth of chamfer appears equal all round; this is an early case of perspective correction, hardly to be expected in vernacular architecture.[5]

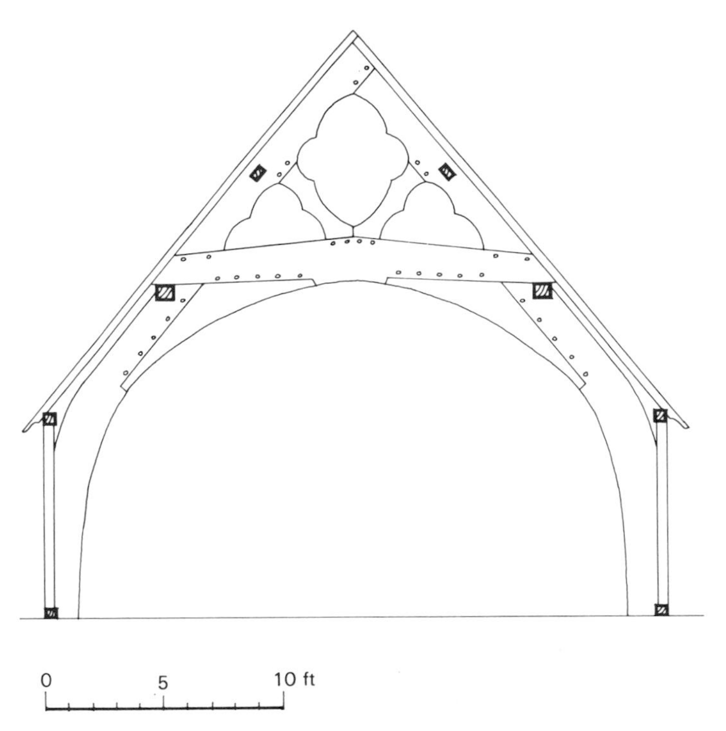

Figure 274 *Amberley Court, near Hereford – base-cruck hall truss*

Severns

1335. Castle Gate, Nottingham. Figures 275 and 276(a), (b) and (c).
(See also Figure 1(a).)

This was a single three-bay solar, originally with a rear hall. Its site at Middle Pavement was over Nottingham's famous sandstone caves, probably inhabited since the Neolithic.

It was dismantled in 1969 and re-erected on its new site almost opposite the Castle entrance, where it could be seen on all four sides. Its end elevation is worth noting for its braces, all visibly attempting to push the frame backwards to compensate for the forward thrust of the jetty. The features of greatest interest are the ground-floor framing designed to span the cellar, or cave, over which it was built, and the rafter roof with jowled crown-posts in the gable trusses and unusual jointing of the collar-plate to them.

The date of the building has been obtained by dendrochronology.[6]

Figure 275 *'Severns', Nottingham – as reconstructed. Courtesy 'Raymonds', Derby*

Figure 276 *'Severns', Nottingham – structural details*

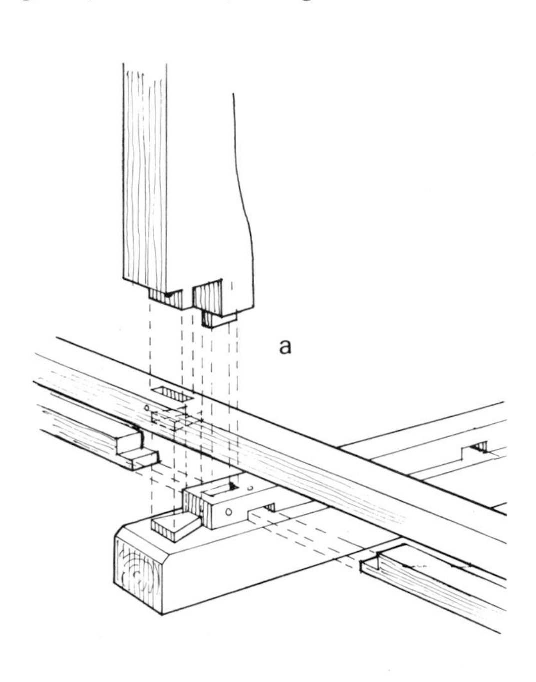

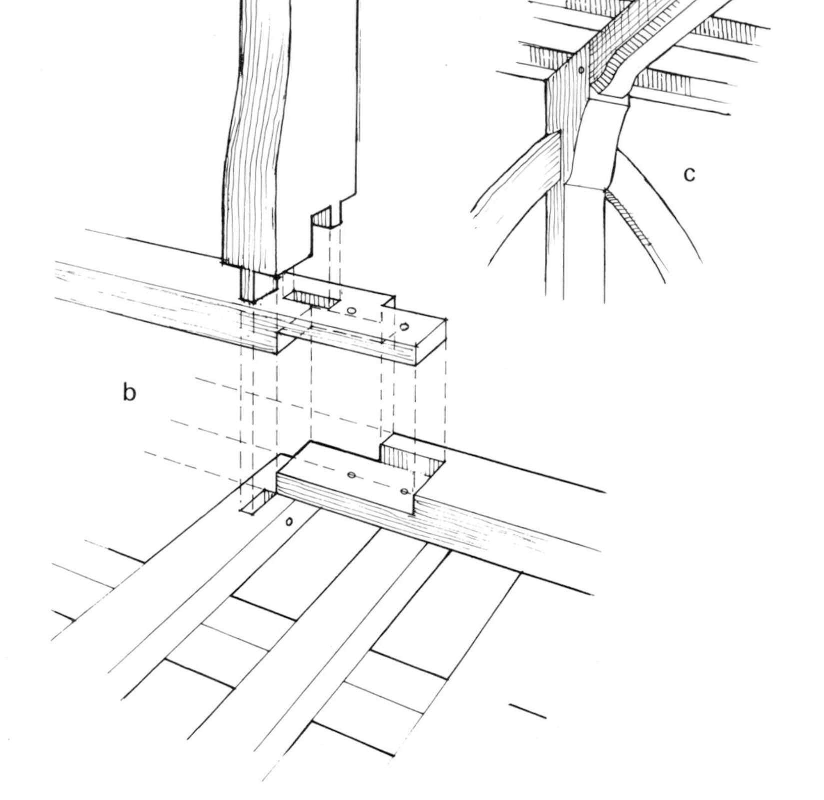

(a) Ground-floor construction showing 12 inch by 12 inch cross-beam spanning caves with sill-beam and post foot
(b) First-floor jetty from inside
(c) Jowled crown-post and collar-plate

White Hart (south wing)

Fourteenth century. Newark, Nottinghamshire. Figure 277.

One of the finest examples of rafter roofs in the midlands was over the former hall in the south wing of this medieval inn. Though the hall was only of two bays – typical of inns, as travellers were presumably important enough to demand service in their rooms – its visual height, emphasized by a very steep roof and 20 foot span, must have been extremely impressive. The unusual double wall-frames are wholly functional, their inner bracing serving to triangulate the joints between the posts, girding-beam and wall-plate, and at the same time reinforce the scarf-joints of the last member, while the outer skin is the characteristic swept-brace design.

The east wing of this complex, also with a rafter roof but different design of crown-post, must be near-coeval.

Bear Steps

Fourteenth century. Shrewsbury, Shropshire. Figure 278(a), (b) and (c).
(See also Figure 73(b).)

This is another complex of buildings of different dates ranging from the early fourteenth century to the mid-sixteenth. The oldest also has a rafter roof with exceptionally short crown-post braced with reversed swept-braces to a steeply cambered tie-beam, a typically upland design.

The adjacent building has an identical roof and a jetty, the upper framing of which has cusped braces in all four corners of each bay, forming a large quatrefoil. The two buildings were hard up against each other, the former 'borrowing' support for its inserted upper floor off the latter's jetty, obviously built a few years earlier and designed to overlook the churchyard.

Figure 277 *White Hart, Newark (south wing) – reconstruction of original hall looking north, bays 1, 2 and 3*

A further feature is a gallery over the little Fish Street at the east end of the four-bay range. But this is later than the fourteenth-century hall.

Lower Norchard

Fourteenth/fifteenth century. Peopleton, Hereford and Worcester. Figure 279.
(See also Figure 73(c).)

This was a typical four-bay peasant house with two-bay hall, the hearth to one side just beyond the passage in the lower and slightly shorter bay. The upper and lower bays, 1 and 4, were floored. Bay 1 was subdivided longitudinally, but off-centre, at ground-floor level, the smaller room being a semi-cellar, probably the dairy; the larger was no doubt the parlour. Bay 4 was conventionally divided into equal-size service rooms. The doorway had a curved door lintel with carving of suspiciously Norman dog-tooth design. The central arch-truss was also moulded but without carving. The close-set pegs of the braces, nine into the collar and eleven into the cruck-blade, were still projecting, giving decoration to the truss. All the swept members were sharply elbowed, including the

Figure 278 *Bear Steps, Shrewsbury*
(a) View from churchyard, the hall on right and Steps central. The left-hand building was originally a guildhall of the sixteenth century

blades, and the upper end wall had the conventional framing of a heavily braced centre-post with shaped, flat triangular door-heads at each side.

The little building was demolished as soon as its owner, a speculative builder, became suspicious that it was interesting. Fortunately a twin still stands in the same village, the only difference being a later cross-wing instead of bay 1 and even more projecting pegs in the arch-truss.

Chorley House

Fifteenth century. Droitwich, Hereford and Worcester (destroyed). Figure 280.
(See also Figures 1(b), 15, 17, 73(a) and 99(a).)

This was one of the greatest halls in Hereford and Worcester, though not typical. Probably it was part of an Augustinian Friary founded in Droitwich in 1331.[7] The hall range was of four bays, each of the end bays being floored. The two middle bays were the hall proper, open

(b) Rafter roof of hall

(c) Fish Street showing gallery on left

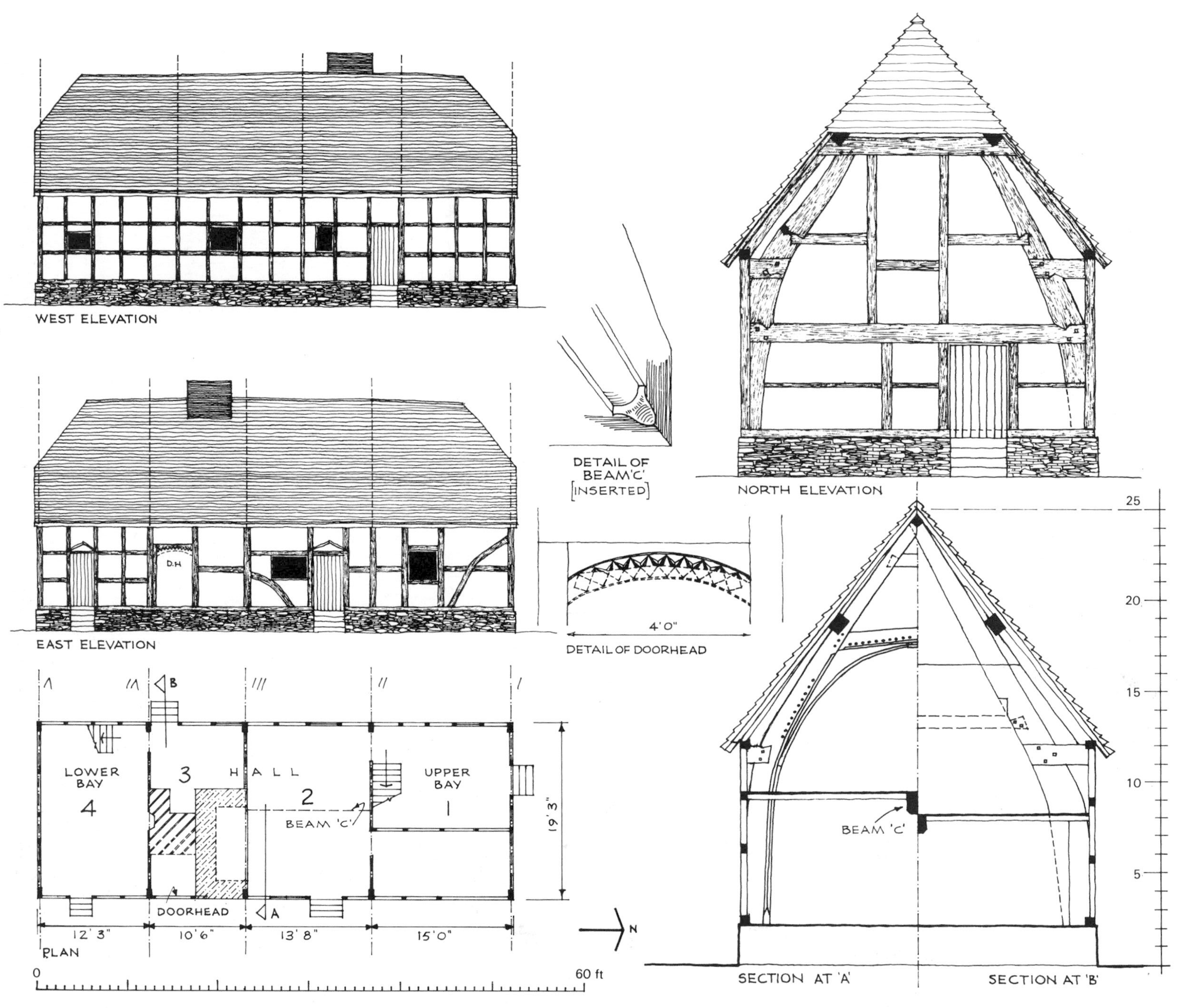

Figure 279 *Lower Norchard, Peopleton, as it stood before demolition. Framing of west elevation is later*

Figure 280 *Chorley House, Droitwich*

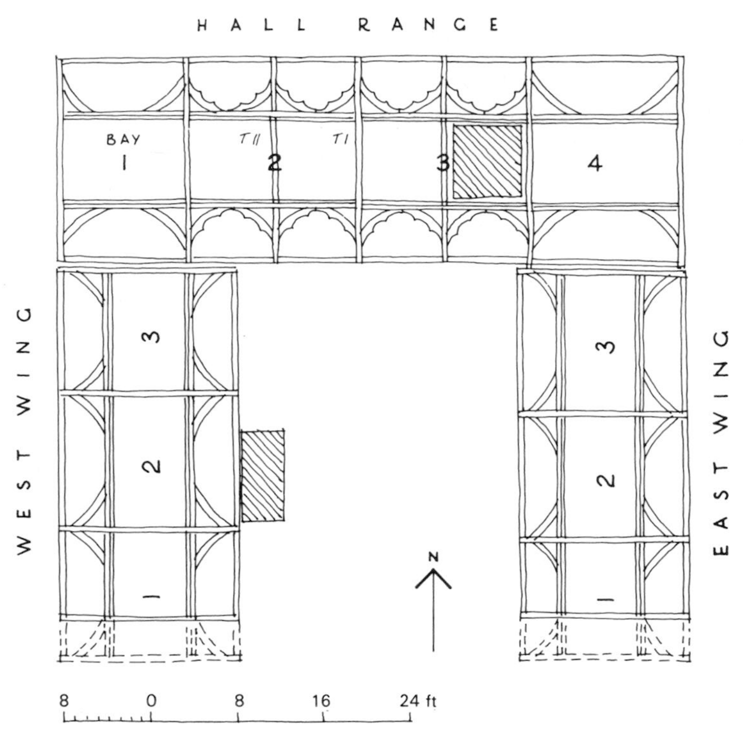

(a) Roof plan showing west wing and east wing curtailed in seventeenth or eighteenth century, the four hall-range bays with cusped wind-braces in the hall itself, and inserted fireplace

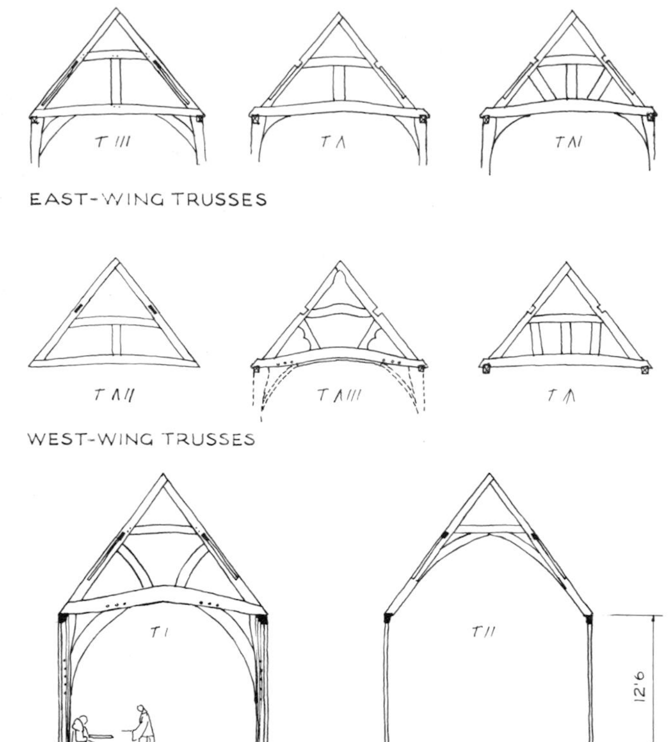

(b) Trusses of east and west wings, and main intermediate truss of hall

to the roof, with the passage occupying half of the lower bay in which the later fireplace had been built. The two bays were of the same length with a broad-timbered central truss and delicate intermediate trusses. The whole structure was a classic example of medieval design, the characteristic upper end wall, swept braces in the wall-frames, elegantly cusped wind-braces, and an ogee door-head in the upper end bay leading into the side wing.

In the side wings not one truss was the same as any of the others. The cusped struts in one of them had been halved from the same log and the cusps followed the natural grain.

After the Dissolution, the building became a fairly grand town house. The hall was floored in a typically Elizabethan manner. The end bays of the side wings were curtailed, probably about 1700 – street widening must already have been a scourge of ancient buildings – and the end frames were replaced by poor timbering. Finally the buildings were subdivided into three separate Victorian houses.

It was listed as seventeenth century throughout, which was of little support in the struggle for its preservation against the threat of commercial development. The latter eventually won after a fight of two years, during which the structure had stood roofless. A scheme for its conversion into a branch library was actually commissioned and sketched, but there was no support from the central department (then the Ministry of Housing and Local Government) and little enough from the County or Droitwich Council. The developer thus had his way, only to be bought out by the Droitwich New Town Development Corporation two years later. Its site is still a car-park.

Shell Cottage

Fifteenth century. Himbleton, Hereford and Worcester. Figure 281.

Shell Cottage, close by the manor house of the same name (see Chapter 7), is surely the smallest medieval post-and-

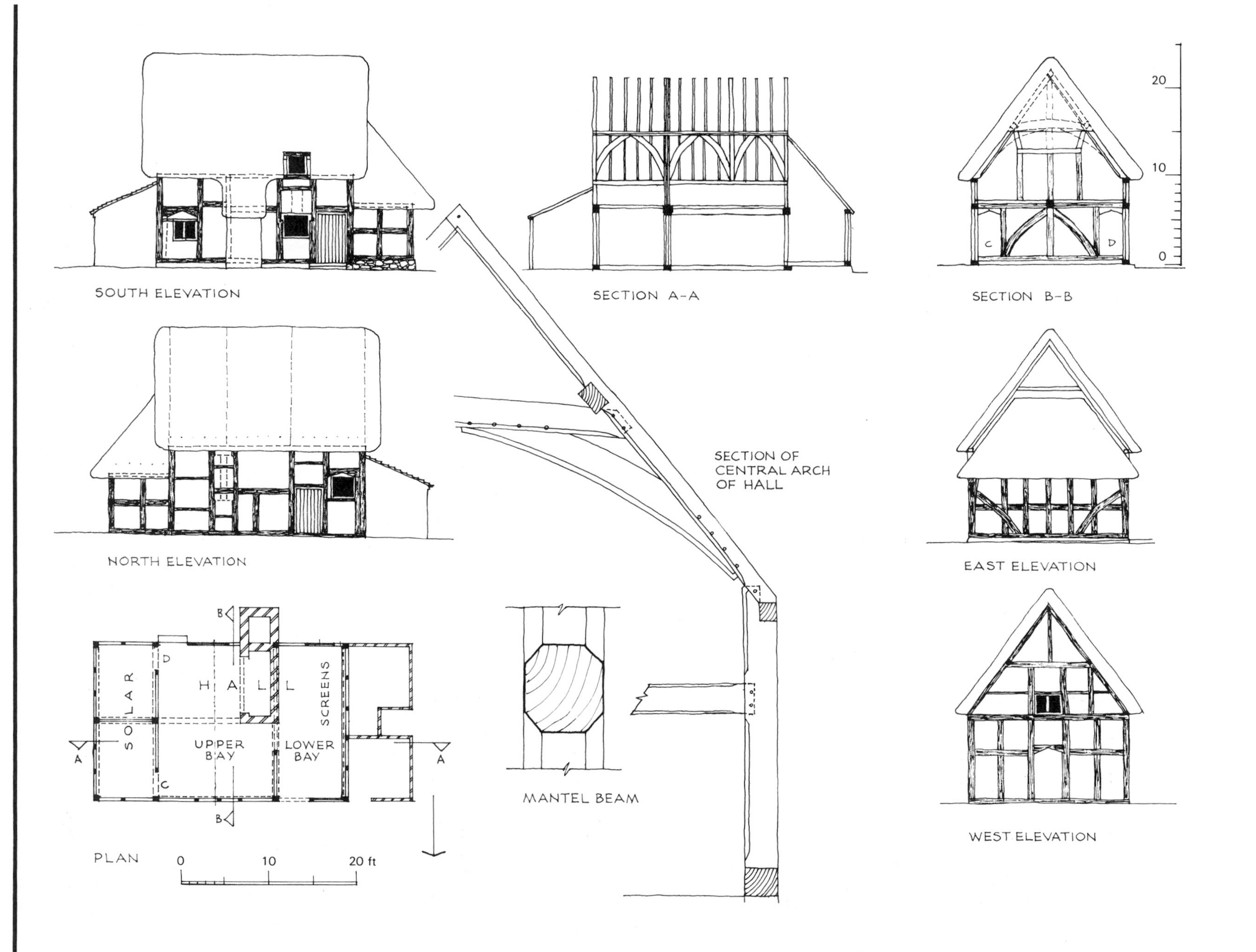

Figure 281 *Shell Cottage, Himbleton*

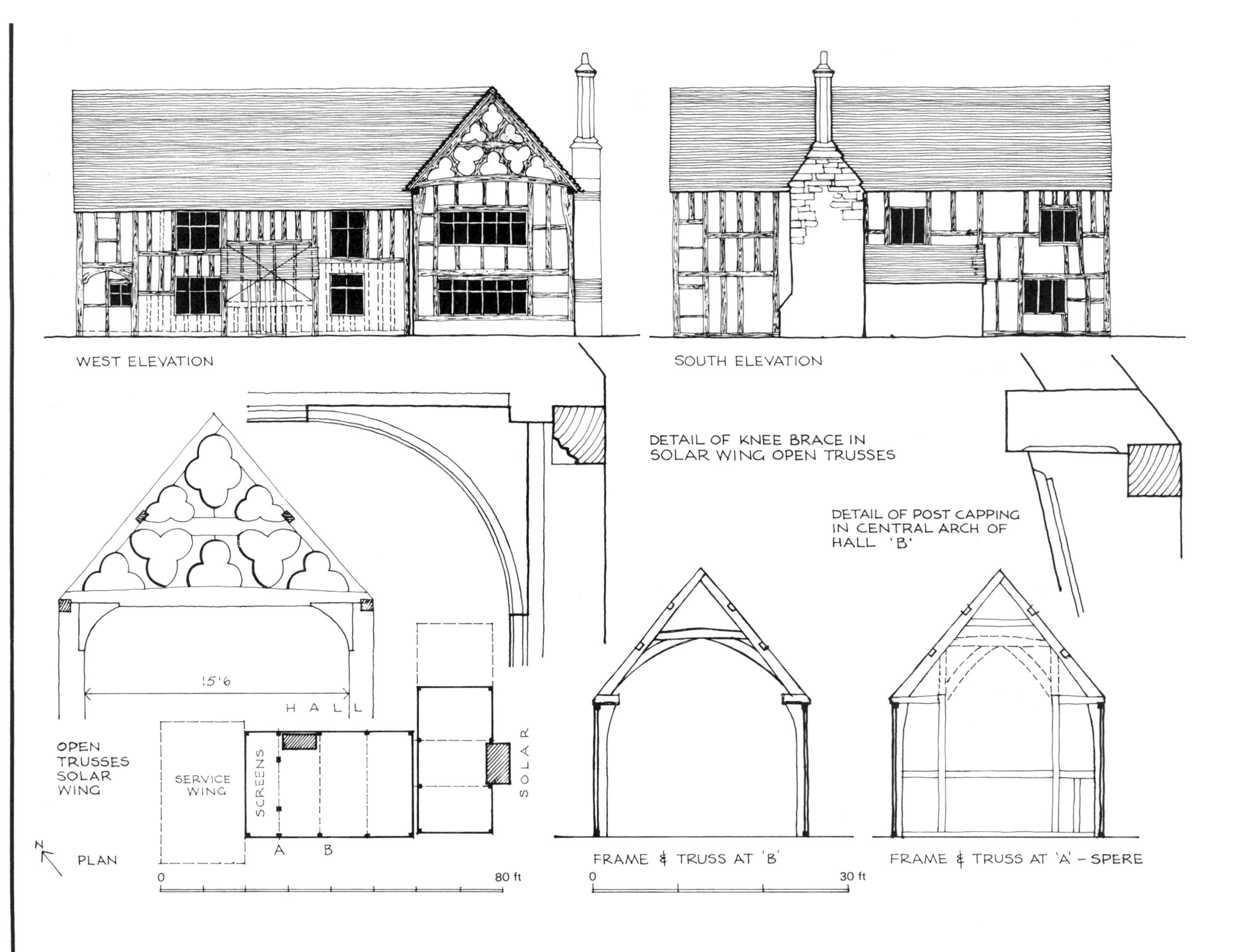

Figure 282 *Throckmorton Court, near Pershore*

truss hall in existence. The two-bay hall is only 22 feet long, the lower bay a mere 8 feet. The upper bay has an intermediate arch-braced open truss, even though it is only 7 feet from the upper end, and there are three pairs of matched and chamfered wind-braces on each side. The main truss is a 'clasp-purlin' type, but the principals are not stepped to a reduced section at the collar. It has a mantel-beam but no tie-beam. The purlins are a single length from truss I to III. There is no ridge-piece and all the visible edges of the main members are chamfered and stopped. The rafters are squared and die-straight. All the original roof members, including most of the rafters, are heavily encrusted with soot from the open hearth.

The upper end wall of the hall is of the standard central post and swept-brace pattern; correctly placed chevron-headed doorways lead into the out-shut solar, with two chambers each measuring precisely 7 feet by 9 feet. The service bay at the other end has gone, but a central doorway with the same style of head suggests that the buttery and pantry were one room. The last features to be noted are the vestiges of diamond mullions in the wall-plate and the mid-height rail in each side wall, showing that the dais end of the hall could match the best with a tall window on each side.

The proximity of this little building to Shell Manor and its architectural perfection may suggest a connection; perhaps the cottage was its dower house.

Throckmorton Court

1500/1550. Hereford and Worcester. Figure 282.

The main elevation reads like a book. Beginning with the four-centred arch door-head announcing the passage, the gallery over the passage is marked by the short girding-beam; then comes the lower bay of the hall, open to the roof, with the girding-beams dividing the wall-frames into three equal tiers; next the upper bay of the hall with typical large window, the mullions of which still partially exist below the upper girding, the lower part having been replaced by brickwork containing the present door; and lastly, another floored bay, the original solar, with single girding at floor level.

An unusual detail to be seen inside is the vestigial tie-beam of the central truss. Either this was cut out when the hall was floored or its stub ends are original – like embryonic hammer-beams. Their chamfering suggests the latter, and this is confirmed by two other examples in Shropshire, Walleybourne Farm and Moat House, Longnor. There is purpose in the stub ends: these permit the tenons of the principals to be aligned with the grains.

The solar wing of three bays and with further single-storey bays at the back, must have been built soon after the hall, combining, as in many farmhouses, the private and service rooms within the one wing. The original service bay at the other end of the hall was probably demolished at the same time. The late features of the solar are that truss knee-braces are merely coved-brackets continuing the ovolo-mould of the posts to the tie-beam; the cusping is shallow and unstructural, reducing the sections of the principals and collar just where they should be at their largest. The last detail to note is another four-centred door-head, marking the upper-floor entrance to the solar. Here, however, it is adjacent to the hall, with the stair, now gone, leading up to it against the rear wall of the hall. The solar fireplace is contemporary with the wing, and the shaft no doubt is a restoration of the original.

Bromsgrove House

Fifteenth/sixteenth century. Avoncroft Museum of Buildings. Figure 283.
(See also Figures 64(a) and (b), 78, 79(c), 81, 91, 92 and 114.)

This was the first venture of the Avoncroft Museum of Buildings, and for lack of funds and trouble with the local authority in its reluctance to give planning permission, took seven years to rebuild.

A typical two-bay hall lying parallel with the street, with two-bay jettied solar wing, it also had on its town site a service bay and three-bay wing behind it.

The hall as originally was of the 'half-floored' type shown by the fact that the cross-beam is tenoned into the wall-posts, and the floor joists are tenoned at one end into the cross-beam and at the other into the girding-beam of the lower end wall. None of this could have been inserted without dismantling both the side walls and the end wall.

The timber-framed chimney stack was no doubt put in when the upper bay of the hall was floored, the first fireplace being an open hearth, probably in the centre of the upper bay. The new fireplace occupied the usual place, backing on to the passage and to one side of the entrance from the passage into the hall. It had a stone back wall and jamb, and some of the joists of the original upper floor were cut out to get the smoke away. Later a brick flue was built at the back of the timber stack.

Other features discernible from the dismantled timbers, though not in the house as it stood in Bromsgrove, were the tall window in the upper bay, the little window in

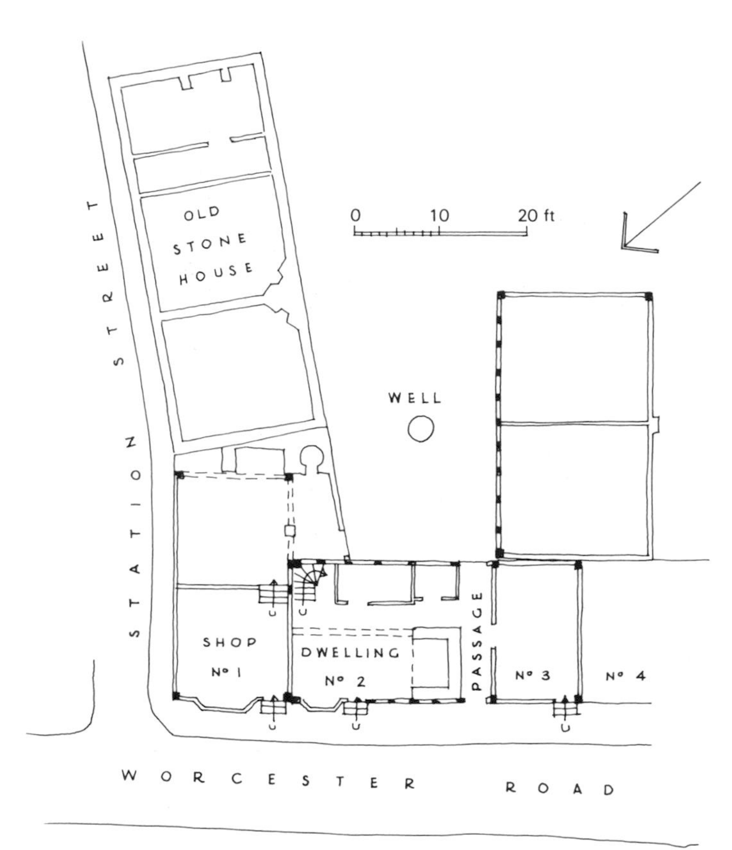

Figure 283 *The Bromsgrove House, Avoncroft Museum of Buildings as it last stood in Bromsgrove, with seventeenth-century stone house and contemporary service wing at rear. The reconstructed house at Avoncroft is the solar cross-wing and hall, but without cellar*

the solar overlooking the main entrance from the street, and the ogee door-head. Internally, shutters were located through the survival of grooves and hinges, and after the building's reconstruction the restored shutters were found to be the means of controlling the smoke by adjusting them differently according to the strength and direction of the wind.

The solar may be earlier than the hall. Its swept brace pattern is especially notable on the front elevation, but also exists in the exposed side wall. The upper floor is one in which the boards are grooved into the floor beams that lie parallel with them, aligned from front to back (see also page 92). These early features are contradicted by such details as trenched purlins and the use of a ridge-piece. Probably the solar is *c.*1500, the hall some fifty years later, and the chimney between 1550 and 1600. The house never looked so rich as it must have done around 1600 when the whole exterior was painted in deep red, and the interior of the upper chamber in red and yellow with black-lined floral pattern. Soon afterwards the solar was rendered and plaster casts were incorporated in the jettied gable. These survived and would have been preserved. Indeed full preparations had been made when the demolition contractor (there was no money for methodical and careful dismantling) felled the entire front on to the street at 6 a.m. on a Sunday morning – allegedly on police instructions.

It has since been shown that the lion rampant on two of the plaques was the arms of a family of dyers who owned the building in the seventeenth century.[8]

White Hart (front range)

1459. Newark, Nottinghamshire. Figures 284(a) and (b), 285(a) and (b).
(See also Figure 82.)

This is the grandest building of the period, recently restored to our drawings by a local firm.[9] Its range of windows, restored against the wishes of the Historic Buildings Council; its wealth of carving and colour, perhaps a little too stark for some tastes after its repainting; its superb purlin roof, again open to the rafters; and its huge and perfectly formed timbers; make it quite unique in England, though it was no doubt typical of the richer inns of its day.

It was also painted internally. The first-floor suite still has two layers of green, gold and red paint on the vertically sliding shutters of the front wall. The pattern at the top floor, scarcely visible, was drawn in black lines and seemed to have been a vine motif. The ground floor had a vine pattern in colour, which has partially survived only on the carriageway partition, and there vandalized by false black timbers painted over it as well as over a great swept brace.

The struggle for the restoration of this building went on for ten years while it was the property of the National Coal Board, and leased to Dorothy Perkins. Only its acquisition by the Nottingham Building Society saved it.

Spon Street

1450/1500. Coventry, Warwickshire. Figures 286, 287, 288 and 289.

Most of the buildings of Coventry[10] are similar to the

Figure 284 *White Hart, Newark (front range)*

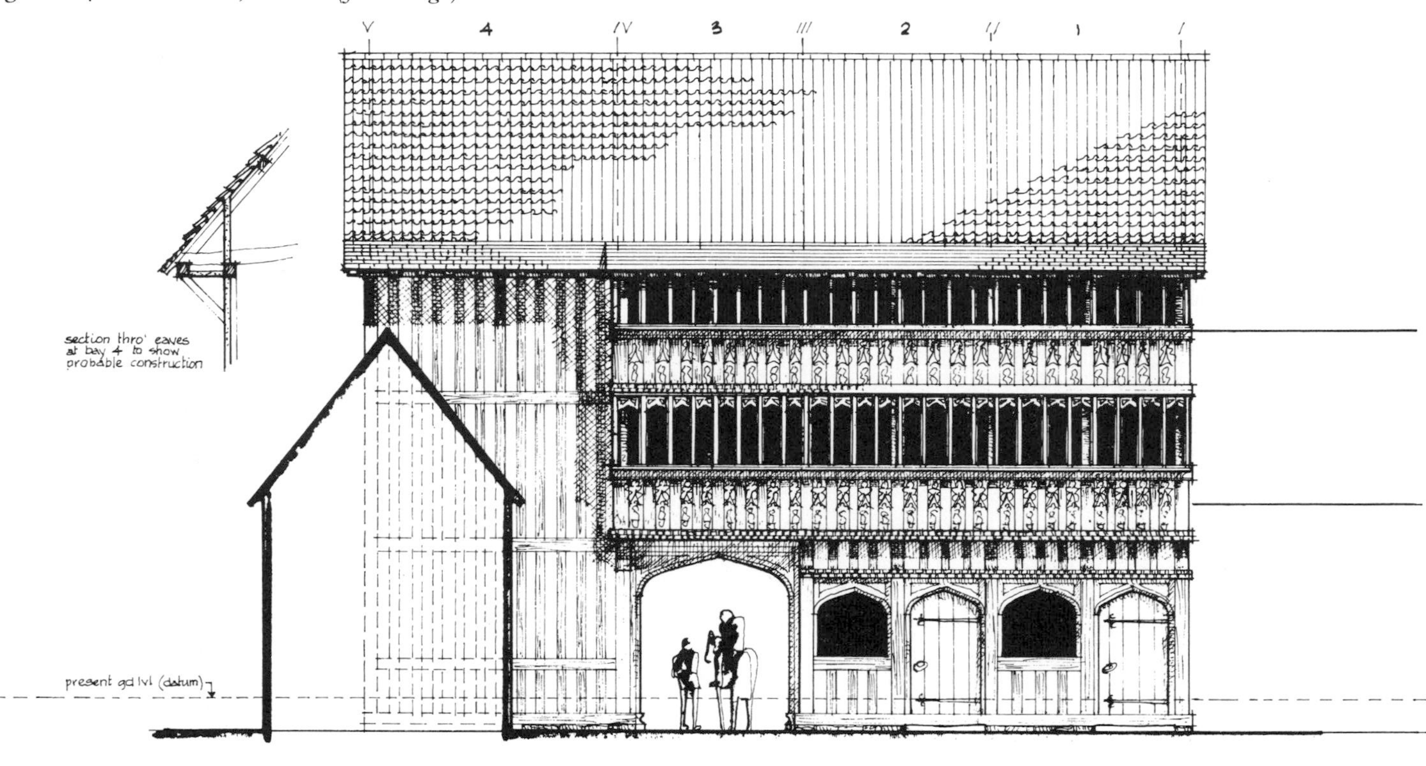

(a) Elevation as originally

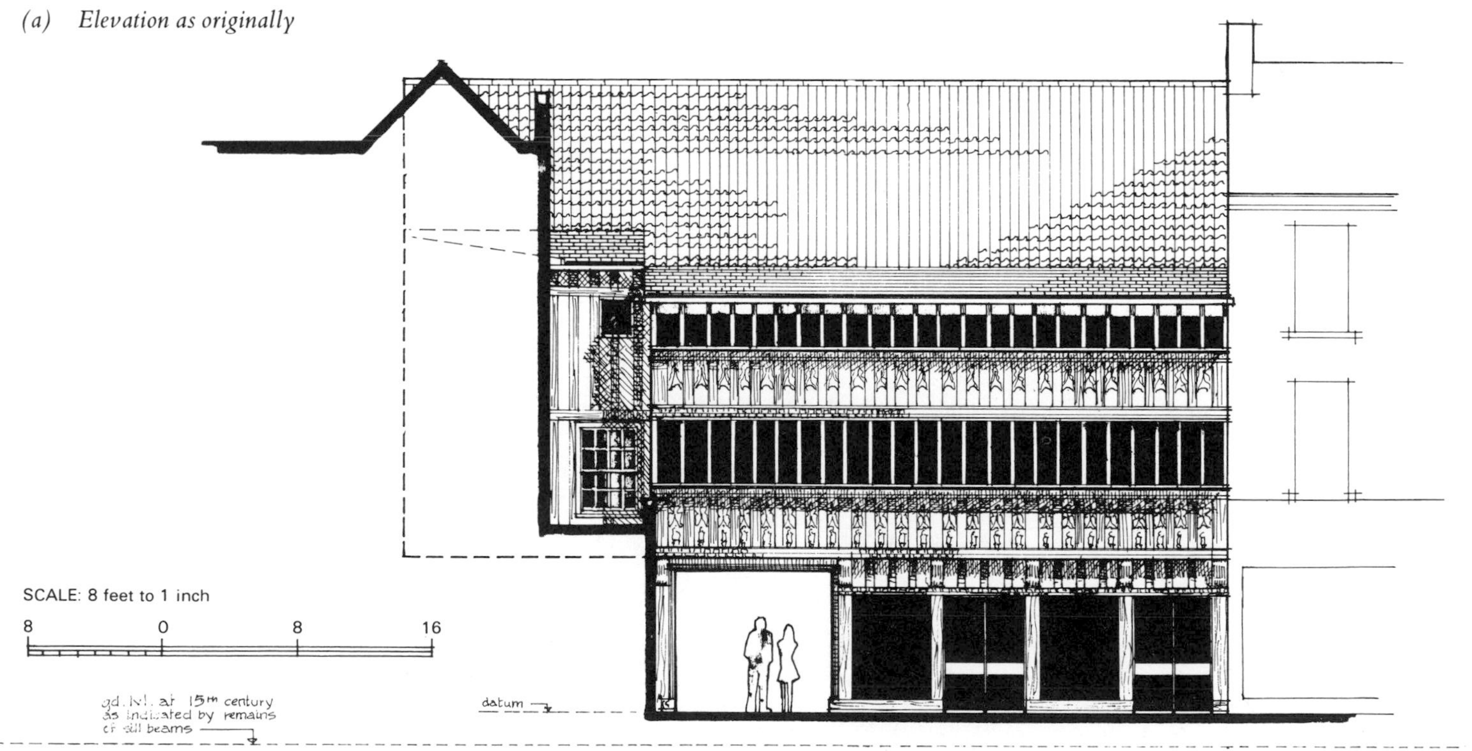

(b) As proposed by us

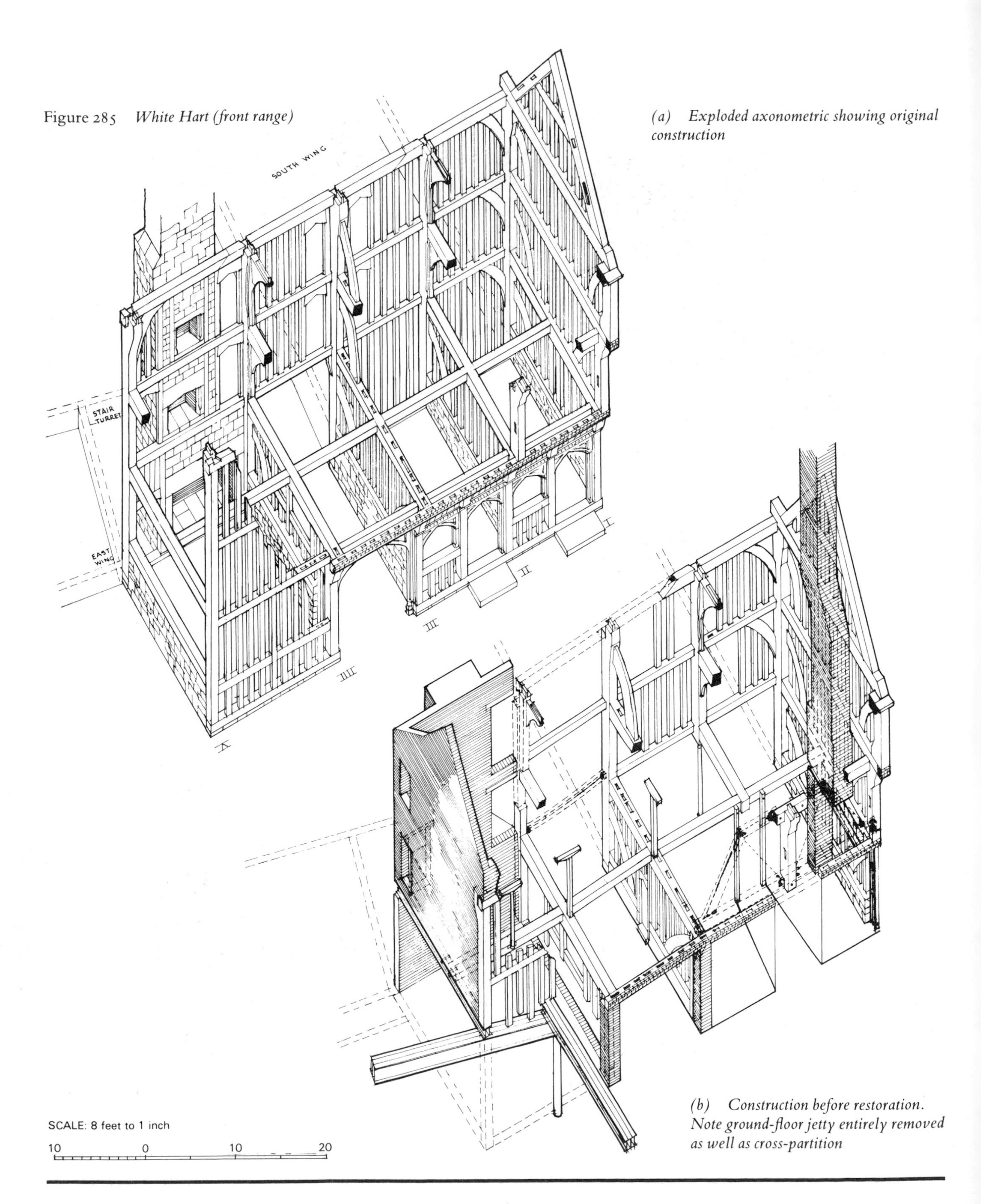

Figure 285 *White Hart (front range)*

(a) Exploded axonometric showing original construction

(b) Construction before restoration. Note ground-floor jetty entirely removed as well as cross-partition

Figure 286 *9 Spon Street, Coventry, showing typical features of fifteenth-century city centre building. Note also plaque on right setting out the brief history of the building and the date of its restoration*

Figure 287 *9 Spon Street, Coventry, looking into the roof-space of the hall in the rear block. Note the trimming of the rafters for the former smoke outlet*

Figure 288 *View of south side of Spon Street with Old Windmill Inn in foreground and 21 Spon Street (formerly 122–3 Much Park Street) next to it. Number 9 is marked by slightly higher roof at the extreme end of Spon Street*

Figure 289 *The north side of the street with number 169 of earlier date (c.1350) in foreground, and the semi-detached 163–5 (formerly 8–10 Much Park Street) beyond the two lower buildings. Courtesy John Greaves Smith*

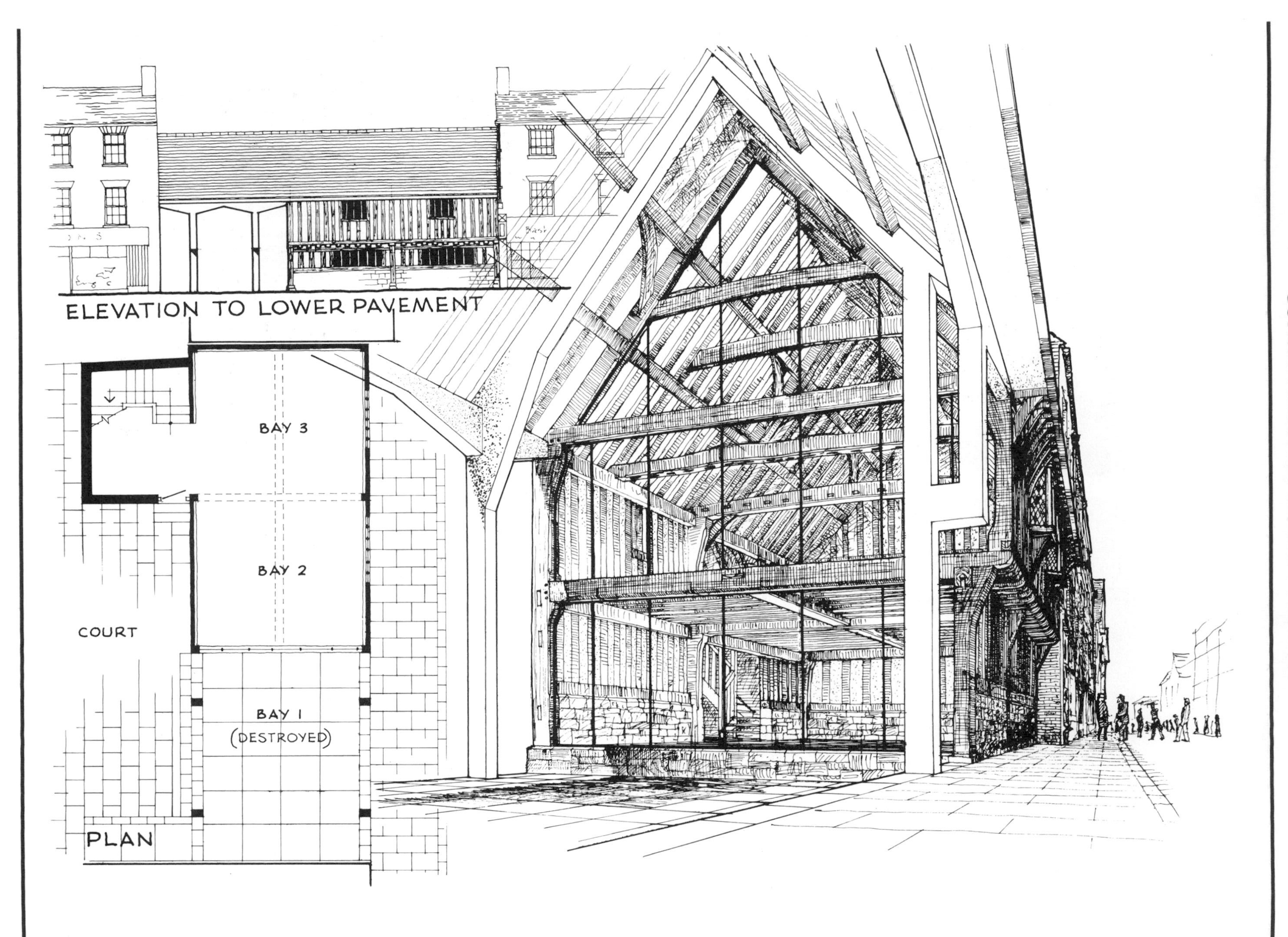

Figure 290 *Peacock Inn, Low Pavement, Chesterfield*

White Hart in date, 1450/1500, though somewhat less grand. They consist of the main upper chamber with shop or workshop below facing the street.

The halls, usually of diminutive size, are at the back and open from ground floor to roof, sometimes the equivalent of three storeys in height. Some are semi-detached. The old city centre buildings rescued from the post-war redevelopment and re-erected in Spon Street are noticeably of larger scale than the indigenous buildings of Spon Street.

Typical features are the jetty, composed of heavy floor beams spanning from back to front; the doorway with four-centred or three-centred arched door-heads; and the small, originally tracery headed windows with a string course at first floor.

Summary

So much for the styles of buildings down to 1550. As we have seen, all was then changed, and the great 'prodigy'[11] houses completely altered domestic architectural style, timber construction no doubt contributing to them, especially in the design of their huge and continuous window ranges – to let in the new light.

Peacock Inn

c.1500. Low Pavement, Chesterfield, Derbyshire. Figure 290.

Typical of the almost innumerable buildings of the sixteenth century is this little three-bay building recently restored by the local authority.[12] Its untypical features are first the raised sill-beam with the posts standing on stone plinths, and stone up to the sill-beam between the posts, and second the plan of the upper floor in which the bays did not accord with the arrangement of rooms. There were only two rooms, a cross-beam and collar being constructed in the middle bay for the sole purpose of providing a partition. The function of the building may have been as at Much Wenlock Guildhall, a courthouse with a little stone-built gaol at the back.

Our proposal was that it should become one of the main entrances into the large commercial development behind the Low Pavement frontages, by replacing its wholly destroyed third bay with a portal of reinforced concrete, following the original plan and profile of the missing bay, and from which the whole interior of the remaining two bays could be seen through a full-height glazed partition.

Ancient High House

1595. Stafford. Figures 291 to 296.
(See also Figures 86, 96, 97(a), (b) and (c), and 105.)

This building, as already noted, is immensely complicated in design and construction, mainly to achieve architectural effect of height and scale. The bay lengths on the attic floor are reduced, so that four gables along the front surmount only three structural bays of the floors beneath them (see Figure 105). The corner jetty also gives an almost breathtaking sense of daring.

The total overhang, measured diagonally on the corner, is over 6 feet, and before the present restoration there was no ground-storey corner post as this had been removed to make way for a typically Victorian angled entrance (Figure 291). The jettied first-floor joists had fractured, and that the whole corner had not collapsed speaks oceans for the strength of the framed floors. The stress over the jettied corner posts at the other end of the north elevation had pulled the whole framework outwards (Figure 292). Elimination or reduction of the intermediate partitions also contributed to the strain.

Without taking the whole building apart, restoration has had to be restricted to restraining the movement rather

Figure 291 *Ancient High House, Stafford – north-east corner showing angled Victorian doorway and absence of ground-storey corner post*

Figure 292 *Ancient High House – restored north-west corner. The lean towards the north can still be seen, particularly at the lower window which is attached to the framework in true vertical and horizontal*

Figure 293 *Ancient High House – detail of north-west corner before restoration*

than correcting it. In the reframing of the whole of the rear elevation, all the post-and-stud tenons have thus had to be cut as parallelograms.

The other most serious defect was rotten timbers (Figure 293), caused by inaccessible rainwater gutters and downpipes and comprehensive patching of decay in nailed-on softwood. Only in this building have we used steel within timber – the flitch system – to avoid the prohibitive expense of replacing corner posts with integral console brackets.

Other notable features of this building are the original windows (Figure 294), many of which had been replaced by Georgian or covered by lath-and-plaster and filled with brick; the staircase (Figure 295), and some extremely rare wallpaper of the eighteenth century (Figure 296).[13]

Bailiff's House

1610. Bewdley, Hereford and Worcester. Figure 297.
(See also Figure 43(a) and (b).)

This house earned fame in 1969 as the first building in England to be dated by dendrochronology. This was done by Veronika Siebenlist, the successor of Professor Huber at Munich who had pioneered dendrochronology in Europe since the war. The tree rings of two tie-beams and one post of this house combined to make a 230 year curve. The agreement of this curve with the South German master curve was 61.1 per cent, enough to remove any doubt about the synchronization. It confirmed the inscribed date, 1610, on the carved lintel of the main doorway as

Figure 294 *Ancient High House – original window rediscovered by stripping*

Figure 295 *Ancient High House – detail of the staircase*

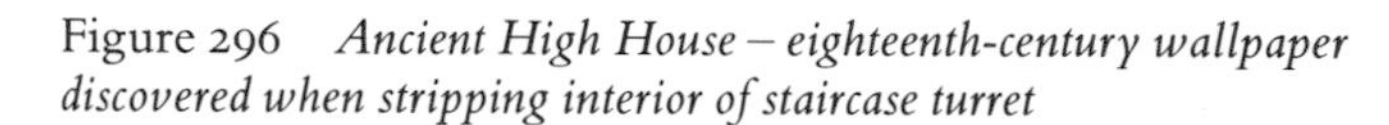

Figure 296 *Ancient High House – eighteenth-century wallpaper discovered when stripping interior of staircase turret*

the date the building was completed, its trees having been felled in 1607, the dendrochronological date.[14]

Since this first breakthrough the science has greatly advanced, and many other buildings, including Bredon Barn, have been dated by the core-bored samples taken by Siebenlist.[15] The explanation for the coincidence of the growth characteristics of the timbers, presumably from the nearby Wyre Forest, and the Spessart oaks of Bavaria is that ground conditions were similar in both regions and the effects of rainfall, temperature, atmospheric pressure and sunlight tend to follow altitude rather than latitude.

Numerous master curves have been worked out for Britain according to regions, and it is the comparison of these with each other and with continental curves that have led to this conclusion, unsuspected in 1969.

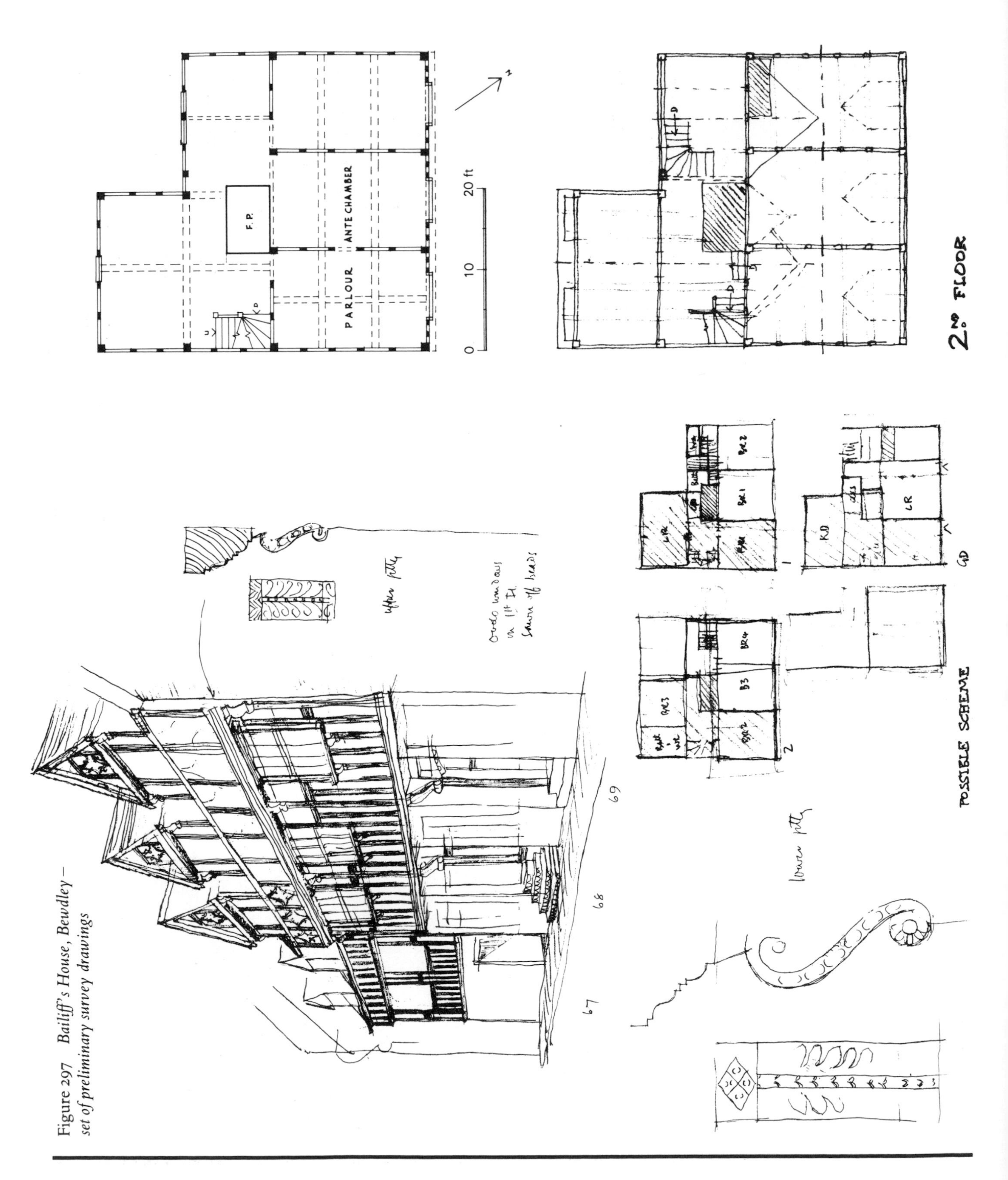

Figure 297 *Bailiff's House, Bewdley – set of preliminary survey drawings*

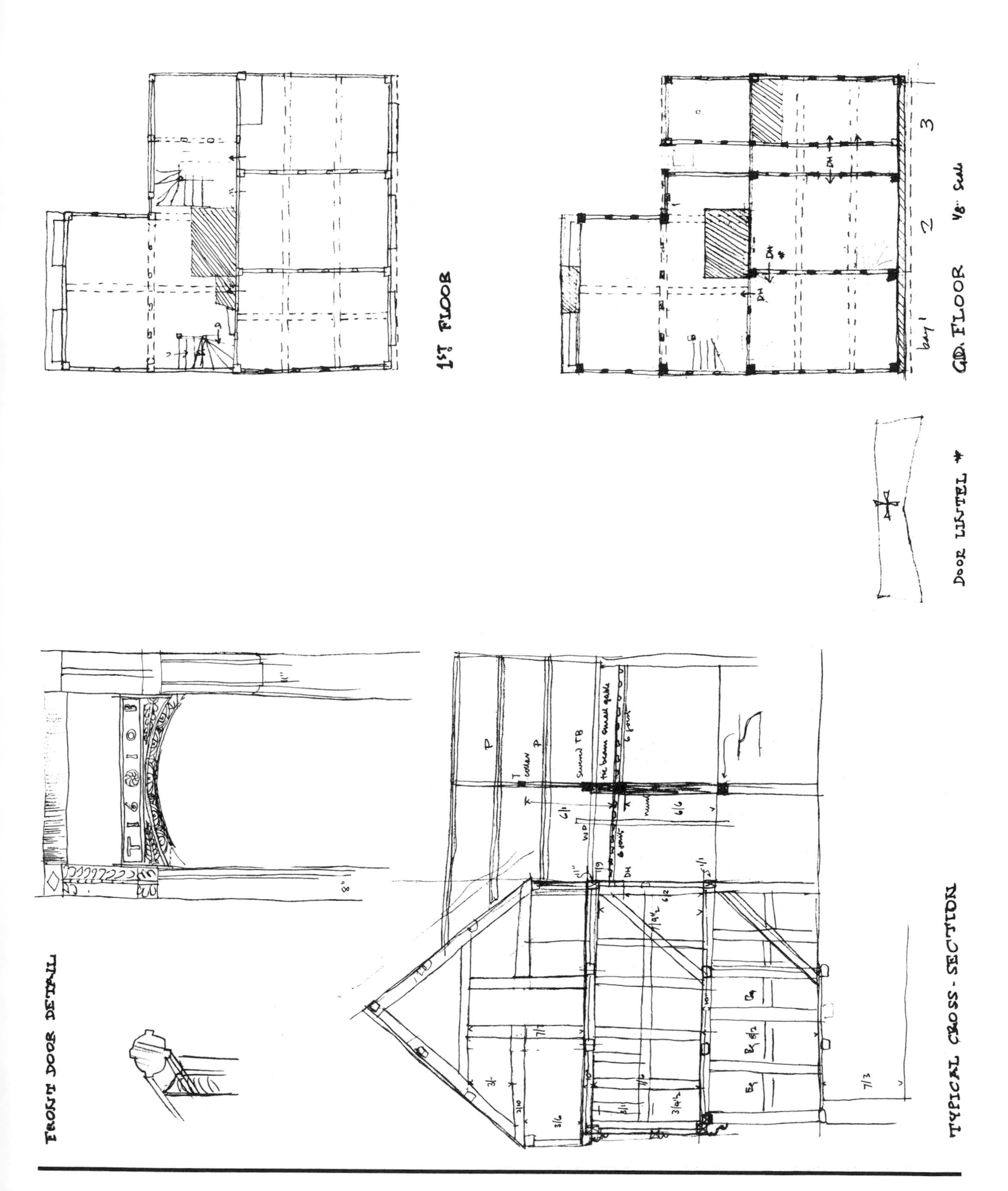
1ST FLOOR
GD. FLOOR
1/8" Scale
DOOR LINTEL
FRONT DOOR DETAIL
TYPICAL CROSS-SECTION

The house is another with decorative timbers and gables over the street. It is typical also in its rear gallery or passage giving access to the upper-floor rooms. Its only concession to the past is its gabled rear wing. It was, of course, built as the 10 Downing Street of Bewdley, and so is of a quality comparable with that of the High House.

Lane House

c.1600. Feckenham, Redditch, Hereford and Worcester. Figure 298(a) and (b).
(See also Figure 110.)

This is the only farmhouse comparable in design and quality to the High House and the Bailiff's House. It has never been painted black-and-white, and still has many original features – a practically straight ogee door-head, original windows integral with the structural frame, an

Figure 298 *Lane House, Feckenham, Redditch*
(a) Main elevation
(b) Detail of door-head and weatherings

Figure 299 *Middlebean Hall Farm – detail of front*

original dormer window, even with original lead glazing, and internally a solid-tread spiral stair. Most important are the 'weatherings' (page 128), almost certainly contemporary with the house.

The plan, already mentioned (page 98), is one that breaks precedent with the old hall plan.

Middlebean Hall Farm

1635. Hereford and Worcester. Figure 299.
(See also Figure 116.)

This house, of two builds, was of the conventional four-bay plan with two central hall (or house-place) bays. It was not medieval but rather of two full-height storeys and roof-space in all but the service bay. In 1635 the front was pushed out 5 feet, a new porch built and the semblance of a long gallery (page 99) added, purely for external show. The first floor was also shallowly jettied with a moulded bressummer on little brackets. All was close-timbered except the porch with its diagonal timbering.

Boring Mill Cottage

1636. Ironbridge Gorge Museum. Figures 300 to 306.

Formerly 'Rose Cottage', this building was condemned by the local authority and then rescued by the Ironbridge Gorge Museum and restored as a house and forge, as originally. The house occupied bay 1, and the forge, with huge fireplace and chimney, bay 4 (Figure 300). Apart from its design and use, its most interesting features were the two dates and certain details that clearly belonged to each.

At first sight '1642', cast in plaster in the external render of the big dormer window, could have been false (Figure 301). But when the render was stripped and the plaque

Figure 300 *Boring Mill Cottage, Ironbridge – before restoration. Work has just started on bay 4*

Figure 301 *Boring Mill Cottage – plaster date plaque, now preserved at the Museum. Courtesy Martin Charles*

Figure 303 *Boring Mill Cottage – original window with attached architrave, a late feature of such windows and very rarely found. Courtesy Martin Charles*

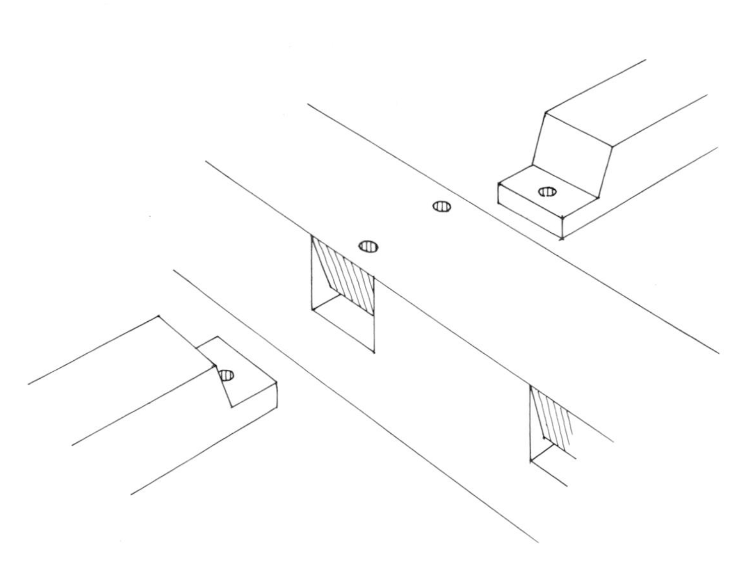

Figure 302 *Boring Mill Cottage – tusk-tenon floor joint*

Figure 304 *Boring Mill Cottage – the inserted fireplace structure as seen from bay 2. Courtesy Martin Charles*

taken down (and preserved), the earlier date '1636' was found, authentically carved on the tie-beam of the dormer. Concerning details, first, the joists of the upper floor, that had originally existed in bay 1 only, were tusk-tenoned into the beams (Figure 302). This joint was the normal in most regions from *c.*1500 or earlier until the mid seventeenth century, when the top, middle or soffit tenon was readopted but without the practically countless variations of medieval versions.[16] The joists, when the plaster ceiling was removed, were also found to be chamfered and stopped, therefore they were intended to be seen.

Figure 305 *Boring Mill Cottage – an original joist cut for the flue structure. Courtesy Martin Charles*

Second, the original house had had ovolo-mould mullion windows, one of which was uncovered in perfect condition (Figure 303). This window looked straight into the little stone building, a position suggesting that the latter was an addition; and this theory was supported by the simple tenon jointing of its ceiling joists. At the same time the condition, not only of the window but also its surrounding timbers, suggested that they had not been exposed to weather for very long.

Finally, a large timber-framed fireplace structure, with fireplaces opening into both bays 1 and 2, had obviously been inserted (Figure 304). This was proved by a number of original joists which had been cut to let the flue through (Figure 305).

Thus the 1636 house had been enlarged by incorporating bay 2 and by the addition of a well-built cellar, an underground one being precluded by the nearby stream. At the same time fireplaces were introduced, the parlour ceiling was plastered and parts of the exterior rendered – all improvements worth commemorating by the new date plaque, 1642 (Figure 306).

Figure 306 *Boring Mill Cottage – restored bay 1 and cellar building. Courtesy Martin Charles*

A POSSIBLE ETHIC FOR THE CONSERVATION OF TIMBER STRUCTURES

Bernard Feilden
ICCROM, Rome, June 1979

The Venice Charter of 1964 is the accepted basis for conservation of architecture, whilst the lesser known Murray Pease report of 1966 deals with conservation of works of art. Conservation is such a rapidly developing field that in some respects neither of these documents goes far enough; nevertheless they must be respected and followed as an invaluable basis for all conservation work.

Wood presents some special problems because it is an organic material, unlike stone, brick, concrete or metals – the other major structural materials. It is food for fungi and insects. It is actively affected by relative humidity, which causes it to expand and contract; moreover, excessive moisture robs it of its compressive strength. It is therefore extremely vulnerable to decay if it is not kept in a proper environment. The resistance to decay in different types of wood varies widely, but under good conditions wood with a medium rating will last well over a thousand years. Under bad conditions, especially where ventilation is inadequate or a poorly maintained building has a leaking roof or other defects allowing water to penetrate, the working life of wood can be short indeed.

Where timber is very plentiful, wood as a building material can form massive log walls, as in the magnificent examples from Kiji in the USSR, or can be used in framed construction, which has produced breathtaking examples of the carpenter's skill, such as Japanese temples and pagodas, the trussed roofs of Gothic churches or cathedrals, the arched ribs of the Sala della Regione in Padova and Vicenza, as well as many bridges and the hammer-beam arched trusses of Westminster Hall and Eltham Abbey in Britain, to mention but a few examples. However, it is in the vast numbers of timber-framed houses, barns and other domestic buildings that some of the most difficult conservation problems are likely to be met.

As a structural material, wood is strong in both compression and tension. Framed buildings are more nearly statically determinate than masonry structures. They depend crucially upon the strength of their joints, many of which have to transmit tension. Unfortunately, the joints are most vulnerable to dampness and to fungus attack followed by beetle attack.

Defects in wooden structural members, whether at the joints or elsewhere, are more critical to a structure's load-bearing capacity than in the case of the other primary building materials.

The universal methodology of conservation requires that objects of any kind should be fully inspected and documented before any intervention is made. The initial inspection should define the object, which in the case of a building includes its setting, and should deal with it as a whole. Such a superficial inspection can indicate what further investigation is necessary. The first requirement of a building is 'firmness', that is, it must stand up and resist all types of loadings. However, we must not forget 'commodity' and 'delight' as well. With a timber-frame building, the more sophisticated the design the more vulnerable it is to decay. As the joints are so vital, after the initial inspection it may well be decided that it is necessary to strip the external plaster and remove internal finishes in order to be assured that the structural condition is sound enough to merit conservation. Whether this detailed inspection is sufficient excuse for total removal of the external plaster in order to reveal the timber framework, is a question outside this particular discussion, as many other values besides the purely architectural ones of structural integrity must be considered. The point is, however, that no valid practical proposals can be submitted for approval until 'firmness' and structural integrity have been assured.

The detailed inspection must not ignore historical or archaeological evidence. It is essential that adequate funds be provided for this first stage of any conservation work as experience shows that full investigation saves greater costs at later stages.

To make valid proposals for approval by a competent authority, the objective of the conservation work must be defined and the method of presentation of the object made explicit.

During all conservation treatments, the following standards of ethics must be rigorously followed:

1. The condition of the object, and all methods and materials used during treatment, must be clearly documented;
2. Historic evidence should be fully recorded and must not be destroyed, falsified, or removed;
3. Any intervention must be the minimum necessary;

4 Any intervention must be governed by unswerving respect for the aesthetic, historical and physical integrity of cultural property.

Interventions should:

1 Be reversible, if technically possible; or
2 At least not prejudice a future intervention whenever this may become necessary;
3 Not hinder the possibility of later access to all evidence incorporated in the object;
4 Allow the maximum amount of existing material to be retained;
5 Be harmonious in colour, tone, texture, form and scale, if additions are necessary, but be less noticeable than original material, while at the same time being identifiable;
6 Not be undertaken by conservator/restorers who are insufficiently trained or experienced, unless they obtain competent advice. However, it must be recognized that some problems are unique and have to be solved from first principles on a trial and error basis.

The presentation of the object at least must not detract from the 'messages' contained within the object, and at best should clarify and make more comprehensible these 'messages' without any distortion or manipulation. The archaeological and structural evidence must be assessed, alternative proposals for conservation must be discussed, and above all no one must form premature conclusions. All aspects must be considered by a multidisciplinary team subject to the structural imperative of 'firmness'.

Without full inspection and documentation this essential process of evaluation cannot be effective. In the evaluation the following values should be considered:

Cultural values
Documentary
Historic
Archaeological and age
Aesthetic
Architectural
Townscape
Landscape and ecological
Use values
Functional
Economic
Social
Political
Emotional values
Wonder
Identity
Continuity

In architectural conservation, problems often arise because the utilization of the historic building, which is economically and functionally necessary, must also respect cultural values.

The cost of conservation may have to be allocated partially to each of the above separate values in order to justify the total to the community. There may be conflicts between some of the values. In certain cases, archaeological values will predominate. In other cases, artistic or historical considerations will prevail, while in yet others practical and economic considerations may modify the scope of conservation. Sound judgement based upon wide cultural preparation and mature sensitivity give the ability to make correct value assessments, and resolve contradictions in a creative way.

A programme for conservation has to be realistic and must also be guided by sound theory, otherwise it will not achieve its objectives. The theory exists to check whether the possible and practical proposals are valid, or whether they should be revised.

In conserving historic buildings we have the great responsibility of either preserving or destroying cultural property, but we must remember that nothing can last forever; there is no 'final solution', and sometimes a timber structure must be deemed to be beyond repair. If it is beyond repair, the design can be reproduced using traditional techniques.

We have seven descending degrees of conservation: prevention, preservation, consolidation, restoration, reproduction, reconstruction, re-evaluation (or adaptive use), which are described as follows.

Prevention of deterioration (or indirect conservation)

Prevention entails protecting cultural property by controlling its environment, thus preventing agents of decay and damage from becoming active. Neglect must also be prevented, as timber buildings are extremely vulnerable.

Therefore, prevention includes control of relative humidity, temperature and light, as well as measures to prevent fire, arson, theft and vandalism. In the industrial and urban environment, it also includes measures to reduce atmospheric pollution, traffic vibrations and ground subsidence due to many causes, particularly the abstraction of water.

Preservation

Maintenance, cleaning schedules and good management aid preservation. Repairs must be carried out when necessary to prevent further decay and to keep cultural property in the same state. Regular inspection of cultural property is the basis of preservation, being the first step in preventive maintenance and repair.

Preservation deals directly with cultural property. Damage and destruction caused by humidity, chemical agents, and all types of pests and micro-organisms must be stopped in order to preserve the object or structure.

Consolidation (or direct conservation)

Consolidation is the physical addition or application of adhesive or supportive materials into the actual fabric of cultural property, in order to ensure its continued durability or structural integrity.

With buildings, when the strength of structural elements has been so reduced that it is no longer sufficient to meet future hazards, the consolidation of the existing material is necessary and new material may have to be added. However, the integrity of the structural system must be respected and its form preserved. No historical evidence should be destroyed. Only by first understanding how a historic building as a whole acts as a 'spatial environmental system' is it possible to make adjustments in favour of a new use, introduce new techniques satisfactorily, or provide a suitable environment for objects of art.

The utilization of traditional skills and materials is of essential importance, as these were employed to create the object or building. However, where traditional methods are inadequate, the conservation of cultural property may be achieved by the use of modern techniques which should be reversible, proven by experience, and applicable to the scale of the project and its climatic environment. In buildings made of perishable materials such as wood, mud, brick or rammed earth, traditional materials and skills should be used for the repair or restoration of worn or decayed parts.

Finally, in many cases it is wise to buy time with temporary measures in the hope that some better technique will evolve, especially if consolidation may prejudice future works of conservation.

Restoration

The object of restoration is to revive the original concept or legibility of the object. Restoration and reintegration of details and features occur frequently and are based upon respect for original material, archaeological evidence, original design and authentic documents. Replacement of missing or decayed parts must integrate harmoniously with the 'whole', but must be distinguishable on close inspection from the original so that the restoration does not falsify artistic or historic evidence.

Contributions from all periods must be respected. All later additions that can be considered as a 'historical document', rather than merely a previous restoration, must be preserved. When a building includes superimposed work of different periods, the revealing of the underlying state can only be justified in exceptional circumstances: when the part removed is widely agreed to be of little interest and when it is certain that the material brought to light will be of great historical or archaeological value, and when it is clear that its state of preservation is good enough to justify the action. Restoration also entails superficial cleaning, but with full respect for the patina of age.

Reproduction

Reproduction entails copying an extant artefact, often in order to replace some missing or decayed, generally decorative, parts to thus maintain its aesthetic harmony. If valuable cultural property is being damaged irretrievably or is threatened by its environment, it may have to be moved to a more suitable environment. A reproduction is thus often substituted in the former location in order to maintain the unity of a site or building.

Reconstruction

The use of new materials for reconstruction of historic buildings and historic town centres may be necessitated by disasters such as fire, earthquake or wars, but reconstruction must be based upon accurate documentation and evidence, never upon conjecture.

The re-erection of fallen stones to create an accurate and comprehensible version of the original structure is a special type of reconstruction called *anastylosis*.

Moving entire buildings to new sites is another form of reconstruction which is justified only by overriding national interest. However, it entails some loss of essential cultural values and the generation of new environmental risks. In some countries wooden buildings are designed to be demountable and movable.

Re-evaluation

The best way of preserving buildings is to keep them in use, a practice which may involve what the French call *mise en valeur*, or modernization and adaptive alteration.

Adaptive reuse of buildings, such as utilizing a medieval convent in Venice to house a school and laboratory for stone conservation, or turning a fine timber barn into a domestic dwelling, is often the only way that historic and aesthetic values can be made economically viable. It is also often the only way that historic buildings can be brought up to contemporary standards by providing modern amenities. Adaptive reuse of timber-framed buildings in historic town centres occurs frequently with the objective of preserving townscape and emotional values.

In practice, interventions may involve some loss of a 'value' in cultural property, but are justified in order to preserve the objects for the future. Conservation involves making interventions at the various scales given above with levels of intensity which are determined by the physical condition, the causes of deterioration, and the probable future environment of the building under treatment. Each case must be considered individually and also as a whole, taking all values into account.

As has been outlined above, the theory of conservation postulates that any intervention be the *minimum* necessary, be reversible if possible or at least not prejudice future interventions, and retains the *maximum* of original material. Traditional methods should be used wherever possible and new techniques should only be used where proved.

The respective roles of architectural conservators and craftsmen need clarification. Craftsmen by their training provide invaluable skills, but the possession of these skills may tend to blind them to the objectives of conservation, one of which is to preserve the maximum amount of existing material. It has been found that craftsmen prefer to use their skill in renewing rather than repairing and this is dangerous. Sadly, because of faulty education, many craftsmen hate history, but it has also been found that if they realize the 'continuity' that exists in their trade and appreciate the skill of past performers, they begin to appreciate historic values.

Their skills need directing into the right channels, and here the architectural conservator has a real role.

To sum up, for successful conservation of wooden buildings it is suggested there must first be inspection and documentation, then structural analysis and preparation of alternative proposals. These proposals should define the objective of the work, so that everyone's efforts are directed towards the same goal; they should also outline how the building is to be presented. Due to the nature of the material in the conservation of timber structures, the ethic imposed by considerations of structural integrity, stability and durability generally takes precedence over other values.

When they realize that the Egyptian carpenter of 1800 BC in Fayum had the same types of tools, with the one exception of the brace and bit, then they will take more interest in history and conservation. This interest would be heightened by the marvellous display of the tools used in the construction of timber buildings as exhibited in the Musé d'Outil in Troyes.

TERMINOLOGY

Architrave Cover-board, generally moulded, for door-frames or for adjoining frames, originally lowest of three courses of classical entablature.

Bargeboards External cover-board, plain, moulded or carved on gable verge.

Battens Small scantlings, not exceeding 2–3 inches.

Beam Large-scantling horizontal member.
Aisle tie-beam Spanning between wall-post or external wall-frames and arcade-post in aisled structures.
Anchor-beam Tenoned through the morticed member, usually post, exposing tongue which is pegged to prevent spread.
Bridging-beam Spanning longitudinally between cross-beams or cross-beam and end wall-frame.
Dragon-beam Jettied over dragon-post set diagonally to floor structure – possibly corruption of 'diagonal'.
Girding-beam Main member at each storey height in unjettied wall-frame (in America, 'side-girt' and 'end-girt').
Hammer-beam Jointed and braced to post, not spanning between posts but instead cut off beyond brace.
Mantle-beam Usually spanning between posts or cruck-blades of central open truss about 6 feet above floor, probably originally for suspending pots over central hearth.
Ring-beam Strictly wall-plate of circular structure, but also member at head of walls of hipped roof structure.
Sill-beam Set on low wall or plinth as footing for wall-frames; 'interrupted' when tenoned into sides of main posts.
Strainer- or straining-beam In aisled structures spanning between arcade-posts some distance below tie-beam.
Tie-beam Spanning between posts or opposite wall-frames, or base member of roof truss.

Bevel Angled or chamfered edge of board, especially in oak panelling.

Bird's mouth or **bird-beak** Normal joint of rafters to wall-plate by which rafter is notched into top surface of plate.

Box-frame Form of construction in which upper floors and roof are supported by external framed walls and internal framed partitions.

Brace Inclined member triangulating joint between vertical and horizontal members or intersecting members.
Arch-brace Swept member generally from post to tie-beam, also from post to principal in open trusses.
Crossover brace Straight or swept members crossing each other in form of St Andrew's cross; also multiple-braces forming diamond-panel pattern.
Knee-brace Short member forming bracket in continuous contact with vertical and horizontal members.
Passing-brace Long straight member, bisecting or half-lapped across other members.
Scissor-brace Straight braces in roof truss, half-lapped across each other.
Wind-brace Swept or straight member triangulating joint between principal and purlin.

Bracket Member in continuous contact between post and beam triangulating joint – may be integral with post.
Console-bracket Voluted top and bottom in the form of the classical console.
Cove-bracket Quadrant shaped either in contact with horizontal and vertical members or separate, tenoned top and bottom and generally filled in between braces in wattle-and-daub.

Butt (see also **Scarf**) Trunk or bole of tree.

Bridle Male part of scarf-joint engaging open mortice.

Buttress (of tree) Spreading of trunk near to ground.

Chamfer Method of finishing edges of timbers.
Chamfer-stop Decorative method of terminating chamfer.

Chase Groove for housing of long edge, flashing, pipes, wiring etc.

Clapboard See **weatherboard**.

Clash Effect of medullary rays exposed by cleaving or sawing oak.

Cob (pisé) Mixture of clay and straw to form walls.

Cogged Jointing method of joists to beams by which joist ends are dropped into beam housings.

Cove Quadrant-shaped overhang.

Cruck Form of portal frame construction with swept or elbowed members extending practically the full height of the building.
Base-cruck Variation of cruck in which members extend to half the height of the roof, carrying trussed superstructure.
False cruck (quasi-cruck) Pair of timbers of cruck form but not trussed or framed as in true crucks.
Jointed-cruck Cruck comprising upper and lower components scarfed together.

Upper-cruck Cruck or cruck form supported on upper cross-beam or, in masonry buildings, with the cruck feet set in walls some distance above floor.
Cruck-spur (spur tie) Horizontal member between cruck-blade and adjacent wall-frame.

Dormer or **dormant** Originally wall-plate; thus dormer window is one built off wall-plate.

Dovetail Joint between top and bottom surfaces of two members (e.g. tie-beam and wall-plate) of which edges fan outwards from neck to prevent slip; also used for cogged joists.
Bare-face dovetail With one edge only.

Dowel Turned peg for centering a member as opposed to structural peg for jointing – see *peg*.

Eaves Overhang of roof-slope.

Fair-face Halved surface of a structural timber or smoother and better finished surface.

Field Raised surface of door- or wall-panel bevelled at edges, also in decorative infill panels consisting of boards fielded to form relief pattern or motifs.

Flitch Method of reinforcing beam by vertically halving and inserting steel plate.

Flittern Young oak pole.

Halved-face See **fair-face**.

In stick Method of stacking timbers out of contact with each other by inserting battens between each layer.

Jetty Cantilevered overhang of one storey over the one below it.

Joist Member of suspended or upper floor.

Jowl Swelling at top of post, generally the buttress of tree up-ended – also formed from swelling at the branch.

Key Wedge or folding wedges used to lock a tabled scarf – see **scarf**.

Lap Method of jointing scarfed or intersecting members.
Half-lap Joint of two members halved in thickness at the joint.
Notched-lap As for half-lap but with notch in one or other edge to prevent withdrawal.
Secret notch-lap As for notched-lap but with notch concealed at back of exposed surface.

Lattice girder Beam made up of top and bottom component connected by open web of diagonal and sometimes vertical timbers.

Mortice Female part of mortice-and-tenon joint.
Dead-mortice Shallow rectangular cavity as for peg at foot of cruck-blades.
Through-mortice Penetrating full depth of member, generally of beam for opposite tusk-tenoned joints, also through posts for anchor-beam.

Mortice-and-tenon Definitive joint of timber-framing interlocking entire structure.

Mullion Upright member of window, normally tenoned into head and sill but not pegged.

Muntin Central upright of a panelled door or panelling.

Nogging Filling in panels with brickwork or compression timbers between joists or studs.

Notch Angled or V-shaped cut in a straight edge.

Out-shut Lean-to addition, generally along the back wall of building.

Palisade Wall of contiguous uprights.

Panel-infill Material or method of filling space between horizontal and vertical or diagonal-framed timbers.

Peg Tapered and pointed wooden pin obtained by cleaving – varying from square in section to irregular polygonal – never turned.

Plate Horizontal member at head of frame or wall, bearing joists of next storey or rafters.
Arcade-plate At head of arcade or arcade-posts.
Collar-plate Spanning between crown-posts or trusses in rafter roof to support collars (also crown-plate or collar-purlin).
Sole-plate (-piece) The bottom member of framed or stud partition in which members are of similar scantling.
Top-plate Generally referring to the arcade-plate.

Ploughed Method of forming groove or chase.

Post Straight vertical timber.
Aisle-post Supporting wall-plate or aisle tie-beam of aisled structures.
Arcade-post Supporting arcade-plate or tie-beam of nave of aisled structure.
Corner post At the external corners of quadrilateral structure.

Crown-post Set on centre of tie-beam to support collar-plate.
Dragon-post At the corner of structure jettied at front and side.
Frank-post Old term probably denoting a main or principal post in a structure.
King-post Set generally centrally on the tie-beam to support ridge of roof.
Prick-post Old term probably denoting secondary post.
Principal post Coinciding with trusses in side walls of structure.
Queen-post Set as a pair symmetrically on tie-beam to support principal rafter or outer ends of collar-beam of truss.
Wall-post Framed into the wall.

Post construction Form of construction using earth-fast posts.

Post-and-truss Form of construction in which posts and trusses form bay-dividing cross-frames.

Purlin Roof beam spanning between trusses, lying in section at right angles to the rafters and supporting them intermediately between plate and ridge.
Clasped-purlin Fixing of purlin by which it is trapped between end of collar and principal rafter.
Collar-purlin See *collar-plate*.
Tenoned-purlins Fixing of purlin by means of tenoning into principal rafter.
Trenched-purlin Halved over back of principal in rectangular notch of equal depth so that members lie roughly in same plane on which common rafters are laid.

Quartered Method of conversion by which log is sawn or cleft into four radial segments.

Racking Tendency of structure under stress to assume form of parallelogram instead of rectangle, resisted by bracing.

Rafter
Common Inclined member directly supporting roof covering.
Compound-rafter Made up of square-section rafters with distance-pieces separating them vertically, the distance-pieces usually supporting the purlins passing through the space between the rafters.
Hip-rafter At corner of hipped roof, receiving commons.
Principal rafter Inclined members of roof truss directly supporting purlins – in rafter roofs, applied to pair of rafters coinciding with the tie-beam.
Sprocket-rafter Attached foot of rafter to decrease slope of roof and form 'bell-mouth' and increase eaves overhang.
Valley-rafter At junction of roof-slopes forming internal angle, receiving feet of commons.
Verge-rafters Outside gable supported on projecting ends of horizontal members.

Rail Short horizontal member of wall-frame.

Rearing Raising from horizontal to vertical.

Reversed assembly Phrase referring to post-head joint in which lateral plate rests on tie-beam.

Ridge Apex of roof.
Ridge-piece or *ridge-pole* Horizontal member spanning between trusses to which heads of common rafters are pegged.

Roof
Gabled With two directions of slope to side walls, end walls being built up to ridge, generally purlin roofs.
Hipped roof With four directions of slope, generally rafter roof but also possible with purlins.
Half-hipped roof Hipped through upper half or roof-slope, generally in purlin roofs above purlin level.
Pentroof Built against a wall, thus sloping in one direction only.
Purlin roof Double roof structure with rafters intermediately supported by purlins.
Rafter roof Single roof structure, without purlins or with purlins functioning only as stiffeners.

Saddle Short timber jointed to head of two uprights to connect them, generally cruck-blades.

Scarf or **splice** Method of jointing members in the same alignment, vertical or horizontal.
Bladed-scarf Half-lap with diminished ends as bare-face tenons housed into other member.
Bridle-scarf Open mortice of one member engaged by tenon of other one, abutting as splayed-scarf (see below).
Half-lapped scarf Each member halved longitudinally either horizontally or vertically without alteration of profile at any point.
Sallied-scarf Each end of joint pointed to restrain side-slip.
Scissor-scarf Each member splay-cut through half its breadth in opposite directions, generally with squint butts (see below).

Splayed-scarf In which each member is splay-cut to fit other without alteration of profile.
Square-butt scarf In which end of each member is cut at right angles to the direction of the scarfed members.
Squint-butt scarf In which the ends are undercut.
Tabled-scarf Each member stepped at centre of splay. Joint may be locked by squint-butts only or by key together with squint-butts.
Vertical-scarf In which timbers are cut for splay or half-lap vertically instead of horizontally.

Scotch Incision in post near head in which to insert temporary prop.

Settlas Local term for brick-supported shelf of slate or wood, round the walls, for milk pans or barrels – also settless, settlus and settle.

Skew Non-rectangular in setting one member or frame to another.
Skew-pegging Random directioning and alignment of pegs in repair joints.

Soffit Underside.

Splice See **scarf**.

Squint In joints – see **scarf**.

Stave Normally cleft upright for wattle-and-daub panels, pointed at top for insertion into hole and at bottom for sliding into groove.

Stud Secondary upright in wall-frame.

Strut Short compressive member, straight or swept.
Queen-strut In roof truss set on tie-beam or collar symmetrically on either side of centre line – always as a pair.

Tenon Male part of mortice-and-tenon joint formed by reducing end of timber and forming shoulders on each side.
Bare-face tenon Timber reduced and shouldered on one side only.
Cock-tenon Vertical tenons of post-head to secure tie-beam.
Fish-tenon See *slip-tenon*.
Haunched-tenon With splayed shoulder, normal for floor joists. See also *tusk-tenon*.
Slip-tenon Detached tenon to engage mortice by which member may be inserted into standing or preformed frame as if tenoned.
Stub-tenon Member reduced and shouldered on all four sides.
Tongued-tenon Tenon penetrating receiving member and projecting on blind side.
Tusk-tenon Normal or bare-face tenon haunched to resist shear.

Through-and-through Method of sawing log into planks – not quartered.

Tong A-frame or straight-bladed cruck, not generally included in cruck category.

Tongue The male part of a groove in tongued-and-grooved boards.

Transome Horizontal component of windows or panelling bisecting mullions.

Truss Framed structure consisting of tie-beam and principal rafters generally with additional secondary members, such as collar and struts, also a total cross-frame from ground level to ridge.
Arch-truss Open truss without tie-beam, consisting of posts, swept braces and principals.
Spere-truss Dividing hall from through-passage.

Verge Edge of roof at gables.

Vice Spiral stair.

Wattle-and-daub Traditional infill material of wall panels.

Wattle groove V-groove in sides and bottom rail of timbers to hold wattle or other infill material.

Weatherboard Overlapping boarding for walls (in America, 'clapboard').

Weatherings Small pentroofs attached to external walls to protect lower part of walls.

Wedge, folding Pair of wedges driven in from opposite direction.

Yoke Short timber connecting pair of uprights just below head, usually of cruck-blades.

NOTES AND REFERENCES

Chapter one Structural types

1 R. A. Cordingley, 'British historical roof-types and their members: a classification', *Transactions of Ancient Monuments Society*, vol. 9 (1961). Names all conceivable variations of timber roofs, but I have not consistently followed his nomenclature, as for instance 'through-purlin' which I have termed 'trenched-purlin'. Several other differences will become clear later.

2 F. W. B. Charles, 'Post-construction and the rafter-roof', *Vernacular Architecture Journal*, vol. 12 (1981).

3 P. Rahtz, *The Saxon and Medieval palaces at Cheddar*, BAR British Series 65 (1979). Apart from helpful discussion for which I am most grateful, the author had no part in the design of the reconstruction. For that I take sole responsibility and criticism.

4 C. A. Hewett, 'The barns at Cressing Temple, Essex and their significance in the history of English carpentry', *Journal of Society of Architectural Historians*, vol. XXVI, no. 1 (1967).

5 I am grateful to Elna Mööller for most of my Danish information.

6 J. T. Smith announced this discovery at the VAG Conference of 1980.

7 S. R. Jones, and J. T. Smith, 'The Great Hall of the Bishops Palace at Hereford', *Medieval Archaeology*, no. 7 (1960); C. A. R. Radford, E. M. Jope, and J. W. Tonkin, 'The Great Hall of the Bishops Palace at Hereford', *Medieval Archaeology*, vol. XVII (1973). We examined one of the great posts in 1982, and there is no doubt that it penetrates the floor of the present building, but excavation would be needed to see whether it also penetrated the ground. The fact that the post has no plinth is fairly conclusive evidence that it does so.

8 This was found at the great barn of Canteloup in Normandy during its reconstruction as the abbey church of St Wandrille in 1969.

9 R. T. Mason, *Framed Buildings of the Weald* (Coach Publishing House 1969).

10 W. Horn, and E. Born, *The Plan of St Gall* (University of California Press 1979), vol. 1, p. 180 ff, fig. 127. Horn shows two basic Roman roof types from Vitruvius as interpreted by Barbaro in his translation of 1556. Horn also provides evidence of purlin roofs with king-posts as early as the sixth century.

11 F. W. B. Charles, 'The Guesten Hall roof, Worcester Cathedral', *Transactions of the Ancient Monuments Society*, vol. 18 (1971).

12 J. T. Smith, 'Cruck construction: a survey of the problems', *Medieval Archaeology*, vol. VIII (1964). Gives the clearest definitions of cruck and quasi-crucks, without, however, taking into account their method of erection.

13 W. Horn, and E. Born, *The Barns of the Abbey of Beaulieu at its granges of Great Coxwell and Beaulieu St Leonards* (University of California Press 1965). They compare Great Coxwell with the huge but practically wholly destroyed barn of Beaulieu St Leonards. The drawings clearly show the quasi-crucks of the former among all other features.

14 N. W. Alcock, *Cruck Construction* (CBA 1981). This is the most comprehensive work on crucks and includes not only definitions but all current theories on origins as well as a gazetteer of all known crucks.

15 J. Walton, 'Hog-back tombstones and the Anglo–Danish house', *Antiquity*, vol. XXVIII (1954).

16 I am grateful for W. Horn's permission to anticipate his forthcoming work.

17 M. de Paor, and L. de Paor, *Early Christian Ireland* (Thames & Hudson 1958). They made this important suggestion entirely independently of my study of Irish 'cruck' churches.

18 F. W. B. Charles, 'Medieval cruck-building and its derivatives', *Medieval Archaeology*, Monograph series, no. 2 (1967).

19 G. Webb, *Architecture in Britain – the Middle Ages* (Penguin 1956), p. 12, notes the 'crucks' as 'antae' after classical models of expressing structure on the front wall.

20 Lord Raglan, *The Cruck Truss* (Royal Anthropological Institute 1956).

21 N. W. Alcock *op. cit.*, see J. T. Smith, *op. cit.*, p. 5 ff.

22 S. R. Jones, 'West Bromwich Manor House', *Transactions of the South Staffs. Archaeological/Historical Society* (1975–6).

23 R. T. Mason, 'Fourteenth-century halls in Sussex', *Sussex Archaeological Collection*, vol. XCV (1957).

24 N. Drinkwater, 'The Old Deanery, Salisbury', *Antiquaries Journal*, vol. XLIV, part 1 (1964).

25 W. Horn, and F. W. B. Charles, 'The cruck built barn of Frocester in Gloucestershire', *Journal of Society of Architectural Historians* (1984).

26 J. T. Smith, 'The early development of timber buildings: the passing brace and reversed assembly', *Archaeological Journal*, vol. CXXXI (1974).

27 N. W. Alcock, *op. cit.*

Chapter two Timber

1 W. Harrison, 'An historical description of the Ilande of Britayne with a brief rehearsal of the nature and qualities of the people of England (from Holinshed's chronicles 1577), in F. J. Furnivall (ed.), *New Shakespeare Society*, vol. I (1877).

2 N. D. G. James, *A History of English Forestry* (Blackwell 1981).

3 C. Y. Taylor in reviewing N. D. G. James, *op. cit.*, in *Forestry*, April 1982 writes:

It is interesting to recall, although not brought out in this book, that Evelyn in England and Colbert in France were contemporaneous and each wanted to ensure that his country would have a continuing supply of oak for maritime purposes. Colbert, who had his king's authority, achieved a great deal for French forestry. Evelyn had no official status and could bring about no direct influence. It is noteworthy, too, that whereas tall, clean oak stems were required in France, it was the open grown oak that was wanted in England. The French must feel satisfied with their object of management.

4. J. Boswell, *Life of Johnson* (Oxford University Press 1950).
5. H. L. Edlin, *Woodland Crafts in Britain* (David & Charles 1974), p. 85.
6. W. Harvey, 'Westminster Hall roof', *Journal of the Royal Institute of British Architects* (1922).
7. L. F. Salzman, *Buildings in England Down to 1540* (Oxford University Press 1967), pp. 206 and 251.
8. Identification of timbers in ancient buildings is invariably a source of contention. The expert referred to here is Alexander Scott.
9. V. Parker, *The Making of Kings Lynn* (Phillimore 1971), p. 107.
10. L. F. Salzman, *op. cit.*, p. 239.
11. O. Rackham, W. J. Blair, and J. T. Munby, 'The thirteenth-century roofs and floor of the Blackfriars Priory at Gloucester', *Medieval Archaeology*, vol. XXII (1978), p. 120.
12. R. K. Field, 'Worcestershire peasant buildings, household goods and farming equipment in the late Middle Ages', *Medieval Archaeology*, vol. IX (1965), p. 105 ff.
13. F. W. B. Charles, 'Dendrochronology – the science of dating buildings by tree rings', *Forestry and Home Grown Timber* (1971).
14. L. D. W. Smith, 'Part of the c.1500 estate survey for the Archer property in Tamworth-in-Arden', to appear in *Transactions of Birmingham and Warwickshire Archaeological Society*. I am grateful to Smith for telling me about this document.
15. B. M. Feilden writes that the original beams were larger in section than anything the Royal Navy put into the Rope Walk at Chatham – see *Architects' Journal* (9 March 1983), p. 49.
16. L. F. Salzman, *op. cit.*, p. 238; see also C. A. Hewett, *English Historic Carpentry* (Phillimore 1980), p. 162 ff.
17. L. F. Salzman, *op. cit.*, pp. 200 and 218.
18. S. R. Jones, and J. T. Smith (1960), *op. cit.*, and C. A. R. Radford, *et al*, *op. cit.*
19. B. Hope-Taylor, *Yeavering: an Anglo British Centre of Early Northumbria* (HMSO 1979).
20. Even larger timbers were required for Henry VIII's Nonsuch Palace in 1538 for the wing towers, which required fourteen 'eighty feet timbers' – see H. M. Colvin, *The History of the Kings Work* (HMSO 1982), vol. IV, p. 192.
21. C. A. Hewett (1980), *op. cit.*, p. 144.
22. R. A. Cordingley, *Stokesay Castle, Shropshire* (College Art Association of America 1963), p. 91 ff.
23. P. Smith, *Houses of the Welsh Countryside* (HMSO 1975).
24. Sir Cyril Fox, and Lord Raglan, *Monmouthshire Houses – Part I – Medieval* (National Museum of Wales 1951), p. 38.
25. J. T. Smith, and C. F. Stell, 'Baguley Hall: the survival of pre-conquest building traditions in the 14th century', *Antiquaries Journal*, vol. XL (1960); and H. Schmidt, 'The Trelleborg House reconsidered', *Medieval Archaeology*, vol. XVII (1973), p. 55.
26. W. L. Goodman, *The History of Woodworking Tools* (Bell & Sons 1964), p. 131.
27. H. L. Edlin, *op. cit.*, p. 17.
28. W. Rose, *The Village Carpenter* (Cambridge University Press 1952).

Chapter three Organization and framing

1. J. Harvey, *The Medieval Architect* (Wayland 1972), p. 142.
2. *Ibid*, p. 207.
3. *Ibid*, p. 252.
4. R. K. Field, *op. cit.*
5. E. Gooder, *The Pittancers' Rental 1410–1411* (Birmingham University Extramural Department 1973).
6. W. Horn, and E. Born (1979), *op. cit.*
7. F. W. B. Charles, and K. Down, 'A sixteenth-century drawing of a timber-framed house', *Worcester Archaeological Society*, third series, vol. 3 (1972).
8. L. F. Salzman, *op. cit.*, p. 336, plate 19 – carpenters and their tools, 15th century from J. Van de Gheyn 'Croniques et conquestes de Charlemaine', Vroment et Cie, Brussels, plate 95.
9. Charles J. Venables.
10. E. L. N. Viollet-le-Duc, *Dictionnaire raisonné de l'architecture français du XI au XV siècle* (1867–70). There are slight differences in detail but not in principle between the numbers in the text and those shown by Viollet-le-Duc.
11. O. Rackham, 'Grundle House: on the quantities of timber in certain East Anglian buildings in relation to local supplies', *Vernacular Architecture Journal*, no. 3 (1972). Comparing his results with these, Grundle House is considerably larger than our example, and the problem of selecting timbers suitable for building have not been given the emphasis which I consider would have been necessary.
12. F. W. B. Charles, 'Scotches, lever sockets and rafter-holes', *Vernacular Architecture Journal*, vol. 5 (1974); see also K. W. E. Gravett, *Vernacular Architecture Journal*, vol. 8 (1977).

Chapter four Historical changes

1. W. Horn, 'On the origins of the medieval bay system', *Journal of Society of Architectural Historians*, vol. XVII (1958).
2. M. Wood, *The English Medieval House* (Dent 1965), p. 19.
3. G. Beresford, 'The medieval clay-land villages: excavations at Goltho and Barton Blount', *Medieval Archaeology*, Monograph series, no. 6 (1975).

4 J. West, *Village Records* (MacMillan 1962). I was privileged to help West in his field research in Worcestershire, trying to locate houses which he had already noted from inventories.
5 S. R. Jones (work in preparation). I am grateful for the information he has given me.
6 E. C. Rouse, '16th and 17th century domestic paintings', *Oxoniensia*, vol. 37 (1972). I am grateful for his advice on the painting of various styles and dates.
7 W. J. Smith – I am grateful for his calling my attention to a number of references to paintings, for example, from Sir Thomas Kytson's household accounts of August/September 1574, quoted by Nathaniel Lloyd, *A History of the English House* (The Architectural Press 1949), p. 80:

> For plastering and whitening the fore front of my master his house in Coleman Street and the courte, with the blackening of the timber work.

From C. B. Andrews (ed.), *Torrington Diaries* (Metheun 1954):

> 1790 June 13 – Stockport. . . . All the houses of this town were formerly built of oaken timber; this, now, in general has given way to brick, tho, this inn is striped and barr'd with as much black timber, as would build a man of war.

And,

> 1790 June 21 – Bramhall Hall. . . . My reception was very civil into this oldest of all the striped houses; black and white, flourish'd into as many devices as a boy wou'd draw upon his kite.

In view of all these we must say that there are notable exceptions to nineteenth-century blackening, but J. Nicholls, *The Progress and Public Processions of Queen Elizabeth* also records:

> Upon the occasion of Queen Elizabeth's visit to Worcester it was ordered by the Common Council of the city 'that every inhabitant within the liberties of the citie shall forthwith whitlyme and colour their houses with comely colours'.

8 P. A. Nicklin kindly gave me a copy of her *Aspects of the History of Bewdley* which included this quotation.
9 F. W. B. Charles, 'Timber-framed houses in Spon Street, Coventry', *Birmingham and Warwickshire Archaeological Society*, vol. 89 (1978–9), p. 98.
10 W. Harrison, *op. cit.*, p. 113 ff.
11 F. W. B. Charles, 'Chester House Library, Knowle, Warwickshire', *Architects' Journal*, vol. 165, no. 17 (1977), p. 782.
12 J. Moxon, *Mechanik Exercises* (Praeger 1970).
13 S. R. Jones, 'Chamfer stops – a provisional mode of reference', *Vernacular Architecture Journal*, vol. 2 (1971)
14 H. M. Colvin, *A Guide to the Sources of English Architectural History* (Pinhorns Handbooks 1967).
15 A. L. Cummings, *The Framed Houses of the Massachusetts Bay Area 1625–1725* (Harvard University Press 1979).
16 C. Lines, 'Fairy-tale home in the woods', *Warwickshire and Worcestershire Life* (1968).

Chapter five Preliminary survey

1 T. Smith, *Traditions of the Old Crown House in Der-yat-end* (Henry Wright 1863).
2 S. R. Jones (1975–6), *op. cit.*

Chapter six Structural survey and repairs

1 It is also extremely difficult to make sure that the damp timber has been wholly eliminated before filling. Unless this is done, the process of decay through wet rot continues more rapidly than before.
2 A. R. Powys, *Repair of Ancient Buildings* (SPAB 1981), p. 120.
3 I am grateful to R. Murdoch of the Lead Development Association for his confirmation of this.
4 This material holds its shape after compression for a limited period and then expands to fill irregularities of the surfaces to which it is applied.
5 It is not possible to give a single reference. Recommended books are: J. L. Martin, B. Nicholson, and N. Gabo, *Circle: international survey of constructive art* (Faber & Faber 1937); W. Herzogenrath (ed.), *Baühaüs – an abridged edition of the catalogue for the exhibition '50 jahre baühaüs* (Institut für auslandsbeziehunger 1975).
6 W. Morris, *On Art and Socialism* (John Lehmann 1947).
7 The firm of P. V. Firminger is gratefully acknowledged for their advice and help in many departments of our larger jobs, and for their consistent financial fairness and control working to 'fixed-price' contracts.

Chapter eight Cheylesmore

1 I am grateful to Margaret Tomlinson for her help in unravelling many problems, especially concerned with Cheylesmore and the construction of the city wall, also the earliest records of Cheylesmore. Cheylesmore as a royal manor is also briefly mentioned in H. M. Colvin, *The History of the Kings Works* (HMSO 1982), vol. 11, p. 909.
2 P. B. Chatwin, 'Early Coventry', *Birmingham and Warwickshire Archaeology Society*, vol. 53 (1928). He is author of several articles on Warwickshire houses (see Birmingham Archaeological Society Transactions), and he informed the Royal Commission of this discovery.
3 M. Rylatt, *City of Coventry: Archaeology and Development* (Coventry Museum 1977). The remains of the Old Stone House in Much Park Street have been preserved, the site

having been excavated by G. G. Astill in 1971. There was also much stone in the foundations of 7 Much Park Street discovered by A. Hannan. Astill and Hannan are preparing a joint report on 7 Much Park Street and the Stone House. See also F. W. B. Charles (1978–9), *op. cit.*, p. 99.

Chapter nine Much Wenlock Guildhall

1 I am grateful to V. H. Deacon and S. Mullins for the help they have given in researching the documents of the Royal Burgh of Wenlock.

Chapter ten The Wellington

1 J. Charge, 'The raising of the Old Wellington Inn and Sinclair's Oyster Bar', *The Structural Engineer* (1972).

Chapter eleven Buildings

1 C. A. Hewett, 'Siddington Barn', *Country Life* (1971), and 'Siddington Barn', *Archaeological Journal*, vol. 129 (1972).
2 I am grateful to Michael Dawes for his hospitality and free run of his house to explore its architectural history.
3 C. J. Bond, 'The estates of Evesham Abbey: a preliminary survey of the medieval topography', *Vale of Evesham Historical Society Research Papers* (1973).
4 S. R. Jones (1975–6), *op. cit.*
5 I am grateful to R. Furneaux Jordan for pointing out this feature and its significance at an Attingham Conference.
6 W. G. Simpson has in preparation a paper on his dendrochronological discoveries at this building and others in the Nottinghamshire area with which we have been concerned.
7 I am grateful to J. Hetherington for his research into the documentary evidence of this building.
8 I am grateful to Pat Hughes and Jennifer Costigan for their research into inventories through which the owners of this house and their trades have been traced to the seventeenth century.
9 Guy St John Taylor and his assistant, Philip Siddall, with Ove Arup, who had already helped us in our work, restored this building in 1981 and it was a privilege to be able to advise them.
10 F. W. B. Charles (1978–9), *op. cit.*, and 'The timber-framed buildings of Coventry – 169 Spon Street', *Birmingham and Warwickshire Archaeological Society*, vol. 86 (1974).
11 J. Summerson, *Architecture in Britain – 1530–1830* (Penguin 1953), p. 30 ff.
12 P. Dixon and P. Borne first surveyed this building when it was due for demolition together with practically the whole of Low Pavement in 1974. Their findings and the public protest that followed saved it. Our report (January 1975) established the practicability of restoring it *in situ* and Feilden and Mawson were then called in to work out a conservation scheme with the developers' architects, Elsom, Pack and Roberts, the local authority taking responsibility for the Peacock.
13 A. L. Cummings first recognized the significance of the wallpaper illustrated and proceeded to a close scrutiny of many others including fragments which he dated *c.*1720. Both he and the Victoria and Albert Museum are now in possession of photographic records and it is hoped that the originals will be preserved without further damage *in situ*.
14 F. W. B. Charles (1971), *op. cit.*
15 J. Fletcher has reported to the National Trust on his analysis of these samples and will be publishing the results of samples selected by himself after the fire of 1981.
16 C. A. Hewett, *The Development of Carpentry 1200–1700* (David & Charles 1969). Gives examples of these and many other joints and is the authority to refer to. However, this should always be with a degree of caution as regards dating since the field of his research is based on, if not confined to the County of Essex, and there is no county that is not in some respect unique in carpentry technique – especially joints.

BIBLIOGRAPHY

Addy, S. O., *The Evolution of the English House*, Allen & Unwin 1933

Airs, M., *The Making of the English Country House 1500–1640*, The Architectural Press 1975

Alcock, N. W., *Cruck Construction*, CBA 1981

Alcock, N. W., and Barley, B. W., 'Medieval roofs with base-crucks and short principals', *Antiquaries Journal* (1973)

Alford, D. W. E., and Barker, T. C., *A History of the Carpenters Company*, Allen & Unwin 1968

Allen, E., *Stone Shelters*, Massachusetts Institute of Technology 1971

Andrews, C. B. (ed.), *The Torrington Diaries*, Methuen 1954

Armstrong, J. R., *Traditional Buildings Accessible to the Public*, Wakefield 1979

Atkinson, T. D., *Local Style in English Architecture*, Batsford 1947

Ayres, J., *The Shell Book of the Home in Britain*, Faber & Faber 1981

Barker, P., *Techniques of Archaeological Excavation*, Batsford 1977

Barley, M. W., *The English Farmhouse and Cottage*, Routledge & Kegan Paul 1961

Barley, M. W., *The House and Home*, Vista Books 1963

Beresford, G., 'The medieval clay-land villages: excavations at Goltho and Barton Blount', *Medieval Archaeology*, Monograph series, no. 6 (1975)

Beresford, M., and Hurst J. G., *Deserted Medieval Villages*, Lutterworth Press 1972

Berger, R., *Scientific Methods in Medieval Archaeology*, University of California Press 1970

Bjerknes, K., and Liden, H. L., *The Stave Churches of Kaupauger*, Fabritius Forlag 1975

Bond, C. J., 'The estates of Evesham Abbey: a preliminary survey of the medieval topography', *Vale of Evesham Historical Society Research Papers* (1973)

Borne, P., and Dixon, P. W., *The Peacock Inn, Chesterfield, Derbyshire*, 1975

Boswell, J., *Life of Johnson*, Oxford University Press 1950

Bowyer, J., *Handbook of Building Crafts in Conservation*, Hutchinson 1981

Bowyer, J., *Vernacular Building Conservation*, The Architectural Press 1980

Bramwell, M., *The International Book of Wood*, Mitchell Beazley 1976

Briggs, M. S., *A Short History of the Building Crafts*, Oxford University Press 1925

Briggs, M. S., *Everymans Concise Encyclopaedia of Architecture*, Dent 1959

British Archaeological Association Conference Transactions for year 1976, *Medieval Art and Architecture at Ely Cathedral*, BAA 1979

Brunskill, R. W., *Vernacular Architecture*, Faber & Faber 1970

Brunskill, R. W., *Vernacular Architecture of the Lake Counties*, Faber & Faber 1974

Brunskill, R. W., *Traditional Buildings of Britain*, Gollancz 1981

Brunskill, R. W., *Traditional Farm Buildings of Britain*, Gollancz 1982

Brunskill, R. W., and Taylor A. C., *English Brickwork*, Ward Lock 1977

Buchanan, C., *Traffic in Towns*, HMSO 1963

Cantacuzino, S., *Architectural Conservation in Europe*, The Architectural Press 1975

Carson, C., *Impermanent Architecture in the Southern American Colonies*, University of Chicago 1981

Charge, J., 'The raising of the Old Wellington Inn and Sinclair's Oyster Bar', *The Structural Engineer* (1972)

Charles, F. W. B., and Horn, W., 'The cruck built barn of Middle Littleton in Worcestershire, England', *Journal of Society of Architectural Historians*, vol. xxv, no. 4 (1966)

Charles, F. W. B., 'Medieval cruck-building and its derivatives', *Medieval Archaeology*, Monograph series, no. 2 (1967)

Charles, F. W. B., 'Ancient timber structures', *Forestry*, vol. 41, no. 1 (1968)

Charles, F. W. B., 'Restoration of ancient timber-framed buildings', *Forestry and Home Grown Timber* (1968)

Charles, F. W. B., 'Timber-framed buildings in Worcestershire', in N. Pevsner, *Buildings of England, Worcestershire*, Penguin 1968

Charles, F. W. B., 'Problems of restoring ancient timber-framed buildings', *Forestry and Home Grown Timber* (1970)

Charles, F. W. B., 'Repair and restoration of old timber-framed buildings', *The Estates Gazette*, vol. 216 (1970)

Charles, F. W. B., 'Timber-frame tradition', in R. Berger, *Scientific Methods in Medieval Archaeology*, University of California Press 1970

Charles, F. W. B., 'Dendrochronology – the science of dating by tree rings', *Forestry and Home Grown Timber* (1971)

Charles, F. W. B., 'The Guesten Hall roof, Worcester Cathedral', *Transactions of the Ancient Monuments Society*, vol. 18 (1971)

Charles, F. W. B., and Down, K., 'A sixteenth-century drawing of a timber-framed house', *Worcester Archaeological Society*, third series, vol. 3 (1972)

Charles, F. W. B., and Horn, W., 'The cruck built barn of Leigh Court, Worcestershire, England', *Journal of Society of Architectural Historians*, vol. XXXII, no. 1 (1973)

Charles, F. W. B., 'The timber-framed buildings of Coventry 169 Spon Street', *Birmingham and Warwickshire Archaeological Society*, vol. 86 (1974)

Charles, F. W. B., 'Scotches, lever sockets and rafter-holes', *Vernacular Architecture Journal*, vol. 5 (1974)

Charles, F. W. B., 'Chester House Library, Knowle, Warwickshire', *Architects' Journal*, vol. 165, no. 17 (1977), p. 782

Charles, F. W. B., 'Timber-framed houses in Spon Street, Coventry', *Birmingham and Warwickshire Archaeological Society*, vol. 89 (1978–9)

Charles, F. W. B., 'The timber-frame tradition and its preservation', *Association for Studies in the Conservation of Historic Buildings Journal*, vol. III (1979)

Charles, F. W. B., 'Post-construction and the rafter-roof', *Vernacular Architecture Journal*, vol. 12 (1981)

Chatwin, P. B., 'Early Coventry', *Birmingham and Warwickshire*

Archaeology Society, vol. 53 (1928)
Chesher, V. M., and F. J., *The Cornishman's House*, Bradford Barton 1968
Christie, H., *The Stave Churches of Nes*, Fabritius Forlag 1979
Cocke, T., Findlay, D., Halsey, R., Williamson, E., and Wilson, G., *Recording a Church: an illustrated glossary*, CBA 1982
Colvin, H. M., *A Guide to the sources of English Architectural History*, Pinhorns Handbooks 1967
Colvin, H. M., *The History of the Kings Works*, HMSO 1982
Cook, O., and Smith, E., *English Cottages and Farmhouses*, Thames and Hudson 1954
Cook, O., and Smith, E., *The English House through Seven Centuries*, Nelson 1968
Cordingley, R. A., 'British historical roof-types and their members: a classification', *Transactions of Ancient Monuments Society*, vol. 9 (1961)
Cordingley, R. A., *Stokesay Castle, Shropshire*, College Art Association of America 1963
Crossley, F. H., *Timber Building in England from early times to the end of the 17th century*, Batsford 1951
Cummings, A. L., *The Framed Houses of the Massachusetts Bay Area 1625–1725*, Harvard University Press 1979
Davey, A., Heath, B., Hodges, D., Milne, R., and Palmer, M., *The Care and Conservation of Georgian Houses*, Paul Harris publishing with Edinburgh New Town Conservation Committee 1978
Denyer, S., *African Traditional Architecture*, Heinemann Educational 1978
Drinkwater, N., 'The Old Deanery, Salisbury', *Antiquaries Journal*, vol. XLIV, part 1 (1964)
Edlin, H. L., *Woodland Crafts in Britain*, David & Charles 1974
Elliott, J. S., *Concerning the Manor House, Dowles and its Demesne*, Herald Press 1917
Emmison, F. G., *Elizabethan Life*, Essex County Council 1976
Evans, G. E., *The Pattern Under the Plough*, Faber & Faber 1966
Evelyn, J., *Sylva, or a Discourse of Forest Trees*, The Royal Society 1664
Feilden, B. M., *Conservation of Historic Buildings*, Butterworth 1982
Feilden, B. M., *Introduction to Conservation*, UNESCO 1979
Field, R. K., 'Worcestershire peasant buildings, household goods and farming equipment in the late Middle Ages', *Medieval Archaeology*, vol. IX (1965)
Fiodorov, B., *Architecture of the Russian North*, Aurora Art 1976
Fitchen, J., *The New World Dutch Barn*, Syracuse University Press 1968
Fletcher, J. M., 'Tree-ring chronologies for the 6th to 16th centuries for oaks of Southern and Eastern England', *Journal of Archaeological Science*, vol. 4 (1977)
Fletcher, J., *Dendrochronology in Europe*, BAR International Series 51, 1978
Fletcher, J. M., and Spokes, P. S., 'The origin and development of crown-post roofs', *Medieval Archaeology*, vol. VIII (1964)
Forrester, H., *The Timber-framed Houses of Essex*, The Tindal Press 1959
Fox, Sir Cyril, *The Personality of Britain*, National Museum of Wales 1959
Fox, Sir Cyril, and Raglan, Lord, *Monmouthshire Houses – Part I – Medieval*, National Museum of Wales 1951
Fox, Sir Cyril, and Raglan, Lord, *Monmouthshire Houses – Part II – Sub-Medieval*, National Museum of Wales 1953
Fox, Sir Cyril, and Raglan, Lord, *Monmouthshire Houses – Part III – Renaissance*, National Museum of Wales 1954
Gooder, E., *The Pittancer's Rental 1410–1411*, Birmingham University Extramural Department 1973
Goodman, W. L., *The History of Woodworking Tools*, Bell & Sons 1964
Gravett, K., *Timber and Brick Building in Kent*, Phillimore 1981
Hadfield, M., *Discovering English Trees*, Shire 1970
Hall, R., *A Bibliography of Vernacular Architecture*, David & Charles 1972
Harding, J. M., *Four Centuries of Charlwood Houses*, The Charlwood Society 1976
Harris, R., *Discovering Timber-framed Buildings*, Shire 1978
Harrison, W., 'An historical description of the Ilande of Britayne with a brief rehearsal of the nature and qualities of the people of England (from Holinsheds Chronicles 1577), in F. J. Furnivall (ed.), *New Shakespeare Society* vol. I (1877)
Harvey, J., 'Conservation of old buildings, a select bibliography', *Transactions of the Ancient Monuments Society* (1972)
Harvey, J., *Conservation of Buildings*, John Baker 1972
Harvey, J., *The Medieval Architect*, Wayland 1972
Harvey, J., *Sources for the History of Houses*, British Records Association 1974
Harvey, W., 'Westminster Hall roof', *Journal of the Royal Institute of British Architects* (1922)
Hauglid, R., *Norwegian Stave Churches*, Dreyers Forlag 1970
Hayes, R. H., and Rutter, J. G., *Cruck framed buildings in Ryedale and Eskdale*, Scarborough and District Archaeological Society 1972
Herzogenrath, W. (ed.), *Baühaüs – an abridged edition of the catalogue for the exhibition '50 jahre baühaüs*, Institut für auslandsbeziehunger 1975
Hewett, C. A., 'Structural carpentry in medieval Essex', *Medieval Archaeology*, vol. VI–VII (1962–3)
Hewett, C. A., 'The barns at Cressing Temple, Essex and their significance in the history of English carpentry', *Journal of Society of Architectural Historians*, vol. XXVI, no. 1 (1967)
Hewett, C. A., 'Siddington Barn', *Country Life* (1971)
Hewett, C. A., 'Siddington Barn', *Archaeological Journal*, vol. 129 (1972)
Hewett, C. A., *Church Carpentry*, Phillimore 1974
Hewett, C. A., *The Development of Carpentry 1200–1700*, David & Charles 1969
Hewett, C. A., *English Historic Carpentry*, Phillimore 1980
Hickin, N. E., *The Woodworm Problem*, Hutchinson 1963
Hickin, N. E., *The Dry Rot Problem*, Hutchinson 1965
Hickin, N. E., *The Conservation of Building Timbers*, Hutchinson 1967
Hickin, N. E., *The Insect Factor in Wood Decay*, Hutchinson 1968

Hilton, R. H., *A Medieval Society*, Weidenfeld and Nicholson 1966

Hope-Taylor, B., *Yeavering: an Anglo British Centre of Early Northumbria*, HMSO 1979

Horn, W., 'On the origins of the medieval bay system', *Journal of Society of Architectural Historians*, vol. XVII (1958)

Horn, W., and Born, E., *The Barns of the Abbey of Beaulieu at its granges of Great Coxwell and Beaulieu St. Leonards*, University of California Press 1965

Horn, W., and Born, E., *The Plan of St. Gall*, University of California Press 1979

Horn, W., and Charles, F. W. B., 'The cruck built barn of Frocester in Gloucestershire', *Journal of Society of Architectural Historians* (1984)

Hoskins, W. G., 'The Rebuilding of Rural England', *Past and Present*, vol. IV (1953)

Howard, F. E., and Crossley, F. H., *English Church Woodwork*, Batsford 1933

Innocent, C. F., *The Development of English Building Construction*, David & Charles 1971

Insall, D., *The Care of Old Buildings Today*, The Architectural Press 1973

Insall, D. W., *Conservation in Action: Chester's Bridgegate*, HMSO 1982

Isham, N., and Brown, A., *Early Connecticut Houses*, Dover 1965

Jagger, M., and Hughes, M., *Hampshire's Heritage*, Hampshire County Planning Department 1980

James, N. D. G., *A History of English Forestry*, Blackwell 1981

Jenkins, J. G., *Traditional Country Craftsmen*, Routledge & Kegan Paul 1969

Jones, B., *English Furniture at a Glance*, The Architectural Press 1971

Jones, S. R., 'Tir-y-coed – a fifteenth century farmhouse in the parish of Melverley, Salop', *Transactions of Shropshire Archaeological Society*, vol. LVI, part 2 (1959)

Jones, S. R., 'Chamfer stops – a provisional mode of reference', *Vernacular Architecture Journal*, vol. 2 (1971), p. 12

Jones, S. R., 'West Bromwich Manor House', *Transactions of the South Staffs. Archaeological/Historical Society* (1975–6)

Jones, S. R., and Smith, J. T., 'The Great Hall of the Bishops Palace at Hereford', *Medieval Archaeology*, no. 7 (1960)

Jones, S. R., and Smith, J. T., 'The houses of Breconshire – Part I – the Builth district', *The Brecknock Society in Bryncheinog*, vol. IX (1963)

Jones, S. R., and Smith, J. T., 'The houses of Breconshire – Part II – the Hay and Talgath district', *The Brecknock Society in Bryncheinog*, vol. X (1964)

Jones, S. R., and Smith, J. T., 'The Wealden houses of Warwickshire and their significance', *Transactions of the Birmingham Archaeological Society*, vol. LXXIX (1964)

Jope, E. M., *Studies in Building History*, Odhams 1961

Kelly, J. F., *Early Domestic Architecture of Connecticut*, Dover 1963

Leland, J., *Itinerary of Britain 1583*, Centaur Press 1964

Lines, C., 'Fairy-tale home in the woods', *Warwickshire and Worcestershire Life* (1968)

Lloyd, N., *A History of the English House*, The Architectural Press 1949

MacEwen, M., *Further Landscapes*, Chatto & Windus 1976

Mainstone, R. J., *Development of the Structural Form*, Allen Lane 1975

Martin, J. L., Nicholson, B., and Gabo, N., *Circle: international survey of constructive art*, Faber & Faber 1937

Mason, R. T., 'Fourteenth-century halls in Sussex', *Sussex Archaeological Collection*, vol. XCV (1957)

Mason, R. T., *Framed Buildings of the Weald*, Coach Publishing House 1969

Mercer, E., *English Vernacular Houses*, HMSO 1975

Mercer, H. C., *Ancient Carpenters' Tools*, Horizon Press 1929

Morris, W., *On Art and Socialism*, John Lehmann 1947

Morris, W., *Selected Writings and Designs*, Penguin 1962

Morris, W., *Three Works: A dream of John Ball, The Pilgrims of Hope, News from Nowhere*, Lawrence & Wishart 1973

Moxon, J., *Mechanik Exercises*, Praeger 1970

Nicholls, J., *The Progress and Public Processions of Queen Elizabeth*, J. Nicholls 1788–1821

de Paor, M. and L., *Early Christian Ireland*, Thames & Hudson 1958

Parker, V., *The Making of Kings Lyn*, Phillimore 1971

Peate, I. C., *The Welsh House*, Hugh Evans & Sons 1944

Peters, J. E. C., *Development of Farm Buildings in Western Lowland Staffordshire*, Manchester University Press 1969

Peters, J. E. C., *Discovering Traditional Farm Buildings*, Shire Publications 1982

Pevsner, N., *Buildings of England*, vols 1–46, Penguin 1951–76

Pevsner, N., *Pioneers of Modern Design*, Penguin 1960

Pinto, E. H., *The Craftsman in Wood*, Bell & Sons 1965

Portman, D., *Exeter Houses*, University of Exeter 1966

Powys, A. R., *Repair of Ancient Buildings*, SPAB 1981

Rackham, O., 'Grundle House: on the quantities of timber in certain East Anglian buildings in relation to local supplies', *Vernacular Architecture Journal*, no. 3 (1972)

Rackham, O., *Trees and Woodland in the British Landscape*, Dent 1976

Rackham, O., Blair, W. J., and Munby, J. T., 'The thirteenth-century roofs and floor of the Blackfriars Priory at Gloucester', *Medieval Archaeology*, vol. XXII (1978)

Radford, C. A. R., Jope, E. M., and Tonkin, J. W., 'The Great Hall of the Bishops Palace at Hereford', *Medieval Archaeology*, vol. XVII (1973)

Raglan, Lord, *The Cruck Truss*, Royal Anthropological Institute 1956

Rahtz, P. A., 'Building and rural settlement', in D. M. Wilson, *The Archaeology of Anglo-Saxon England*, Methuen 1976

Rahtz, P., *The Saxon and Medieval Palaces at Cheddar*, BAR British Series 65, 1979

RCHM Inventory *Herefordshire I–III*, HMSO 1931–4

RCHM (England) *Monuments Threatened or Destroyed: a select list 1956–62*, HMSO 1963

Roe, F. G., *English Cottage Furniture*, Phoenix House 1961

Rose, W., *The Village Carpenter*, Cambridge University Press 1952
Rouse, E. C., '16th and 17th century domestic paintings', *Oxoniensia*, vol. 37 (1972)
Rylatt, M., *City of Coventry: Archaeology and Development*, Coventry Museums 1977
Salzman, L. F., *Buildings in England down to 1540*, Oxford University Press 1967
Schmidt, H., 'The Trelleborg House reconsidered, *Medieval Archaeology*, vol. XVII (1973)
Smith, J. F., *A Critical Bibliography of Building Conservation*, Mansell 1978
Smith, J. T., 'Medieval roofs: a classification', *Archaeological Journal*, vol. CXV (1960)
Smith, J. T., 'Cruck construction: a survey of the problems', *Medieval Archaeology*, vol. VIII (1964)
Smith, J. T., 'Timber-framed building in England', *Archaeological Journal*, vol. CXXII (1966)
Smith, J. T., 'The early development of timber buildings: the passing brace and reversed assembly', *Archaeological Journal*, vol. CXXXI (1974)
Smith, J. T., and Stell, C. F., 'Baguley Hall: the survival of pre-conquest building traditions in the 14th century', *Antiquaries Journal*, vol. XL (1960)
Smith, L. D. W., 'Part of the *c.*1500 estate survey for the Archer property in Tamworth-in-Arden', to appear in *Transactions of Birmingham and Warwickshire Archaeological Society*
Smith, P., *Houses of the Welsh Countryside*, HMSO 1975
Smith, T. *Traditions of the Old Grown House in Der-yat-end* (Henry Wright 1863)
Stokes, M. A., and Smiley, T. L., *An Introduction to Tree-Ring Dating*, University of Chicago Press 1968
Summerson, J., *Architecture in Britain – 1530–1830*, Penguin 1953
Taylor, A. C., *The Pattern of English Building*, Faber & Faber 1972
Thompson, E. P., *William Morris, Romantic to Revolutionary*, Merlin Press 1977
Turner, R., *The Smaller English House 1500–1939*, Batsford 1952
Vernacular Architecture. Journal of the Vernacular Architecture Group published by the Group annually from vol. I (1970)
Victoria History of the Counties of England, *Shropshire VIII*, Oxford University Press 1968
Victoria History of the Counties of England, *Warwickshire III–V*, Oxford University Press 1969
Viollet-le-Duc, E. L. N., *Dictionnaire raisonné de l'architecture français XI au XV siècle* (1867–70)
Walton, J., 'Hog-back tombstones and the Anglo-Danish House' *Antiquity*, vol. XXVIII (1954)
Ward, P., *Conservation and development in historic towns and cities*, Oriel Press 1968
Webb, G., *Architecture in Britain – the Middle Ages*, Penguin 1956
West, J., *Village Records*, MacMillan 1962
West, T., *The Timber-framed House in England*, David & Charles 1971
Wood, M., *The English Medieval House*, Dent 1965
Wood-Jones, R. B., *Traditional Architecture – The Banbury Region: Minor Domestic Architecture before 1600*, Manchester University 1964

INDEXES

Page references in italics refer to illustrations.

Subject index

Buildings index